PROCEEDINGS
OF AN
INTERNATIONAL CONFERENCE
ON

# PHASE TRANSFORMATIONS IN

# FERROUS ALLOYS

# PHASE TRANSFORMATIONS IN FERROUS ALLOYS

PROCEEDINGS OF AN INTERNATIONAL CONFERENCE COSPONSORED BY THE FERROUS METALLURGY COMMITTEE OF THE METALLURGICAL SOCIETY OF AIME AND THE PHASE TRANSFORMATIONS TA OF THE MATERIAL SCIENCE DIVISION OF THE AMERICAN SOCIETY FOR METALS, HELD OCTOBER 4-6, 1983, IN PHILADELPHIA, PENNSYLVANIA.

EDITED BY

## A. R. MARDER
BETHLEHEM STEEL CORPORATION
BETHLEHEM, PENNSYLVANIA 18016

AND

## J. I. GOLDSTEIN
LEHIGH UNIVERSITY
BETHLEHEM, PENNSYLVANIA 18015

*A Publication of The Metallurgical Society of AIME*

**A Publication of The Metallurgical Society of AIME**
420 Commonwealth Drive
Warrendale, Pennsylvania 15086
(412) 776-9000

Printed in the United States of America.
Library of Congress Catalogue Number 84-61582
ISBN NUMBER 0-89520-481-9

# TABLE OF CONTENTS

# PREFACE

This book contains the proceedings of an International Conference on Phase Transformations in Ferrous Alloys held in Philadelphia, Pennsylvania, USA, October 4-6, 1983. The conference was co-sponsored by the Ferrous Metallurgy Committee of The Metallurgical Society of the American Institute of Metallurgical Engineers and the Phase Transformations TA of the Material Science Division of the American Society for Metals.

Twenty-three years ago, in Philadelphia, on October 19, 1960, the Ferrous Metallurgy Committee sponsored the symposium, Decompositon of Austentite by Diffusional Processes. The published proceedings of that meeting has since become a classic in the field of phase transformations. In the Foreword of that book, the editors, Drs. V.F. Zackay and H.I. Aaronson, wrote, "Although research in the decomposition of austenite has now been in progress for more than a century, the stage of exploration is still characterized more by prospects than by accomplishments." Since that time a great deal of work has been done to enhance our knowledge of ferrous phase transformations, and therefore the Ferrous Metallurgy Committee believed that it was appropriate to review and update our progress.

There have also been tremendous changes in the steel industry since 1960. At that time, new research centers were being established at many steel companies, and the industry was in a period of prosperity. As a result of the research conducted on ferrous phase transformations, several products and processes were commercially developed over the past 20 years. Notably, high strength low alloy steels (HSLA), quench and tempered plate steel, controlled rolling, heat treated rail steel, dual phase alloys and, most recently, continuous heat treating of sheet steel.

The steel industry is now undergoing a period of uncertainty with steel production becoming more international and retrenchment and merger the byword for U.S. Steelmakers. Research activity has followed this same cycle and perhaps it is timely that we review where we have been so that we know where we are going in the field of ferrous phase transformations.

The organizing committee, consisting of Drs. H.I. Aaronson, J.M. Rigsbee, J.I. Goldstein and A.R. Marder chose to divide the conference into four separate sessions as follows:

I.   Mechanical Properties
II.  Martensite Transformations
III. Diffusional Transformations
IV.  Bainite and Deformed Austenite

Along with invited papers, researchers were encouraged to submit short "research in progress" reports, that are also included in these proceedings.

The organizers are especially indebted to our session chairmen who worked along with the members of the organizing committee to produce an efficient, yet lively meeting. They were: G.S. Ansell, Rennselaer Polytechnic Institute, W.S. Owen, Massachusetts Institute of Technology, W.C. Leslie, University of Michigan, and R.F. Hehemann, Case Western Reserve University. All the papers in this proceedings underwent a peer review and we wish to acknowledge the efforts of the anonymous reviewers, many of whom were conferees. In addition, the encouragement of the Ferrous Metallurgy Committee is greatly appreciated as is the help of the TMS/AIME staff, including John Ballance and Elizabeth Luzar.

Finally, we would like to thank Janis Kowalik, a graduate student for efficiently running the slides at the meeting and R.S. Laubach for developing the index.

# Phase Transformations in Iron Meteorites

J. I. Goldstein

Department of Metallurgy and Materials Engineering
Lehigh University
Bethlehem, PA   18015

Iron meteorites (Fe-5 to 25 wt% Ni alloys) were slow cooled from 1300°C at the slowest rates known to man (~ 1 to 100°C/million years). Among the phases which form on cooling are Widmanstätten ferrite, martensite, decomposed martensite, Fe C, (FeNi) P and ordered FeNi at low temperatures. These ferrous phase transformations have been studied by optical microscopy, TEM, EPMA, AEM and SEM. Laboratory alloys have been heat treated to study most of these transformations. In addition computer models of the growth of Widmanstätten ferrite have been developed to explain the diffusion gradients which are still present in these meteorites and to calculate cooling rates for these unique alloys. The development of these phase transformations are discussed with relation to pertinent phase diagrams.

Iron meteorites are fragments of naturally produced solid material that have survived passage, from interplanetary space, through the Earth's atmosphere and have landed on the surface of the Earth. Meteoritic material probably originates in the belt of asteroids between Mars and Jupiter. The iron meteorites are composed mainly of iron and nickel with small amounts of cobalt, phosphorus, sulfur and carbon. The nickel content varies from 5 to 60 wt% although in the vast majority of cases it lies in the range between 5 and 12%. The structures of these meteorites, Figures 1 and 2, were developed during slow cooling, over millions of years, of these samples in their parent bodies. The regular octahedral pattern which is observed is called the Widmanstätten pattern. This microstructure was discovered first in iron meteorites in the early 1800's and bears the name of one of the first scientists to study meteorites, namely Widmanstätten. This pattern is found in hundreds of iron meteorites. The two photomicrographs (Figures 1 and 2) illustrate the ferrite, $\alpha$, (bcc) precipitates which form in the matrix austenite, $\gamma$, (fcc). Almost all iron meteorite samples were single crystal austenite before the Widmanstätten pattern formed ( $\leq$ 800°C) and some of the single crystals were more than 1 meter in size.

With the slow cooling which prevailed in asteroidal bodies, the phases which form in iron meteorites should be very close to their equilibrium composition. The purpose of this paper is to examine the phase transformations which occur in the iron meteorites. Particular attention will be paid to the pertinent phase diagrams and the process of diffusion controlled growth of the $\alpha$ phase. The effect of third elements such as P on the nucleation and growth process will be discussed. We will also examine several low temperature phase transformations which also occur in ferrous materials such as the formation of martensite, and the decomposition of martensite. In addition we will discuss the ordering of alloy compositions close to 50-50 FeNi.

Figure 1 - The Widmanstätten pattern in the Mt. Edith iron meteorite (9.6 wt% Ni). It has a cooling rate of ~ 200°C/my. The large rounded precipitates are troilite, (FeS) and the lamellar precipitates are schreibersite, $(FeNi)_3P$. The section is 20 cm long and 15 cm wide. (Photograph courtesy of the Div. of Meteorites, U.S. National Museum).

Figure 2 - The Widmanstätten pattern in the Bristol iron meteorite (8.1 wt % Ni). It has a cooling rate of ~ 5000°C/my. Shocked deformation bands are observed in the ferrite and rather extensive regions of transformed austenite are found between the ferrite plates. The field of view is 0.8 x 0.6 cm.

## Non Equilibrium in Iron Meteorites

In three dimensions the $\alpha$ ferrite occurs as an interpenetrating arrangement of plates that are oriented parallel to the faces of an octahedron and the Widmanstatten pattern arises because plates of $\alpha$ grow with habit planes approximately parallel to the (111) octahedral planes of the parent austenite $\gamma$ (Figures 1 and 2). An electron microprobe trace for Ni taken at right angles to the $\alpha$ growth front is illustrated in Figure 3 for the Grant meteorite. Such a Ni profile with major Ni gradients in austenite shows that the meteorite is not in total equilibrium at the final growth temperature.

The observed diffusion profiles may be explained by the nucleation and diffusion-controlled growth of ferrite, $\alpha$, when an originally homogeneous parent $\gamma$ is slowly cooled through the two phase $\alpha + \gamma$ region of the Fe-Ni equilibrium diagram (1) (Figure 4). As the meteorite cools down to about 450°C, the Ni content of both the $\alpha$ and $\gamma$ phases increases and the amount of $\alpha$ phase increases as the amount of $\gamma$ phase decreases. Below about 450°C the Ni content in the $\alpha$ phase will decrease. Since the rate of diffusion of Ni at any temperature is about two orders of magnitude faster in bcc $\alpha$ than in fcc $\gamma$, chemical gradients in $\alpha$ will be much flatter than the gradients in $\gamma$.

Assuming that at any stage of growth, local equilibrium is maintained at the $\alpha/\gamma$ growth interface, one can understand how the Ni gradients in $\alpha$ and $\gamma$ develop. Ni is rejected by the growing $\alpha$ phase, which passes into the residual $\gamma$. Because of the slow rates of diffusion the Ni gradient builds up in the $\gamma$. As the temperature decreases Ni builds up in the $\gamma$ near the $\alpha/\gamma$ interface. Because of faster diffusion in bcc $\alpha$, the Ni gradient in $\alpha$ does not build up until low temperatures, less than 500°C. The Ni depletion in the $\alpha$ phase near the $\alpha/\gamma$ boundary is caused by the decreasing Ni solubility in the $\alpha$ phase below 450°C (Figure 4). Although iron meteorites cooled over a period of millions of years, total equilibrium was not achieved. However the Ni gradients which developed in $\alpha$ and $\gamma$ can be explained by considering the Widmanstatten pattern growth in terms of our understanding of diffusion controlled growth. We will return later in this paper to consider the microstructures which form in the $\gamma$ phase when the Ni concentration gradients in the $\gamma$ phase are frozen in and the meteorite is cooled to low temperatures.

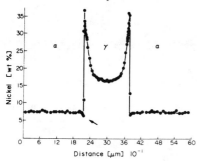

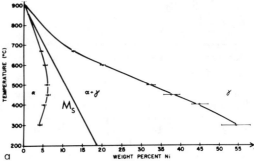

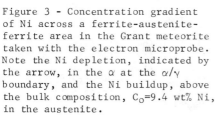

Figure 3 - Concentration gradient of Ni across a ferrite-austenite-ferrite area in the Grant meteorite taken with the electron microprobe. Note the Ni depletion, indicated by the arrow, in the $\alpha$ at the $\alpha/\gamma$ boundary, and the Ni buildup, above the bulk composition, $C_0=9.4$ wt% Ni, in the austenite.

Figure 4 - Fe-Ni binary phase diagram determined by Romig and Goldstein (1). $M_S$ is the martensite start temperature.

## Development of the Widmanstätten Pattern in Iron Meteorites

Most discussions of Widmanstätten pattern nucleation and growth consider the reaction as occurring in a binary FeNi alloy. However it has been shown experimentally that ferrite, $\alpha$, will not nucleate at either grain boundaries or in the matrix grains of austenite, $\gamma$ (2). Goldstein and Doan (3) were able to produce experimentally, for the first time, a Widmanstätten pattern in Fe-Ni alloys containing as little as 0.1 wt% P during cooling of these alloys through the $\alpha + \gamma$ region of the phase diagram. Figure 5 shows the type of structure obtained in a 9.8 wt% Ni, 0.3 wt% P alloy slow cooled to 650°C. The P present in the iron meteorite and in particular the phosphide, $(FeNi_3)P$, phase present in the iron meteorite (4), acts as the nucleating agent for the $\alpha$ phase. The effect of P on the $\alpha + \gamma$ phase boundaries of the binary diagram is to increase the Ni content of the $\alpha$ and to decrease the Ni content of the $\gamma$ (1).

Figure 5 - A light micrograph showing the microstructure of an 89.9 wt% Fe-9.8 wt% Ni-0.3 wt% P alloy slowly cooled to 650°C. 1% nital etch. Ferrite in a Widmanstatten pattern is formed within the austenite during cooling (Goldstein and Doan (3)).

Narayan and Goldstein (4,5) have experimentally grown intragranular $\alpha$ in Fe-Ni-P alloys containing between 5 and 10 wt% Ni and 0 and 1.0 wt% P and have examined the nucleation and growth process of these precipitates using analytical electron microscopy (AEM) techniques. Figure 6 shows a light micrograph of intragranular $\alpha$ precipitates in a 6.88 wt% Ni, 0.49 wt% P alloy which was cooled from 790 to 650°C at a rate of 5°C/day. Arrows point to some of the $\alpha$ precipitates and some grain boundary $\alpha$ is also visible. The general shape of the $\alpha$ precipitates during growth can be described as cylindrical. Figure 7 is a TEM micrograph from the same alloy and shows an intragranular $\alpha$ crystal in a matrix that has transformed to martensite during the quench to room temperature. Figure 8 shows the Ni profile obtained across an $\alpha / \gamma$ interface in the same alloy. The results of the Narayan and Goldstein study (6) showed that $\alpha$ ferrite nucleates intragranularly with little undercooling. In addition the measured concentration profiles indicate diffusion controlled growth of $\alpha$ with interfacial equilibrium maintained at the $\alpha / \gamma$ interface.

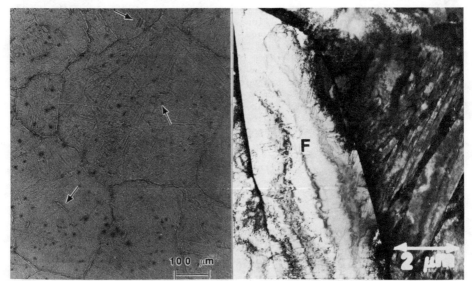

Figure 6 - A light micrograph showing a typical microstructure for
alloys exhibiting intragranular
ferrite precipitation.  Arrows point
to some of the ferrite precipitates.
Grain boundary ferrite is also visible.  The specific alloy described
here contains 6.88 wt% Ni, 0.49 wt%
P and was cooled from 790 to 650°C
at a rate of 5°C/day.

Figure 7 - A TEM micrograph showing a
ferrite crystal which has nucleated
intragranularly in a matrix that has
been transformed to martensite.  The
alloy contains 6.88 wt% Ni and 0.49 wt%
and was cooled from 790°C to 650°C at a
rate of 5°C/day.

A model to simulate the bulk diffusion controlled growth of $\alpha$ phase was
also developed (6).  A numerical method of lines (NMOL) technique was used
for the solution of the system of partial differential equations (7).  The
NMOL technique and the Murray-Landis (8) variable grid spacing transformation were combined to solve the problem of diffusional growth of ferrite.
The numerical model calculates the growth of ferrite in the ternary system
Fe-Ni-P during continuous cooling and generates the concentration profiles
of Ni and P in the ferrite and austenite phases as a function of temperature
and time.  The computation was based on the following assumptions:

1.  The growth of ferrite is controlled by bulk diffusion of Ni
    in austenite.
2.  Interfacial equilibrium of Ni and P occurs at the $\alpha/\gamma$
    interface at all times during the growth process.
3.  Ferrite nucleates and grows as a cylindrical precipitate.
    The morphology of the precipitate changes to platelike as
    impingement of the precipitates occurs.
4.  There is no P gradient in either phase and no Ni gradient
    in ferrite.

The computer model was applied to various samples including the alloy
described in Figures 6 and 7.  The bulk compositions of the alloys and the
cooling rates measured in the laboratory were used as inputs to the computer
model.  Figure 8 shows the excellent agreement of the calculated Ni profiles
with the measured data from the AEM.

The same computer model should be applicable to the study of the cooling history of the iron meteorites. Instead of simulating growth of $1 \mu$ m sized precipitates cooling at 5°C/day, one can consider $\alpha$ precipitates in iron meteorites in a size range from 10 $\mu$m to 1 cm. Calculations for various iron meteorites yield cooling rates from $500°C/10^4$ years to $1°C/10^4$ years (6). Most of the iron meteorites cool in a range from $50°C/10^4$ to $1°C/10^4$ years. These estimated cooling rates are very slow but are two orders of magnitude faster than rates determined from previous computer models (9–12). To accommodate the revised cooling rates predicted by this recent study, meteorite parent bodies need only be a few kilometers in diameter assuming that the iron meteorites are present in the center of the body.

### Microstructure of Retained Austenite and Low Temperature Phase Transformations

The characteristic M-shaped composition profile (Figure 3) arises because there is a higher concentration of Ni in the austenite adjacent to the $\alpha$ ferrite plates from which the Ni was rejected. This high Ni concentration in the austenite falls with distance from the $\alpha/\gamma$ interface. Depending on the amount of diffusion that has occurred, the lowest value of the Ni content of the austenite varies from the bulk meteorite composition to almost the equilibrium composition predicted from the phase diagram. Figure 9 is an optical micrograph showing the microstructures developed in the Edmonton iron meteorite between 15 and 50 wt% Ni along the M shaped composition profile in austenite. Region 4 has the lowest Ni and region 1 has the highest Ni. Figure 10 shows the detailed Ni composition profile in regions k, 1, 2, and 3 of the taenite phase using the AEM from the Carlton iron meteorite. Up to $\sim 20\%$ Ni the untransformed $\gamma$ will transform to martensite $\alpha_2$ (see the $M_s$ curve in Figure 4). The microstructure of the

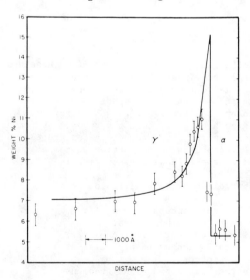

Figure 8 - Ni composition profile measured across a ferrite/austenite interface in a 6.88 wt% Ni and 0.49 wt% P alloy that was cooled from 790°C to 650°C at a rate of 5°/day. The solid line is the Ni profile calculated by the numerical model.

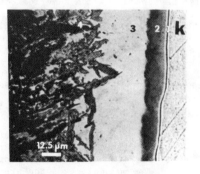

Figure 9 - Decomposition of $\gamma$ into distinct regions with decreasing Ni content away from the $\alpha/\gamma$ boundary of the Edmonton iron meteorite (14). k indicates the $\alpha$ ferrite phase. Region 4 is martensitic.

martensitic area in the austenite of a similar meteorite is shown in Figure 9, region 4. Over geological time the martensite, which may be either lath or lenticular, decomposes on a sub-micron scale to $\alpha + \gamma$ as shown in Figure 11. The total reaction is given by $\gamma \rightarrow \alpha_2 \rightarrow \alpha + \gamma$. Region 3, from $\sim 25$ to 30 wt% Ni, contains retained austenite, $\gamma$.

In the composition range 30-40% Ni the untransformed $\gamma$ breaks down on a fine scale to a duplex mixture of low and high Ni phases which invariably border the Widmanstätten $\alpha$ plates (region 2 in Figure 9). This phase mixture etches deeply and is often termed "the cloudy zone" or "cloudy border." Electron microscope observations (14,15) reveal that this region appears to consist of a structure of essentially single crystal austenite globules surrounded by a honeycomb network of single crystal ferrite, as shown in Figure 12 from the Estherville meteorite. As displayed in Figure 10 the composition profile oscillates wildly as the duplex cloudy border is encountered. Individual analyses taken in the two phases of the cloudy border of the Estherville meteorite are shown in Figure 13. As expected the high Ni phase is $\gamma$, but the low Ni phase contains $\sim 23$ wt% Ni and therefore cannot be $\alpha$. At this composition the low Ni phase would be expected to be martensitic $\alpha_2$. Convergent beam patterns taken from individual phases in the cloudy border show that the high Ni $\gamma$ exhibits superlattice spots characteristic of the $Ll_0$ structure. The mechanism for the formation of the cloudy zone is not known. Order-disorder or spinodal decomposition may be responsible for the formation of this unusual structure.

Within the last few years there has been increasing evidence that the highest Ni regions (region 1 in Figure 9, with 48-52 wt% Ni) close to the $\alpha/\gamma$ interface are ordered with the FeNi $Ll_0$ superlattice structure. Initial observations were made in meteorites using Mössbauer spectroscopy (16), and

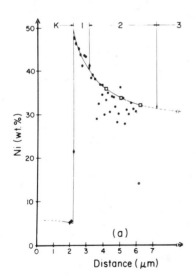

(a)

Distance ($\mu$m)

Figure 10 - Ni composition profile across regions 1, 2 and 3 of the taenite phase of the Carlton iron meteorite. The cloudy zone is region 2.

Figure 11 - A SEM micrograph of decomposed austenite in the Weaver Mts. meteorite (13) showing the breakdown of lath martensite $\alpha_2$ to $\alpha + \gamma$. The platelets are $\gamma$ phase and the highly etched black regions are $\alpha$ phase.

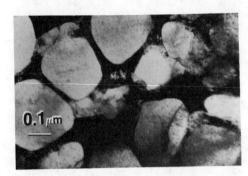

Figure 12 - A TEM micrograph of cloudy border showing a honeycomb structure in the Estherville meteorite.

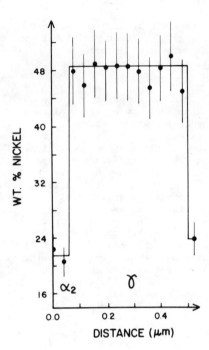

Figure 13 - Ni composition profile within the $\alpha_2$ and $\gamma$ regions of the cloudy border in the Estherville meteorite.

polarized light microscopy (17) to observe the anisotropy associated with the tetragonality of the $Ll_o$ structure.  More direct evidence (18,19) has also been produced in terms of the observations of anti-phase boundaries and the observation of superlattice spots which permit dark field images to be formed in the TEM, as shown in Figure 14.

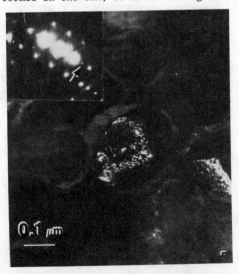

Figure 14 - Superlattice-dark field image of the ordered FeNi superlattice in the cloudy border region of the Estherville meteorite.  The superlattice reflection used is the arrowed reflection in the diffraction pattern.  The superlattice image shows the fine scale of the ordered regions within the $\gamma$ phase.

Neither the cloudy border, nor the order-disorder transformation are reproducible in the laboratory, although the latter can be induced by irradiation (20). Both microstructures are sensitive to reheating and are rapidly destroyed above 300-400°C, emphasizing the metastability of the cloudy zone and the low critical temperature ( ~ 320°C) of the ordering transformation. In view of the long times apparently required for both these transformations, their absence (or presence on an exceptionally fine scale, ~ 10 nm) is an indication of substantial shock-induced heating, followed by relatively rapid cooling to ambient temperatures.

## Summary

The iron meteorites are of much historical interest to the metallurgist particularly because of the presence of the Widmanstätten pattern. This pattern is formed in the large FeNi single crystals and is often visible to the unaided eye. However the iron meteorites yield unexpected information about low temperature phase transformations and the major effect of minor elements in solid solution in the FeNi single crystals. Finally these samples have upon cooling undergone several phase transformations well known to the ferrous metallurgist, namely--Widmanstätten precipitation and growth, martensite formation, and low temperature tempering of martensite. Studies of the iron meteorites have not only contributed information on the cooling history of these unusual materials but have allowed the study of equilibrium and non-equilibrium phase transformations under slow cooling conditions which cannot be reproduced in the laboratory.

## Acknowledgments

The author wishes to thank Dr. A. D. Romig (Sandia Laboratories), Dr. C. Narayan (IBM Research Laboratories), Dr. D. B. Williams (Lehigh University), and Dr. R. C. Clarke (Smithsonian Institution) for their help and collaboration with the author on various phases of iron meteorite research. The research was supported by NSF Grant EAR 7900995 and NASA Grant NGR 39-007-043.

## References

1. A. D. Romig, Jr. and J. I. Goldstein, "Determination of the Fe-Ni and Fe-Ni-P Phase Diagrams at Low Temperatures (700-300°C)" Met. Trans., 11A (1980) p. 1151.
2. N. P. Allen and C. C. Earley, "The Transformations $\alpha \rightarrow \gamma$ and $\gamma \rightarrow \alpha$ in Iron Rich Binary Iron-Nickel Alloys," J. Iron Steel Inst., 166 (1950) p. 281.
3. J. I. Goldstein and A. S. Doan, Jr., "The Effect of Phosphorus on the Formation of the Widmanstatten Pattern in Iron Meteorites," Geochim. et Cosmochim. Acta, 36 (1972) p. 51.
4. C. Narayan and J. I. Goldstein, "Nucleation of Intragranular Ferrite in Fe-Ni-P Alloys," accepted Met. Trans. (1984).
5. C. Narayan and J. I. Goldstein, "Growth of Intragranular Ferrite in Fe-Ni-P Alloys," accepted Met. Trans. (1984).
6. C. Narayan and J. I. Goldstein, "A Major Revision of Iron Meteorite Cooling Rates," submitted Geochim. et Cosmochim. Acta (1984).
7. W. E. Schiesser, "DSS Version 2, Introductory Programming Manual," Lehigh University and Naval Air Development Center (1976).
8. W. D. Murray and F. Landis, "Numerical and Machine Solutions of Transient Heat-Conduction Problems Involving Melting or Freezing.

Part I - Method of Analysis and Sample Solutions," Trans. ASME, 81 (1959) p. 106.

9.  J. A. Wood, "The Cooling Rates and Parent Planets of Several Iron Meteorites," Icarus, 3 (1964) p. 429.

10. J. I. Goldstein and R. E. Ogilvie, "The Growth of the Widmanstatten Pattern in Metallic Meteorites," Geochim. et Cosmochim. Acta, 29 (1965) p. 893.

11. J. I. Goldstein and J. M. Short, "The Iron Meteorites, Their Thermal History and Parent Bodies," Geochim. et Cosmochim. Acta, 31 (1967) p. 1733.

12. J. Willis and J. T. Wasson, "Cooling Rates of Group IVA Iron Meteorites," Earth and Plan. Sci. Lett., 40 (1978) p. 141.

13. P. M. Novotny, J. I. Goldstein and D. B. Williams, "Analytical Electron Microscope Study of Eight Ataxites," Geochim. et Cosmochim. Acta, 46 (1982) p. 2461.

14. L. S. Lin, J. I. Goldstein and D. B. Williams, "Analytical Electron Microscopy Study of the Plessite Structure in Four III CD Iron Meteorites," Geochim. et Cosmochim. Acta, 43 (1979) p. 725.

15. E. R. D. Scott, "The Nature of Dark Etching Rims in Meteoritic Taenite," Geochim. et Cosmochim. Acta, 37 (1973) p. 2283.

16. J. F. Albertsen, G. B. Jensen and J. M. Knudsen, "Structure of Taenite in Two Iron Meteorites," Nature, 273 (1978) p. 453.

17. E. R. D. Scott and R. S. Clarke, Jr., "Identification of Clear Taenite in Meteorites as Ordered FeNi," Nature, 281 (1979) p. 360.

18. S. Mehta, P. M. Novotny, D. B. Williams and J. I. Goldstein, "Electron-Optical Observations of Ordered FeNi in the Estherville Meteorite," Nature, 284 (1980) p. 151.

19. P. M. Novotny, "An Electron Microscope Investigation of Eight Ataxite and One Mesosiderite Meteorites," MS Thesis, Lehigh University (1981).

20. K. Benusa, E. P. Butler, D. B. Williams and J. I. Goldstein, "Ordering in Fe-Ni Alloys Induced by Irradiation in the High Voltage Electron Microscope," 7th International Conference on HVEM (1983), in press.

# STRUCTURE–PROPERTY RELATIONSHIPS IN FERROUS TRANSFORMATION PRODUCTS

A. R. Marder
Research Department
Bethlehem Steel Corporation
Bethlehem, PA   18016

A review of the structure-property relationships in ferrous transformation products shows that the mechanical properties are influenced by
specific morphological features.  The dominant microstructural feature that
contributes to strength is the high angle boundary, whereas for toughness,
it is the frequency and magnitude of change in the orientation of the
ferritic component of the transformation product.  The grain boundary in
ferrite, the cementite/ferrite interface in pearlite and the packet boundary
in martensite and bainite are all examples of high angle boundaries that
control strength.  The ferrite grain, a group of colonies in pearlite, and
the martensite or bainite packet are examples of structural features that
can produce a change in ferrite orientation and are a major influence on
toughness.  Composite multiphase microstructures affect properties in different ways depending upon the combination of transformation products used.
Strength is characterized to a first approximation by empirical equations
that take into account the proportional contribution of each individual
phase.  Finally, continuous yielding occurs in multiphase structures containing martensite, since this is the only transformation product that will
generate sufficient mobile dislocations to prevent discontinuous yielding.

## I  INTRODUCTION

An evaluation of the structure–property relationships in transformed ferrous alloys involves many strengthening mechanisms.  The principle effects to be considered are:  (1) solid solution strengthening in ferrite and martensite, (2) dispersion hardening in pearlite, bainite, tempered martensite and ferrite, (3) dislocation strengthening in cold worked bainite and martensite, and (4) hardening by high angle boundaries in all the transformation products.  Several excellent reviews have been written on strengthening mechanisms in steel (e.g., refs. 1,2) and only a brief review of these strengthening effects will be given in this paper.  The underlying commonality is that some or all of these effects are active in each ferrous transformation product, therefore it is the main purpose of this review paper to characterize the contribution that the structural features of each transformation product makes toward mechanical properties.

### A.  Solid Solution Strengthening

Solid solution strengthening is found in both substitutional and interstitial solid solutions.  The amount of substitutional solid solution strengthening is primarily dependent upon the difference in size between the solute and solvent atoms.  The substitutional solute atoms cause a symmetrical distortion of the solvent lattice which leads to small strengthening effects.  In general, substitutional solutes in dilute solid solutions, e.g., such as ferrite, increase strength linearly, figure 1.

Interstitial solutes, such as carbon and nitrogen in ferrite, introduce an asymmetrical lattice distortion which produces a strengthening effect that is 10–100 times that of substitutional elements.  Interstitial strengthening has been shown to be proportional to the square root of the solute concentration.  A further increase in strength occurs when these interstitial solutes interact with dislocations.  Interstitial solid solution strengthening by carbon is a major contributor to the strength of as quenched martensite.

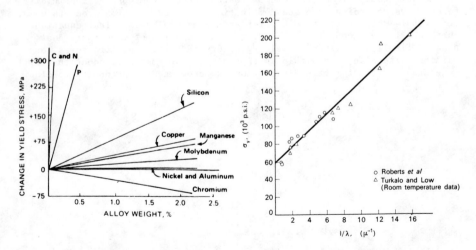

Fig. 1.  Solid solution hardening effects in low carbon ferritic steel.[2]

Fig. 2.  Yield strength of spheroidized controlled by interparticle spacing.[4]

## B.  Dispersion Hardening

Most steels usually contain a dispersion of one or more carbides which vary over a wide range of sizes from coarse pearlitic carbide structures to an ultra-fine dispersion of alloy carbides in quenched and tempered steels.  Thus, a wide variation in mechanical properties can be obtained.  The strengthening produced by a dispersed phase can be approximately determined from the theory of dispersion strengthening which[3] assumes an array of undeformable spherical particles.  Orowan showed that the yield stress ($\sigma_y$) was inversely proportional to the interparticle spacing ($\lambda$):

$$\sigma_y = \sigma_s + \frac{2T}{b\lambda}$$

where $\sigma_s$ is the yield strength of the matrix, T is the line tension of a dislocation and b is the Burgers vector.  Ashby[5] modified the equation to take into account the radius r of the particles.  These relationships can be applied in the simple case of spheroidal carbides in a ferrite matrix, figure 2, and in the less ideal case they can also provide approximations where the dispersion ranges from irregular polyhedra to fine rods or plates.

## C.  Dislocation Strengthening

The flow stress ($\sigma_f$) increases as the density of dislocations ($N_f$) increases[6]:

$$\sigma_f = \sigma_o + K\sqrt{N_f} \qquad [2]$$

where $\sigma_o$ is the flow stress due to other strengthening mechanisms and K is a constant incorporating the shear modulus and the Burgers vector, figure 3.  The work hardening rate depends upon the rate at which $N_f$ increases with strain and therefore is very dependent on dislocation distribution and interaction.  Dislocations may be introduced by cold working, quenching strains, differences in thermal expansion between particles and the matrix, volume changes accompanying precipitation, and strains produced during low temperature transformations.

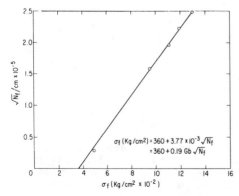

Fig. 3.  Dislocation density vs flow stress in vacuum melted iron.[6]

## D.  Grain Boundary Strengthening

It has been established that the yield strength ($\sigma_y$) of a ferritic steel increases with decreasing grain size [7]:

$$\sigma_y = \sigma_i + K_y d^{-1/2} \qquad [3]$$

where $\sigma_i$ is the friction stress needed to move dislocations, $K_y$ is a constant and d is the grain diameter.  This equation is the basis for the strengthening of ferrite, figure 4, and will be discussed below. Similar types of effects can be seen for the high angle boundaries found between packets of martensite and bainite and for the ferrite/cementite interface.

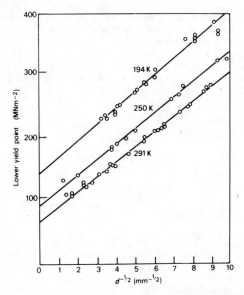

Fig. 4 [7]  Dependence of the lower yield stress of mild steel on grain size.

In evaluating the structure/property relationships in ferrous transformation products, consideration must be given to the following mechanisms: solid solution strengthening, dispersion hardening, dislocation strengthening and gain boundary strengthening.  Some or all of these effects are found in characterizing the mechanical properties of ferrous transformations.

## II    FORMATION OF FERROUS STRUCTURES

An understanding of the structural features that control the mechanical properties of ferrous transformations can be understood by first studying the formation of these structures.  In this way, the individual "building blocks" of each structure can be identified and their contribution to properties can then be evaluated.

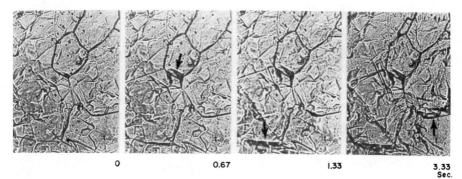

|     |      |      |            |
|-----|------|------|------------|
|  0  | 0.67 | 1.33 | 3.33 Sec.  |

Fig. 5. Cinephotomicrographic sequence showing the formation of ferrite allotromorphs (arrows) on austenite grain boundaries in pure iron at 825°C (cooling rate 12°C/s) (x 125).[8]

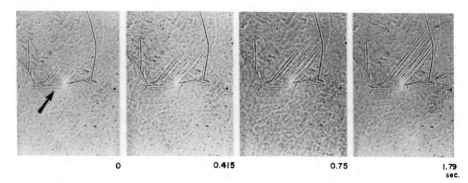

|     |       |      |            |
|-----|-------|------|------------|
|  0  | 0.415 | 0.75 | 1.79 sec.  |

Fig. 6. Cinephotomicrographic sequence showing the growth of Widmanstatten ferrite (arrow) from Fe-0.8C austenite at 617°C (cooling rate of 10°C/s) (x 250).[8]

## A.   Ferrite

The formation of three different types of ferrite have been identified using the hot stage microscope[8]:  (1) Allotriomorphs, as seen (arrows) in fig. 5, that develop on the austenite grain boundary, (2) Widmanstatten ferrite growth as shown in fig. 6, and (3) A phase front generated from grain boundaries that can not be seen in still photographs but is faintly observable in cinephotomicrographs.  Only the relationship between the equiaxed structure and properties will be discussed in this paper.  Fig. 7a shows that the relevant structure of ferrite is the ferrite grain which is usually related to the prior austenite grain.

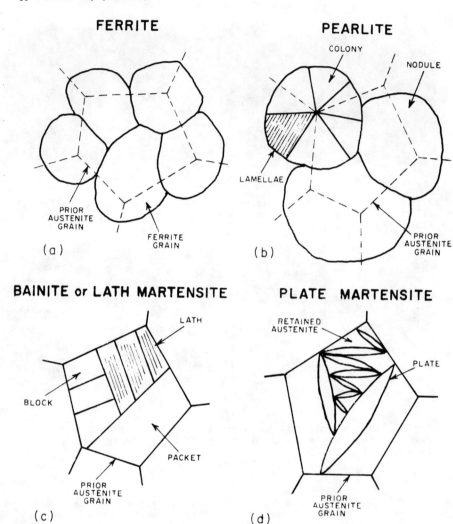

Fig. 7.  The basic microstructural features of ferrous transformation products.

B.  Pearlite

The pearlite transformation lacks surface relief, and therefore is not easily discernable in the hot stage microscope without the use of phase contrast objectives.  Fig.8 shows a growth sequence of a pearlite nodule at a triple point.  The nodule maintains its spherical nature as it grows in three different austenite grains and the growth rate is maintained by changing the ferrite/cementite orientation within the austenite grain, fig. 9, producing colonies.  Each colony contains lamellae of carbide in a single ferrite orientation.  Thus, the building blocks that can affect the properties of pearlite are:  (1) prior austenite grains, (2) pearlite nodules, (3) colonies, and (4) the ferrite/cementite lamellae, fig. 7b.

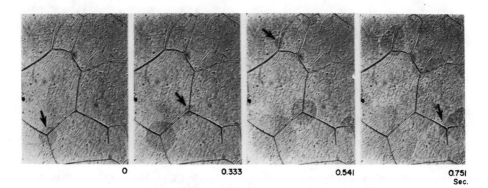

Fig. 8. Cinephotomicrographic sequence of growth of a pearlite nodule (arrow) at the junction of three austenite grains; cooling rate 4800°F and transformation temperature 1090°F. Magnification about 250 times.[9]

Fig. 9. Cross-section views of the microstructure of pearlite nodules in partially transformed hot-stage specimens, showing nodule forming both at specimen surface and interior.[9]   Picral.


C. Lath Martensite and Bainite

Hot stage microscopy shows that the formation of lath martensite, which is found in Fe–C alloy with C<0.6, occurs by two types of nucleation: (a) a side-by-side nucleation of laths or groups of laths that gives the effect of a phase front moving through the austenite matrix, and (b) nucleation of non-adjacent plates which partition the parent austenite, fig. 10.[10]  A similar type of formation is found for bainite, fig. 11.[8] These adjacent parallel laths, which may be separated by either low or high angle boundaries and retained austenite, make up a packet. Several packets may be found in a single austenite grain.[11]  Within each packet, several blocks of laths with the same orientation may be found, Fig. 12.[12] Thus, in both lath martensite and bainite, the prior austenite grain, the packet, block and ferrite lath can all contribute to the properties of this microstructure, fig. 7c.

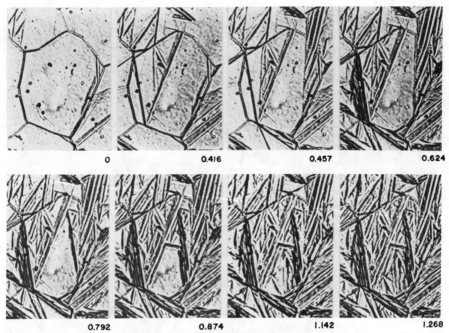

Fig. 10.  Cinephotomicrographic sequence showing the formation of lath martensite in a low-alloy steel at 330°C (cooling rate of 17°C/s) (x227).(8)

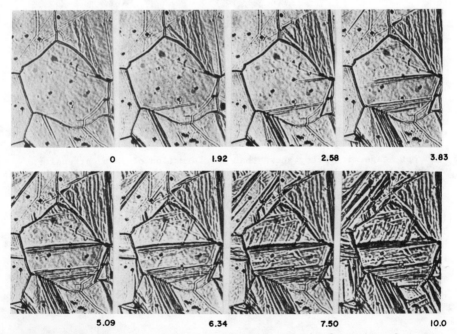

Fig. 11.  Cinephotomicrographic sequence showing the formation of bainite in a low-alloy steel at 550°C (cooling rate 6.2°C/s)(x227).(8)

Fig. 12.   Blocks of parallel laths within a packet of martensite.[12]

## D.   Plate Martensite

Plate martensite is found in quenched Fe-C alloys with C>0.6 and forms by individual nonparallel plates that vary significantly in size, fig. 13.

Fig. 13.   Transformation of a Fe-1.86 C alloy as it is sub-critically cooled.   (a) 75°F (23.9°C); (b) -75°F (-60°C).   Nital etch.   Magnification 500 times.[11]

Nucleation occurs at prior austenite grain boundaries as well as at other plates, with the parent austenite being continually partitioned.   Thus the structural features that can contribute to the properties of plate martensite are:   (1) prior austenite grain boundaries, (2) martensite plates, and (3) prior austenite, fig. 7d.   An additional factor found in very large martensite plates and at very high carbon contents is microcracking,[13] fig. 14.

Fig. 14.  The structure resulting
from impingement of plates microcracking in Fe-1.39 C plate
martensite, magnification 500
times.[11]

### III  THE PROPERTIES OF FERROUS TRANSFORMATION PRODUCTS

#### A.  Ferrite

The control of yield strength by ferrite grain size is probably the
most documented structure/property relationship.  The emperical equation
that relates lower yield strength and grain size is given by the Hall-Petch
relationship[14,15]:

$$\sigma_y = \sigma_i + K_y \, [d_g^{-1/2}(fr) + d_c^{-1/2} \frac{(1-fr)}{2}] \qquad [3]$$

where d is the grain diameter, $\sigma_y$ is the yield stress, $\sigma_i$ is the
friction stress opposing the movement of dislocations in the grains, and
$K_y$ is a constant.  This relationship was shown to hold over a range of
grain sizes from 0.35 to 400 µm (ASTM 19.5 to 0)[16,17].  Although the
values of $\sigma_o$ and $K_y$ can change depending upon alloy content, the exponent of d remains -1/2.  When a substructure exist in these ferrite grains,
equation 3 can be modified to take into account the cell size[18]:

$$\sigma_y = \sigma_i + K_y \, [d_g^{-1/2}(fr) + d_c^{-1/2} \frac{(1-fr)}{2} \qquad [4]$$

where $d_g$ is the grain size, $d_c$ is the cell size and fr is the fraction
of recrystallized grains.

Strengthening by grain size refinement is the only strengthening
mechanism that also increases the toughness.  By decreasing grain size the
ductile-to-brittle-impact transition temperature decreases.  Petch[19]
suggested the following equation for the transition from fibrous to cleavage
fracture:

$$\beta T_c = \ln B - \ln \left| \frac{4qG\gamma'}{K^*} - K^* \right| - \ln d^{1/2} \qquad [5]$$

where $T_c$ is the fracture transition temperature, $K^*$, $\beta$ and B are constants, q is a triaxiality factor (1/3 for a Charpy V-notch), G is the shear modulus, $\gamma'$ is the effective surface energy, and d is the grain diameter. Thus, the mechanical properties of ferrite are uniquely improved by the ferrite grain size structure which is additive to the other strengthening mechanism such as solid solution strengthening and dispersion strengthening.

## B.  Pearlite

Previous results[20] have shown that the austenitizing temperature determines the austenite grain size, whose boundaries provide nucleation sites for pearlite nodules.  For a constant transformation temperature, which dictates the number of nucleation sites, nodule size can be directly related to prior austenite grain size.  On the other hand, lowering the transformation temperature reduces the interlamellar spacing, which is independent of prior austenite grain size or nodule size.  Few studies have been made on colony size but it can be expected that the size of colonies would be controlled by the same variables that control nodule size.  The following is a review of those structural features that control the various mechanical properties of pearlite.

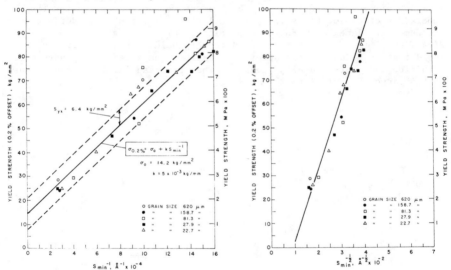

Fig. 15.  Effect of minimum interlamellar spacing on yield strength:
(a) $S^{-1/2}$ and (b) $S^{-1}$.[20]

Strength:

Several investigations have shown that refining the interlamellar spacing of pearlite increases yield strength[20-26] and it has also been shown that the yield strength of pearlite is independent of prior austenite grains size and nodule diameter.[20]  When the yield stress is plotted against the interlamellar spacing, according to a Hall-Petch type equation, the data gives a linear relationship for the exponent of spacing either -1/2 or -1, fig. 15.  A typical result is as follows:[26]

$$\sigma_y = -170 + 0.25 \, S^{-1/2} \qquad [6a]$$

$$\sigma_y = 140 + 4.6 \times 10^{-5} \, S^{-1} \qquad\qquad [6b]$$

where the yield stresses and interlamellar spacing, S, are in units of $MN/m^2$ and m, respectively. The Hall-Petch type relationship, eq. [6a], is typical of hardening caused by grain boundaries and subboundaries. However, for these pearlite results, a theoretically unacceptable negative friction stress is obtained.[25] The inverse linear dependence of strength on interlamellar spacing, eq. [6b], typical of hardening resulting from the formation of a cellular structure in ferrite,[27] gives a positive friction stress. Langford[28] has developed an equation describing the dependence of the yield strength on interlamellar spacing using two functions of S.

$$\sigma_y = \sigma_o + K_1 S^{-1/2} + K_2 S^{-1}$$

where $K_1$ and $K_2$ are constants. The derivation of this equation is based on the concept of a change in the free path for the movement of dislocations, which causes a change in the hardening mechanism. The total deformation is divided into the work required for the formation of a dislocation pile-up (proportional to $S^{-1/2}$) and the work required for the generation of dislocations (proportional to $S^{-1}$) in the space defined by the cementite lamellae. For fine pearlites, with S ranging from 20-450 nm,[20,21,25,26] the $S^{-1}$ relationship holds. However, it was found that equation 7 satisfactorily described the strength dependence for both fine and coarse lamellar pearlite when the mean lamellar spacing was in the range of 70-1250 nm.[24]

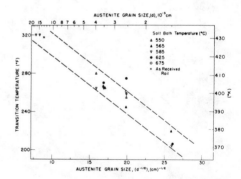

Fig. 16.  Charpy transition temperature vs prior austenitic grain size.[21]

Toughness:

Fracture toughness of pearlite is directly related to the prior austenite grain size with a finer grain size producing a lower transition temperature, fig.16.[21,29,30] Pearlite colony size and interlamellar spacing are not effective microstructural parameters in the control of toughness.[21] Bernstein and co-workers[31,32] have demonstrated by fractographic analysis of brittle cleavage fracture of eutectoid steel, that the fracture facets usually consist of a number of pearlite colonies. The cleavage fracture is capable of propagating across colony boundaries which has led to the concept of an effective grain size or pearlite block size for cleavage fracture.[33] It was found that the effective grain size is controlled by the prior austenite grain size.[31] Since adjacent colonies have nearly the same ferrite orientation, it is not difficult for these fracture facets to pass through colony boundaries.

Ferrite in adjacent colonies will have a common orientation because they grow from the same prior austenite grain.[34]  However, colonies nucleated from different parent austenite grains will not have the same ferrite orientation.  The crack path will change only when a different ferrite orientation is encountered at a colony boundary.  Thus, for finer prior-austenite grain sizes, the crack path will be changed more frequently and a higher toughness will be obtained.[32]  It was also shown that cementite has no effect on this cleavage process.[32].

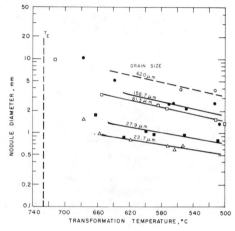

Fig. 17.  Effect of transformation temperature on nodule diameter for prior austenite grain sizes.[20]

The effective grain size that controls toughness is made up of several colonies and is controlled by the prior-austenite grain size[21,31,32]. and it has been shown that prior austenite grain size controls nodule diameter, fig. 17.[20]  These nodules are spherical and grow radially into several austenite grains at the same time, fig. 8, producing different orientations within a nodule, often referred to as colonies or groups of colonies.  This could be how the "effective grain size" originates for the control of toughness.  However, fig. 17 also suggests that transformation temperature will also control the "effective grain size" since the number of nucleation sites for nodules increases as the transformation temperature decreases at any given prior austenite grain size.

Other Properties:

Although it is well established that interlamellar spacing controls strength and that prior austenite grain size significantly influences fracture toughness, other properties are not as easily discernable.  The energy absorbed in ductile tearing, as measured by the upper shelf energy in a Charpy test, is related to interlamellar spacing[21] as is cyclic deformation behavior.[26]  However, it has been reported that work hardening is both dependent[22] and independent[35] of interlamellar spacing.  It was not obvious whether nodule diameter or interlamellar spacing controls ductility as measured by percent reduction in area, although fracture stress is directly related to spacing.[20]  Finally, it has been reported that fatigue crack propogation is both independent[36] and dependent[37] on microstructural factors.

## C.  Martensite

Martensite forms in basically two morphologies, depending predominantly on carbon and/or other alloy additions.  Lath martensite, found in Fe-C alloys with C<0.6%, is made up of packets of parallel laths within a prior austenite grain.  The laths are heavily dislocated and have a habit plane of {557}.[11]  Plate martensite, found in Fe-C alloy with C>0.6%, consists of nonparallel units or plates that vary significantly in size, have a twinned substructure that may also contain dislocations, and have a habit plane that changes from {225}$_\gamma$ to {259}$_\gamma$ as the carbon or alloy content increases.[11]  Thus, these two distinctly different morphologies may be expected to influence properties differently.

Hardness:

The well known martensite hardness relationship with carbon content is seen in Fig. 18, which shows the range of hardness that can be obtained in martensitic structures.  Some of the data may be influencd by the amount of retained austenite, especially above 0.7%, because the $M_f$ is below room temperature.  Liquid helium treatments, the data marked X in fig. 18, give higher hardnesses as a result of more austenite transforming to martensite.  An additional factor of microcracking in the very high carbon alloys may also play a role in lowering the hardness values. Austenite grain size and aging may also influence the maximum hardness. Grain size effects will be discussed in the strength of martensite section, while the effects of aging can be seen in fig. 19.  Aging at room temperature occurs after quenching, which can explain the scatter in reported hardness.

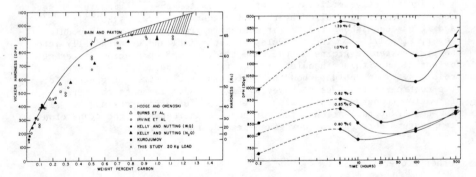

Fig. 18.  The effect of carbon content on the microhardness of martensite.

Fig. 19.  Room temperature aging of Fe-C alloys from 0.80 to 1.39 w/o C.

Light load (17-30 gm) hardness measurements of selected martensite plates can limit the efffect of microcracking and retained austenite. The as quenched hardness of Fe-C martensite is plotted against (w/oC)$^{1/2}$[38,39] in Fig. 20 and shows good agreement with previous results.  A sharp discontinuity exists at the same carbon content where the change in morphology from lath to plate martensite occurs.[40], and it can be seen from these results that carbon content has a greater influence in plate martensite than lath martensite.

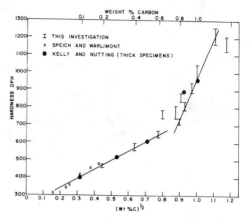

Fig. 20. The effect of carbon on the microhardness of martensite.

Strength:

The factors that have been considered to affect the strength of martensite are:

1. solution hardening by substitutional elements
2. solution hardening by interstitial elements
3. precipitation or segregation of carbon
4. substructure strengthening
5. grain size effects

The solution hardening effect was found to be 35 Pa per % Mn and 21.5 Pa per % Ni.[41] In general interstitial elements, specifically carbon, increase strength by the square root of the carbon content. Speich and Warlimont[42] found that yield strength of low carbon martensite obeyed the following relationship:

$$\sigma_{0.2} \text{ (MPa)} = 413.7 + 17.2 \times 10^5 \text{ (w/o C)}^{1/2} \qquad [8]$$

Roberts and Owen[43] have reported that in a series of Fe-Ni-C alloys, the plate martensite structure had a slightly greater slope in the above equation than the lath martensite. Chilton and Kelly[44] showed similar results, however, they indicated that carbon is more effective in raising the strength when it is out of solution than when it is dissolved interstitially. It has also been reported that the dislocation density of the transformed martensite increases linearly with carbon content which would give rise to a yield strength vs (carbon content)$^{1/2}$ relationship.[41,45]

In a study of the effect of lath width, packet size and prior austenite grain size on the strength of lath martensite in an Fe-0.2C alloy[46], it was shown that packet size was the dominant microstructural feature that controlled the strength, Fig. 21. Regardless of prior austenite grain size, the same distribution of lath widths was produced by the transformation, with the most frequent size being 0.2 μm. Thus lath size had no effect on strength. A similar result was found for Fe-Mn[41,47] and Fe-Ni[41] alloys. The strong effect of packet size was attributed to the interaction of segregated carbon and/or very fine carbides with the packet boundaries[48]. Upon tempering the strong dependency of yield strength on packet size was significantly diminished.

Norstrom[(41)] has developed a comprehensive equation for the yield strength of lath martensite:

$$\sigma_y = \sigma_o + \sigma_i + K_y D^{-1/2} + \alpha Gb[\rho \text{ tot}]^{1/2} \qquad [9]$$

where $\sigma_o$ is the friction stress for $\alpha$–iron, $\sigma_i$ is the solid solution hardening effect, $K_y$ is the Hall-Petch slope for the packet size D, $\alpha$ is a constant, G is the shear modulus, b is the burgers vecter, and $\rho$ tot is the total dislocation density. $\rho$ tot may be expressed as:

$$\rho \text{ tot} = \rho_o + K(\%C) + \frac{\theta}{b}\frac{2}{d} \qquad [10]$$

where $\rho_o$ is the dislocation density within the laths, K is a constant, %C is the carbon content, $\theta$ is the misorientation between laths and d is the lath width. The contribution to strength by the total dislocation density term, $\rho_o$, is made up of a contribution of dislocations in laths, lath boundaries and a carbon-content-controlled dislocation density.

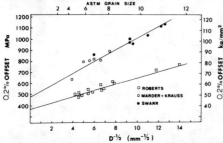

Fig. 21. Hall-Petch plots 0.2 pct offset yield strength vs $D^{-1/2}$ for Fe-0.2C (Marder and Krauss, Swarr) and Fe-Mn (Roberts) as-quenched martensites. D is the packet diameter.[(48)]

Fig. 22. Failure along a grain boundary and a packet boundary in an Fe-0.2 w/o C martensite. Light photomicrograph. Nital etch.

Toughness:

In a study of Fe-Mn alloys, Roberts[(47)] has shown that packet size also influences the impact transition temperature. For every 10 micron decrease in packet size, the transition temperature was decreased 20 C. The fracture of martensite occurs along packet boundaries and prior austenite grain boundaries as seen in Fig. 22. Packet size has also been directly related to the cleavage facets of low carbon tempered martensite fractured below its ductile to brittle transition temperature.[(49,50)]

## D.   Bainite

Two types of bainite, which are characterized by the temperature range in which they form, are commonly found in plain carbon steels. These two structures, called upper bainite and lower bainite, are distinguishable both morphologically and by the orientation relationship between the ferrite and carbide. Upper bainite contains ferrite laths which nucleate side-by-side in packets, figure 7d, and occurs at temperatures above 350°C in steels containing >0.6%C. Carbon is sufficiently mobile to diffuse to the austenite in front of the bainitic laths

so that the carbon enriched austenite can (1) be retained, (2) form high
carbon martensite or (3) form cementite between laths.[2]   The bainitic
ferrite laths become finer and the carbides at the lath boundaries become
more numerous, as the carbon content increases and transformation temperature decreases.  Lower bainite, however, is in the form of plates or
laths with an irrational habit plane, containing carbides within the
laths.  As the transformation temperature is lowered or the carbon
content increased, the ferrite laths become finer and the number of
carbide particles increase.  The lath boundaries within a packet in both
upper and lower bainite are usually low angle boundaries which can
prevent dislocation movement.  On the other hand, packet boundaries, or
the prior austenite grain boundaries, are high angle boundaries which
prevent crack propagation.[2]

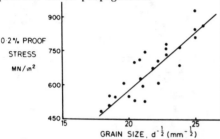

Fig. 23.   Effect of the bainitic
ferrite grain size on the 0.2%
proof stress.[51]

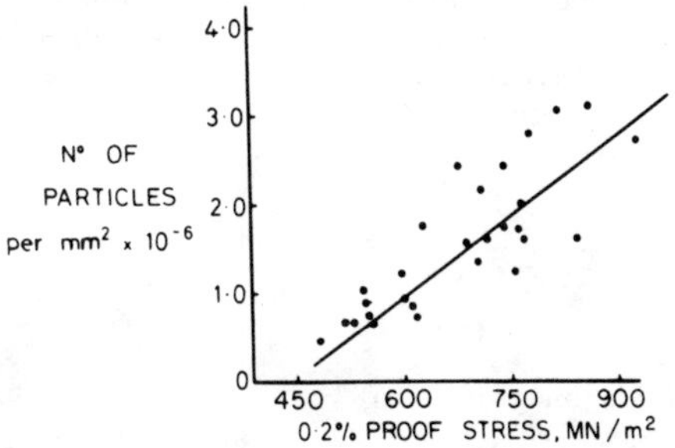

Fig. 24.   Effect of carbide
dispersion on the 0.2% proof
stress.[51]

Strength:

According to Pickering,[2] the increased strength in bainite is
due to the following structural characteristics:

1.  Bainitic lath size which gives a Hall-Petch relationship with
    proof stress, figure 23.

2.  Dislocation density, which increases as the transformation
    temperature decreases.

3.  Carbide disperion, which increases with decreasing trans-
    formation temperature and increasing carbon content, figure 24.

4.  Solid solution strengthening by carbon.

A multiple linear regression analysis gave the following equation
for the 0.2% proof stress:[51]

$$\sigma \ (MPa) = -194 + 17.4d^{-1/2} + 151.1 \ n^{1/4} \qquad [11]$$

Where d is the bainitic lath size in mm and n is the number of carbides
per mm$^2$.  This empirical equation shows a negative constant, which
indicates a threshold carbide distribution below which carbides do not
contribute to strength. As a result, carbides do not contribute to the
strength of upper bainite because they mainly are found at lath
boundaries.  Thus, only in lower bainite and in high carbon upper bainite,
is there a significant contribution from carbide dispersion strengthening.[2]

Other regression analyses include dislocation density but neglect bainitic lath size[52] or show a $d^{-1}$ relationship for lath size and strength.[53] Strength, as measured by hardness, has been related to the bainite packet size in a similar manner to the improvement in lath martensite strength by packet size,[54] however, a definitive study on the effect of packet size on the strength of bainite has not been reported.

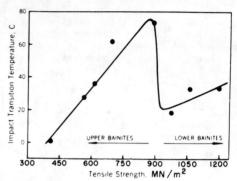

Fig. 25.  Effect of tensile strength and change from upper to lower bainite on impact transition temperature of low-carbon bainitic steels.[51]

Toughness:

It has been observed for bainites of different compositions that upper bainite has a lower impact toughness than lower bainite at comparable strength levels[55], Figure 25.  This is due to the fact that in upper bainite the large interlath carbides or high-carbon martensite areas crack to form a supercritical defect.  The cleavage crack will propagate because it is only obstructed by high angle packet boundaries or prior austenite boundaries[2].  In lower bainite, there is no easy brittle failure initiation since the smaller intralath carbides do not crack.  Once a cleavage crack is initiated, its propagation is obstructed by the many carbides and increased dislocation density.[2]

Naylor and co-workers[54, 56-58] have shown that the transition temperature improves when the packet size decreases, and that the quasi-cleavage facet size is the same as the packet diameter.[54]  An average fracture direction is maintained within a packet because a group of fracture planes, {100} {110} {112} {123}, from differently oriented laths, allow small angle deviations across the laths.[56]  At the next packet the group of fracture planes have a different orientation and therefore high angle deviations are required at packet boundaries.  It was also shown that when lath width decreases the transition temperature is again improved.[57]

Table 1 summarizes this section on the properties of ferrous transformation products.  For strength, it can be seen that dominant microstructural feature is a high angle boundary:  (1) grain boundary in ferrite, (2) cementite/ ferrite interface in pearlite, and (3) packet boundary (including the prior austenite grain boundary) in both martensite and bainite.  The low angle lath width boundary in martensite is considered to be part of the dislocation strengthening contribution.  In bainite, the lath width contribution was determined by a regression analysis that did not consider packet size and the relative contribution of lath width and packet size is yet to be determined.

Similar relationships hold for toughness. Again, the relative contribution for lath width and packet size in both martensite and bainite must be determined, but it appears that the predominant feature is the packet size since it correlates well with the cleavage facet size. The packet can maintain an average fracture direction because all the laths in a packet have approximately the same orientation. For pearlite the fracture unit depends upon a group of pearlite colonies which share a common ferrite orientation. Obviously in ferrite, the ferrite grain size is the structural feature that controls toughness. Therefore, for a given composition, improved toughness will result from reducing the size of the smallest structural unit that maintains a common ferrite orientation.

### TABLE I.  MICROSTRUCTURE AND MECHANICAL PROPERTIES*

| Transformation Product | Controlling Structural Unit | |
|---|---|---|
| | Strength | Toughness |
| Ferrite | Grain Size | Grain Size |
| Pearlite | Interlamellar Spacing | Group of Colonies |
| Martensite (Lath) Bainite | Packet Size (lath width) | Packet Size (lath width) |

* For a given composition

### IV  THE PROPERTIES OF COMBINED MICROSTRUCTURES

Many commercial materials have mixed microstructures and an understanding of the contributing effects of each constituent on the resultant mechanical properties is important from the standpoint of alloy design. The various factors that influence the mechanical properties of multiphase steel alloys can be listed as follows:[59]

1. Volume fraction of each phase:  $f_a$, $f_b$,..., $f_n$

2. Grain size of each phase:  $d_a$, $d_b$,..., $d_n$

3. Shape and distribution of phases.

4. Flow stress ratio of the phases:  $C^* = \dfrac{\sigma_s^a}{\sigma_s^b}$

5. Energy or strength of the interface.

When yield strength is determined by the onset of plastic flow of the soft phase then it will be independent of f and $C^*$. However, when engineering yield strength of the alloy is considered, $\sigma_s$ is dependent on $C^*$ and f. Tamura, et al[60] found that in the region where f is small, $\sigma_s$ is independent of $C^*$. Conversely, for a low $C^*$ the following law of mixtures holds:

$$\sigma_s = \sigma_s^a (1-f_b) + \sigma_s^b f_b \qquad [12]$$

On the other hand when $C^*$ is large, $\sigma_s$ deviates considerably from linearity at large f values.

A. Ferrite & Pearlite

The following yield strength relationship was found to be applicable to the entire range of pearlite contents:[61]

$$\sigma_s \text{ (MPa)} = f_\alpha^{1/3} [35.4 + 58.5 \text{ (w/o Mn)} + 17.4d^{-1/2}] + (1 - f_\alpha)^{1/3}$$

$$[178.6 + 3.9 S_o^{-1/2}] + 63.1 \text{ (w/o Si)} + 425 \text{ (w/o } N_f)^{1/2} \qquad [13]$$

where $f_\alpha$ is the volume fraction of ferrite, d is the grain diameter of the ferrite in mm, $S_o$ is the interlamellar spacing of pearlite in mm, and $N_f$ is the free nitrogen. A similar equation was developed for the tensile strength. An exponent of 1/3 is used for the volume fraction because at low pearlite contents the amount of pearlite does not significantly affect strength. As the pearlite content increases to 100%, the tensile strength becomes increasingly dominant, figure 26, and the effect of interlamellar spacing becomes apparent, figure 27. The effect of increasing amount of pearlite on the components of yield strength, using eq. 13, can be seen in figure 28.[61] In this case, the interlamellar spacing of the pearlite was assumed to be constant.

The 20 ft-1b transition temperature can be represented by the following regression equation:[61]

$$I(°C) = f_\alpha [-46 -11.5d^{-1/2}] + (1 - f_\alpha)[-335 + 5.6 S_o^{-1/2}$$

$$-13.3p^{-1/2} + 3.48 \times 10^6 t] + 48.7 \text{ (w/o Si)} + 762 \text{ (w/o N)}^{1/2} \qquad [14]$$

where the symbols are the same as for eq. 13, p is the pearlite colony size in mm, and t is the pearlite-cementite plate thickness in mm.

This equation shows that as pearlite content increases the impact transition temperature increases, fig. 29. A small pearlite colony size, which in the pure pearlite studies[31] was related to the effective grain size or groups of colonies, decreases the impact transition temperature by impeding crack propogation, figure 30. Similarly, a refined ferrite grain size is beneficial to toughness. Other pearlite morphological features seem to play a role in this regression equation: decreased interlamellar spacing is detrimental but decreased pearlite-cementite plate thickness is beneficial to toughness. Since pearlite spacing and plate thickness are interrelated, there will be an optimum spacing for the best impact properties.

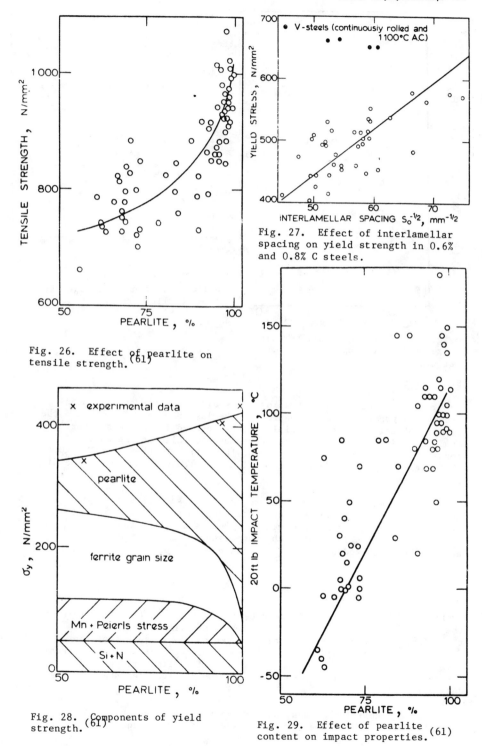

Fig. 26.  Effect of pearlite on tensile strength.[61]

Fig. 27.  Effect of interlamellar spacing on yield strength in 0.6% and 0.8% C steels.

Fig. 28.  Components of yield strength.[61]

Fig. 29.  Effect of pearlite content on impact properties.[61]

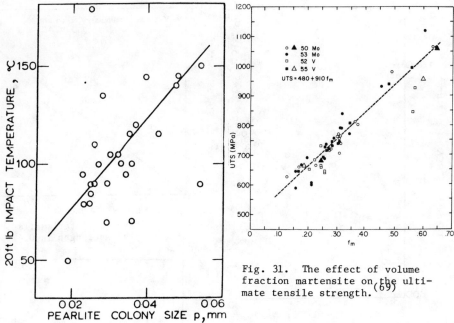

Fig. 30.  Effect of pearlite colony size on impact properties in 0.6% C steels.[61]

Fig. 31.  The effect of volume fraction martensite on the ultimate tensile strength.[69]

## B.  Ferrite & Martensite

Ferrite and martensite alloys, a.k.a. dual phase steels[62-64] have received a great deal of attention in the last few years and the details of this type of material will be covered in the next paper.[65] Dual phase steels can be characterized by the following features: (a) high tensile strength together with a low yield strength or a Y.S./T.S ≤0.6, (b) absence of a yield point elongation, and (c) high ductility and high work hardening rate.[66]

The strength of dual-phase steels is a result of strong martensite acting as the load-carrying constituent and contained within a ductile ferrite matrix.  It has experimentally been shown that the tensile strength increases linearly with increasing volume fraction of martensite and is characterized by the "law of mixtures":[67]

$$\sigma_c = \sigma_\alpha + (\sigma_m - \sigma_\alpha) f_m \qquad [15]$$

where $\sigma_\alpha$ is the stress in the ferrite at the ultimate tensile strain of the martensite, $\sigma_m$ is the tensile strength of martensite, and $f_m$ is the volume fraction of martensite.  It can easily be seen that the previously developed equations for the strength of ferrite and martensite [e.g., eq.(3) and (9)] could be used in eq. 15.  Similar type expressions for the strength of dual-phase steels have been developed that fit the data very well.[67, 68]

A more simplified expression for the tensile strength ($\sigma_{TS}$) of dual phase steel takes the form of a linear regression equation:

$$\sigma_{TS} = \sigma_o + K_m f_m \qquad [16]$$

where $\sigma_o$ and $K_m$ are constants and $f_m$ is the volume fraction martensite. A typical result is found in figure 31. The reported coefficient for martensite strengthening, $K_m$, range between 8.5 and 13.0 MPa/% martensite[69-72] and takes into account carbon content, particle or packet size, dislocation strengthening, and substitutional solid solution strengthening of the martensite. To a first approximation these values are all constant in dual phase steels. The value for $\sigma_o$ varies between 425 and 480MPa[69-72] and includes the solid solution strengthening and grain size effects in ferrite. Again, these values are also constant.

During deformation it has been estimated that the martensite phase carries approximately 30 percent of the load when it occupies 16 per cent of the volume.[73] After comparing the average true stress in martensite at the limit of uniform extension (1414MPa) with the reported yield strength of medium carbon martensite (1138-1654MPa), Szewczyk and Gurland[73] concluded that extensive plastic deformation of martensite would occur at high strains near and in the necked region. It was found that tensile fracture in dual phase steel occurred by void formation after necking at small inclusions and at the ferrite/martensite interface.[69] Furthermore, observable deformation of the martensite was not found until the very high strains that are produced during necking, figure 32, and at no time was martensite found to be cracked.

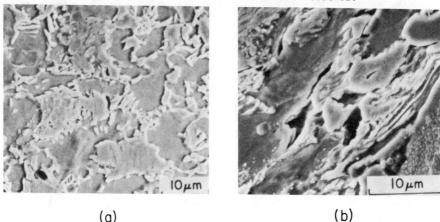

(a)                                    (b)

Fig. 32. Air cooled Mo alloy 50 (a) after necking $\varepsilon$ = 0.215 and (b) after failure $\varepsilon$ = 0.26.[69]

C.  Ferrite & Bainite

Very few studies have been made on the ferrite + bainite microstructure. The few studies that have been conducted have been concerned with the replacement of martensite by bainite in dual phase steels. Rigsbee et al[74] reported that increasing the volume percent bainite increases the ultimate tensile strength and decreases % elongation. The bainite strengthening factor was 1.0 MPa/% bainite as compared to a martensite factor of 8.5 to 13.0 MPa/%, martensite.[69-72]

Recent work[75] has shown that the Y.S./T.S. ratio of ferrite +

bainite is greater than 0.8, a value similar to ferrite + pearlite, which is indicative of discontinuous yielding, but considerably higher than ferrite + martensite structures (0.5). This effect was due to the lack of solute carbon in the F + P and F + B structures in which all the carbon was precipitated as carbide. The F + M dual phase steel had most of the carbon in solution. On the other hand, Sudo et.al.[75] reported that F + B steels had the lowest transition temperature and highest shelf energy in impact testing.

### D.   Martensite & Bainite

The martensite + bainite microstructure has also received little attention. However, unlike the dual phase ferrite structures in which a diffusional ferrite transformation occurs prior to the second phase transformation, the bainite transformation effectively partitions the prior austenite grain before martensite forms,[76] figure 33. It has been found that for optimum impact and tensile properties the bainite constituent should be around 10-25 volume percent.[77,78]

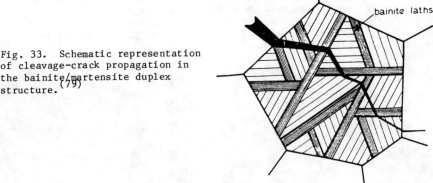

bainite laths

Fig. 33.   Schematic representation of cleavage-crack propagation in the bainite/martensite duplex structure.[79]

Recently,[78] an equation for the strength of martensite + bainite microstructures was proposed, which modifies the law of mixtures concept.

$$\sigma_{0.2} = \sigma_i + kS_m^{-1/2} - (\sigma_i + kS_m^{-1/2} - \sigma_{0.2}^B)V_B \qquad [17]$$

where $\sigma_i$ is the friction stress for martensite, $S_m$ is the average martensite packet size, k is a constant, $V_B$ is the volume fraction bainite and $\sigma_{0.2}^B$ is the 0.2 pct proof stress of bainite. If the bainite is strained to the strength of martensite in the early stages of deformation, $\sigma_{0.2}^B = \sigma_{0.2}^M = \sigma_i + KS_m^{-1/2}$. Eq.(17) can be rewritten as:

$$\sigma_{0.2} = \sigma_i + kS_m^{-1/2} \qquad [18]$$

In general, it was found that tensile data for specimens containing up to 25 volume percent bainite correlated well with eq (18), because the bainite effectively partitions the prior austenite producing a smaller packet size.

The improvement in toughness of these structures is also a result of the partitioning of the prior austenite grain,[77] causing a reduction in the mean free path for cleavage cracks, called the unit crack path.[76]  The impact transition temperature had a better correlation with the unit crack path than with the austenite grain size, thus the initially formed bainite laths partition the austenite grains and reduce the effective grain size for fracture.[79]  In addition, lower bainite provides increased fracture resistance to brittle fracture at lower temperature which further improves the fracture ductility and notch toughness of the martensite + bainite composite.[78]

E.   Ferrite + Bainite + Martensite

It has been reported that additions of martensite to ferrite-bainite steels eliminate the yield point elongation and lower the Y.S./T.S. ratio.[80]  Recently, it was shown that F+B+M structures have tensile properties intermediate between F+M and F+B, figure 34.[81] The effect of additions of bainite on F + M steels containing between 4 and 11% martensite is seen in figure 35.  In agreement with previous results increasing the amount of bainite lowers elongation.[74] However, in samples with up to 20% bainite there does not appear to be any significant effect on the strength or Y.S./T.S. ratio of the F + M alloys.  As reported by Sudo and Iwai,[81] the advantage of the F + B + M structure is better stretch – flangeability than the F + M steel and also a better fatigue strength and toughness.  The deformability of the F + B + M steel is improved by reduction of void formation due to the smaller size of the martensite particles and less cohesion of the ferrite/ bainite interface. Bainite is much softer than the martensite, therefore the bainite can deform almost as easily as the ferrite matrix.

Fig. 34.  Effects of micro-structures on the mechanical properties of Mn-Si-Cr steels in the hot-rolled and controlled-cooled state.[81]

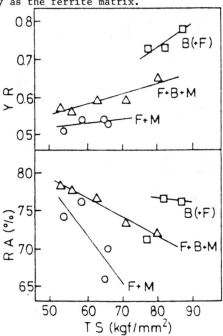

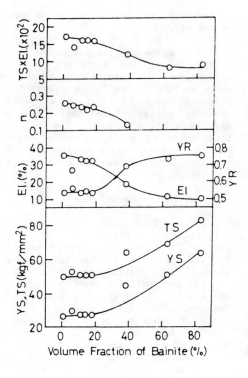

Fig. 35.  Effects of bainite content on the mechanical properties of steel No. 3 which contains martensite content of 4 to 11%.[81]

## F.  Ferrite + Pearlite + Martensite

The effect of various microstructures on the strength of intercritically annealed C-Mn-Si alloys, shows that the properties of the F + P + M alloys are intermediate between F + M and F + P alloys, figure 36.

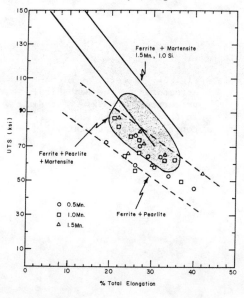

Fig. 36.  The strength ductility relationship in several multiphase alloys.

In all cases the introduction of pearlite produced discontinuous yielding. Although there was no correlation between the Y.S./T.S. ratio and the addition of pearlite, increased martensite content lowered this ratio even in the presence of pearlite, figure 37. Therefore, as shown [82] previously for the dual phase F+ M alloys, increased volume percent martensite introduces mobile dislocations eventually allowing for the onset of continuous yielding at lower flow stresses.  For the small values of pearlite (<20%) evaluated, yield strength was not affected by pearlite content, in agreement with the previously reported results on F + P alloys.[83,84]

Fig. 37.  The effect of martensite on the Y.S./T.S. ratio in ferrite-pearlite-martensite steels.

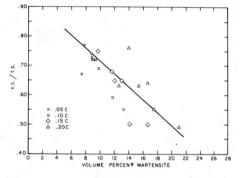

The following regression equations were developed for a similar set of C-Mn-Si alloys:[70]

$$\sigma_{3\%} \ (\text{MPa}) = 273.1 + 13.2 \ (\%M) + 5.2 \ (\%P) \qquad [19a]$$

$$\sigma_{TS} \ (\text{MPa}) = 443.2 + 12.1 \ (\%M) + 4.7 \ (\%P) \qquad [19b]$$

It was concluded from this study that the effects of volume percent martensite and pearlite are additive and little interaction occurs.  From eqs. 19a and 19b, martensite is about 2.5 times as effective in increasing strength as pearlite.  It was also found that martensite had a more deleterrous effect on both uniform and total elongtion than does pearlite on a volume percent basis.  The constant in both equations contains the values for grain size strengthening and solid solution hardening and can be assumed to be similar to the other regression equations.[83,84]

To a first approximation the law of mixtures seems to govern the strength of steels with composite multiphase microstructures .  The strength of these materials can be characterized by empirical equations that take into account the proportional contribution of each individual phase.  In all cases, except for the martensite + bainite alloys, the structure contains a hard phase, e.g., martensite, bainite or pearlite, surrounded by a soft matrix of ferrite.  For an equivalent volume per-cent second phase, as the hardness or strength of the second phase increase, the more difficult it is to deform, and the higher the strength of the composite.  For example, in the highest strength F + M alloy, the second phase martensite only deforms after necking in the tensile test. For martensite + bainite alloys, the softer phase bainite at low volume percent, serves to partition the prior austenite grains prior to the

martensite transfrormation.  The resulting effect is a much smaller
packet size than if bainite did not form first and improved strength and
toughness of the composite.

A low value of the Y.S./T.S. ratio is indicative of continuous
yielding in these composite materials.  Discontinuous yielding was found
in ferrite + pearlite and ferrite + bainite alloys, but continuous
yielding occurred in all composites containing above a critical amount
of martensite.  For example, 4-11% martensite had to be added to ferrite +
bainite steels in order to get continuous yielding, and up to 20%
martensite was added to ferrite + pearlite composites before continuous
yielding occurred.  Thus, enough carbon must be in solution in the
martensite so that the dislocations generated during the transformation
will be mobile enough to produce continuous yielding.

## V   GENERAL SUMMARY

The mechanical properties of individual transformation products and
composite multiphase alloys have been reviewed.  Specific morphological
features of each transformed microstructure influence properties.  For
strength the dominant microstructural feature is the high angle boundary:
(1) the grain boundary in ferrite, (2) the cementite/ferrite interface in
pearlite, and (3) the packet boundary (including the prior austenite grain
boundary) in both martensite and bainite.  For toughness, the structural
feature that maintains a common ferrite orientation is most important:
(1) the grain in ferrite, (2) the group of colonies in pearlite, and
(3) the packet of martensite or bainite.

Combinations of transformation products to form composite multiphase microstructures can influence mechanical properties in various
ways.  Strength is characterized by empirical equations that take into
account the proportional contribution of each individual phase, i.e.,
to a first approximation a modification of the law of mixtures.  In most
cases these composites consist of a hard phase surrounded by a matrix
of soft ferrite.  Strength at equivalent volume percent second phase
increases with the hardness of the second phase.  Continuous yielding
only occurs in multiphase alloys that contain martensite because;
(1) the martensite transformation generates a high density of dislocations and (2) the carbon can be contained in solid solution in the
martensite, thus reducing the carbon pinning of dislocations generated
during transformation.

## ACKNOWLEDGEMENTS

I am grateful to my colleagues A. O. Benscoter, E. T. Stephenson,
B. L. Bramfitt, and G. Krauss for their helpful participation in some
of the research reported in this paper.  Many thanks to Isabella Hartz
for her careful typing of the manuscript.

REFERENCES

1.  E. Hornbogen, "Strenthening Mechanisms in Steel," in Steel-Stengthening Mechanisms, Climax Molybdenum Co., Zurich, 1969, p. 1.
2.  F. B. Pickering, Physical Metallurgy and the Design of Steels, Applied Science Publishers Ltd., London, 1978, Ch. 1.
3.  E. Orowan, Internal Stresses in Metals and Alloys, Institute of Metals, London, 1948, p. 451.
4.  W. R. Tyson, Acta Met, 1963, Vo. 11, p. 61.
5.  M. F. Ashby, Acta Met, 1966, Vol. 14, p. 679.
6.  A. S. Keh, Direct Observations of Imperfections in Crystals, Interscience, New York, 1962, p. 213.
7.  N. J. Petch, Fracture, Proc. of Swampscott Conf., J. Wiley, 1959, p. 54.
8.  B. L. Bramfitt, A. O. Benscoter, J. R. Kilpatrick, and A. R. Marder, Metallography - A Practical Tool for Correlating the Structure and Properties of Materials, ASTM STP 557, 1974, p. 43.
9.  B. L. Bramfitt and A. R. Marder, Met Trans, Vol. 4, 1973, p. 2291.
10. A. R. Marder and G. Krauss, Trans ASM, Vol. 62, 1969, p. 957.
11. G. Krauss and A. R. Marder, Met Trans, Vol. 2, 1971, p. 2343.
12. J. M. Marder and A. R. Marder, Trans ASM, Vol. 62, 1969, p. 1.
13. A. R. Marder, A. O. Benscoter and G. Krauss, Met Trans, Vol. 1, 1970, p. 1545.
14. E. O. Hall, Proc. Phys. Soc., Vol. 64B, 1951, p. 747.
15. N. J. Petch, J. Iron Steel Inst., Vol. 174, 1953, p. 25.
16. W. B. Morrison, Trans ASM, Vol. 59, 1966, p. 824.
17. R. L. Miller, Met Trans, Vol. 3, 1972, p. 3047.
18. B. L. Bramfitt and A. R. Marder, Processing and Properties of Low Carbon Steel, J. M. Gray, ed., AIME, 1973, p. 191.
19. N. J. Petch, Phil. Mag., Vol. 3, 1958, p. 1089.
20. A. R. Marder and B. L. Bramfitt, Met Trans, Vol. 7A, 1976, p. 365.
21. J. M. Hyzak and I. M. Bernstein, Met Trans, Vol. 7A, 1976, p 1217.
22. T. Takahashi and M. Nagumo, Trans Jap. Inst. Metals, Vol 11, 1970, p. 113.
23. T. Gladman, I. McIvor, and F. Pickering, J. Iron Steel Inst., Vol. 210, 1972, p. 916.
24. Z. Mikulec, I. Gottwaldova, and J. Mrovec, Kovove Materialy, Vol. 16, 1978, p. 600.
25. J. G. Sevillano, Strength of Metals and Alloys, Vol. 2, Pergamon Press Ltd., Oxford, England, 1980, p. 819.
26. A. Sunovoo, M. E. Fine, M. Meshii, and D. H. Stone, Met Trans, Vol. 13, 1982, p. 2035.
27. A. W. Thompson, Met. Trans, Vol. 8, 1977, p. 833.
28. G. Langford, Met Trans, Vol. 8, 1977, p. 861.
29. J. H. Gross and R. D. Stout, Weld J., Vol. 30, 1951, p. 4815.
30. J. Fluegge, W. Heller, E. Stolte and W. Dahl, Arch. Eisen, Vol. 47, 1976, p. 635.
31. Y. J. Park and I. M. Bernstein, Met. Trans., Vol. 10A, 1979, p. 1653.
32. D. J. Alexander and I. M. Bernstein, Met Trans., Vol. 13A, 1982, p. 1865.
33. T. Takahashi, M. Nagumo, and Y. Asano, Journ. Jap. Inst. Met., Vol. 42, 1978, p. 716.
34. R. J. Dippenaar and R. W. K. Honeycombe, Proc. Roy Soc., Vol A333, 1973, p. 455.
35. B. Karlsson and G. Linden, Mater Sci. Eng., Vol. 17, 1975, p. 153.
36. S. Nisida, T. Urashima, K. Sugino and H. Masumoto, Strength of Metals and Alloys, Vol. 2, 1CSMA5, 1979, p. 1255.
37. G. T. Gray, III, A. W. Thompson, J. C. Williams, and D. H. Stone, Canadian Met. Q, Vol. 21, 1982, p. 73.

38.  P. M. Kelly and J. Nutting, Proc. Roy. Soc., Vol. 259A, 1960, p. 45.
39.  G. R. Speich and H. Warlimont, J. Iron Steel Inst., Vol. 203, 1968,
     p. 385.
40.  A. R. Marder and G. Krauss, Trans ASM, Vol 60, 1967, p. 651.
41.  L. A. Norstrom, Scand. J. Metallurgy, Vol. 5, 1976, p. 159.
42.  G. R. Speich and H. Warlimont, J. Iron Steel Inst., Vol. 206, 1968,
     p. 385.
43.  M. J. Roberts and W. S. Owen, Physical Metallurgy of Martensite and
     Bainite, ISI Spec. Rep. 93, 1965, p. 53.
44.  J. M. Chilton and P. M. Kelly, Acta Met, Vol. 16, 1968, p. 637.
45.  M. Kehoe and P. M. Kelly, Scripta Met, Vol. 4, 1970, p. 473.
46.  A. R. Marder and G. Krauss, Strength of Metals and Alloys, Vol. III,
     2nd Int. Conf., ASM, Asilomar, Calif. 1970, p. 822.
47.  M. J. Roberts, Met Trans, Vol. 1, 1970, p. 3287.
48.  T. Swarr and G. Krauss, Met Trans, Vol 7A, 1976, p. 41.
49.  T. Inou, S. Matsuda, Y. Okamura and K. Aoki, Trans Jap. Inst. Metals,
     Vol. 11, 1970, p. 36.
50.  S. Matsuda, T. Inoue, H. Mimura and Y. Okamura, Trans Iron Steel
     Inst. Jap., Vol.12, 1972, p. 325.
51.  R. W. K. Honeycombe and F. B. Pickering, Met Trans, Vol. 3, 1972,
     p. 1099.
52.  M. E. Bush and P. M. Kelly, Acta Meta, Vol. 19, 1971, p. 1363.
53.  D. W. Smith and R. F. Hehemann, J. Iron Steel Inst., Vol. 209,
     1971, p. 476.
54.  J. P. Naylor and P. R. Krahe, Met. Trans., Vol. 5, 1974, p. 1699.
55.  K. J. Irvine and F. B. Pickering, J Iron Steel Inst., Vol. 201,
     1963, p. 518.
56.  J. P. Naylor and P. R. Krahe, Met Trans, Vol. 64, 1975, p. 594.
57.  J. P. Naylor and R. Blondeau, Met Trans, Vol. 7A, 1976, p. 891.
58.  J. P. Naylor, Strength of Metals and Alloys, 4th Int. Conf., Vol. 2,
     Nancy, France, 1976, p. 503.
59.  Y. Tomota and I. Tamura, Trans Iron Steel Inst. Jap., Vol. 22, 1982,
     p. 665.
60.  I. Tamura, Y. Tomota, A. Akao, Y. Yamaoka, M. Ozawa and S. Kanatani,
     Trans Iron Steel Inst. Jap., Vol. 13, 1973, p. 283.
61.  T. Gladman, I. D. McIvor and F. B. Pickering, J. Iron Steel Inst.,
     Vol. 210, 1972, p. 916.
62.  Formable HSLA and Dual Phase Steels, A. T. Davenport, ed., TMS/AIME,
     New York, 1979.
63.  Structure and Properties of Dual-Phase Steels, R. A. Kot and
     J. W. Morris, eds., TMS/AIME, Warrendale, PA, 1979.
64.  Fundamental of Dual-Phase steels, R. A. Kot and B. L. Bramfitt, eds.,
     TMS/AIME, Warrendale, PA, 1981.
65.  J. M. Rigsbee, Phase Transformations in Ferrous Alloys, A. R. Marder
     and J. I. Goldstein, ed., TMS/AIME, Warrendale, PA, 1984, p.
66.  A. R. Marder, Fundamentals of Dual-Phase Steels, R. A. Kot and
     B. L. Bramfitt, eds., TMS/AIME, Warrendale, PA, 1981, p. 145.
67.  J. Y. Koo, M. J. Young and G. Thomas, Met Trans, Vol. 11A, 1980,
     p. 852.
68.  G. R. Speich and R. L. Miller, Structure and Properties of Dual-Phase
     Steels, R. A. Kot and J. W. Morris, ed., TMS/AIME, Warrendale, PA,
     1981, p. 113.
69.  A. R. Marder, Met Trans, Vol. 13A, 1982, p. 85.
70.  S. S. Hansen and R. R. Pradhan, Fundamentals of Dual Phase Steels,
     R. A. Kot and B. L. Bramfitt, eds., TMS/AIME, Warrendale, PA, 1981,
     p. 113.
71.  W. R. Cribb and J. M. Rigsbee, Structure and Properties of Dual Phase
     Steels, R. A. Kot and J. W. Morris, eds., TMS/AIME, Warrendale, PA,
     1979, p. 91.
72.  J. H. Bucher, E. G. Hamburg and J. F. Butler, ibid., p. 346.

73. A. F. Szewezyk and J. Gurland, Met Trans, Vol. 13A, 1982, p. 1821.
74. J. M. Rigsbee, J. K. Abraham, A. T. Davenport, J. E. Franklin and J. W. Pickens, Structure and Properties of Dual-Phase Steels, R. A. Kot and J. W. Morris, eds., TMS/AIME, Warrendale, PA, 1979, p. 304.
75. M. Sudo, S. Hashimoto and S. Kambe, Trans. Iron Steel Inst. Jap., Vol. 23, 1983, p. 303.
76. H. Ohtani, F. Terasaki and T. Kunitake, Trans Iron Steel Inst. Jap., Vol. 12, 1972, p. 118.
77. T. Kunitake, F. Terasaki, Y. Ohmori and H. Ohtani, Iron and Steel, Vol. 45, 1972, p. 647.
78. Y. Tomita and K. Okabayashi, Met Trans, Vol. 14A, 1983, p. 485.
79. Y. Ohmori, H. Ohtani, and T. Kunitake, Metal Science, Vol. 8, 1974, p. 357.
80. M. Sudo, M. Higashi, H. Hori, T. Iwai, S. Kambe and Z. Shibata, Trans Iron Steel Inst. Jap., Vol. 21, 1981, p. 820.
81. M. Sudo, and T. Iwai, ibid, Vol. 23, 1983, p. 294.
82. A. R. Marder, Met Trans, Vol. 12A, 1981, p. 1569.
83. F. B. Pickering and T. Gladman, Metallurgical Development in Carbon Steel, Iron and Steel Inst. Special Report 81, London, 1963, p. 10.
84. J. D. Grozier and J. H. Bucher, J. Materials, Vol. 2, 1967, p. 393.

# NUCLEATION OF AUSTENITE DURING INTERCRITICAL ANNEALING OF A

## COMMERCIAL LOW-ALLOY STEEL

S. W. Thompson*, G-S. Fan** and P. R. Howell*

*Metallurgy Program
Department of Materials Science and Engineering
College of Earth and Mineral Sciences
The Pennsylvania State University
University Park, PA 16802, U.S.A.

**Department of Mechanical Engineering
Nanjing Institute of Technology
The People's Republic of China

The nucleation of austenite in an hot-rolled, continuously-cooled steel containing 1.5 wt.% Mn-0.3 wt.% Si-0.11 wt.% C has been investigated using transmission electron microscopy. The results indicate that austenite nucleates predominately at ferrite/pearlite interfaces, pearlite/pearlite interfaces and cementite at ferrite grain boundaries. In addition, austenite nucleated in the ferritic matrix after longer annealing times, and was often retained upon cooling to room temperature. Grain boundary austenite was Widmanstätten in nature, while intragranular austenite displayed a rod-like morphology; both morphologies of austenite were related to the ferrite by the Kurdjumov-Sachs orientation relationship. For this steel composition, the results suggest that cementite particles are of primary importance for austenite nucleation while reduction in interfacial energy is of secondary importance.

Interest in dual-phase steels, that are produced by intercritically annealing ferrite/pearlite (and/or carbide) aggregates, should be accompanied by renewed interest in austenite nucleation. This is true since austenite formation during intercritical annealing is incomplete (i.e. does not go to 100% austenite) and, therefore, the distribution of austenite is an important factor in determining dual-phase properties.

Although Walldow (1) seems to have been the first to have noticed the effect of the starting structure on the location of austenite nucleation sites, later studies (2,3) provided firm evidence for the active nucleation sites in accord with accepted solid-solid nucleation theory (4). A more recent study (5) of dual-phase steels supports this earlier work. As part of an overall examination of the formation of austenite during intercritical annealing, the nucleation of austenite has been investigated using transmission electron microscopy (TEM).

## Experimental Methods

An hot-rolled, continuously-cooled steel plate was obtained from United States Steel in Monroeville, PA, with the following composition (in wt.%): Fe- 0.11C- 1.48Mn- 0.30Si- 0.002P- 0.002S- 0.008Al. Intercritical heat treatments were conducted at 740°C for times ranging from 30 s. to 5 hr. and then quenched in water. Thin foil TEM specimens were prepared from 3 mm. diameter discs, and examined in a Phillips EM 300 operating at 100 kV.

## Results and Discussion

A preliminary investigation of the as-received plate has been conducted, and the important observations are: (I) the microstructure consisted of alternate bands of proeutectoid ferrite and pearlite; (II) the proeutectoid ferrite contained a high dislocation density, and evidence for recovery was extensive; (III) the pearlite regions exhibited complex structures and pearlitic cementite often adopted a discrete, rather than lamellar, morphology; (IV) cementite precipitation occurred at ferrite grain boundaries, and to a lesser extent, in the ferritic matrix. From the evaluation of the starting structure, we may suggest several possible nucleation sites for austenite during intercritical annealing. These sites are: (A) ferrite/pearlite interfaces; (B) pearlite/pearlite interfaces; (C) grain boundary cementite/ferrite interfaces; (D) cementite/ferrite interfaces within pearlite colonies; (E) intragranular cementite/ferrite interfaces; (F) ferrite grain boundaries; (G) imperfections in the ferritic matrix; (H) the ferritic matrix (i.e. homogeneous nucleation).

Of the eight possible nucleation sites, it was found that there are three predominant sites while one other became active after longer annealing times. Figures 1-3 illustrate the three major sites: figure 1 is an example of nucleation at a ferrite/pearlite interface (site A), while figure 2 shows nucleation at pearlite/pearlite interfaces (site B). Figure 3 shows nucleation at a ferrite grain boundary (most likely in conjunction with cementite - site C). From these and other observations, it can be concluded that all three are very potent sites for the nucleation of austenite. Also, it is of interest to note that in figure 1 the austenite is allotriomorphic, while figure 3 is an example of the developement of a Widmanstätten morphology: this suggests that an orientation relationship is developed between the austenite and ferrite.

While the three nucleation sites mentioned were predominant, another site became evident after examining specimens annealed for longer times.

Figure 4 shows austenite which had nucleated in the ferritic matrix and has been retained to room temperature (this is substantiated by the stacking faults observed in one of the particles).  From this and other observations, it can be suggested that the austenite adopts a rod-like morphology.  In addition, selected area diffraction pattern analyses indicated that the orientation relationship developed between austenite and ferrite is that proposed by Kurdjumov and Sachs (6) for both the Widmanstätten and rod-like austenite.

Classical nucleation theory (4) has been employed qualitatively in an attempt to rationalize why the observed sites were active.  These and other sites have been schematically depicted by Speich and Szirmae (3) and by Paxton (7). Roberts and Mehl (8) emphasized the need for a cementite particle to nucleate austenite, and it is generally accepted that austenite nucleation at ferrite grain boundaries is associated with cementite (5), although direct observation is infrequent (3).  This is a reasonable deduction since in the present investigation it has been found that:  (I) at short annealing times, all the precipitates at grain boundaries were austenite (no grain boundary cementite was detected after 1 min. at temperature); (II) many ferrite grain boundaries are free from both austenite and cementite; (III) it is proposed that it would be difficult for the interfacial free energy to be sufficiently reduced at a grain boundary for times of about 1 min., and so the driving force for nucleation must be large (i.e. cementite must be present) for nucleation to occur.  Therefore, sites A,B and C are associated with cementite and have a similar volume free energy term ($\Delta$Gv) that is large.  Sites A through C are all located at the triple junction between a cementite particle and two ferrite grains, which gives a large reduction in the surface free energy term.  Therefore, the sites are equivalent geometrically and as far as the free energy of

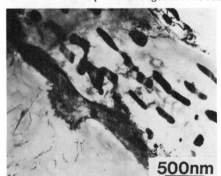

Nucleation at site A - see text. (TEM)

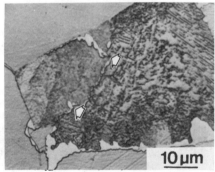

Nucleation at site B (arrowed).
Optical Micrograph.

Nucleation at site C. (TEM)

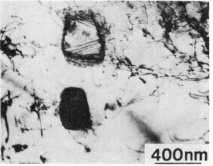

Nucleation at site D. (TEM)

formation for a critical nucleus ($\Delta G^*$) is concerned. The large values of the steady-state nucleation rates at sites A-C, coupled with the observed fast dissolution rate of pearlite (see also Speich, Demerest and Miller (9)) suggests that these sites are expected to be eliminated from the microstructure within the first minute at temperature. Note that it is assumed that site C is associated with a cementite particle. Hence, eliminating this site does not imply that all ferrite grain boundaries are consumed by austenite. In fact, some boundaries are precipitate free even after five hours at temperature.

After pearlite dissolution, further austenite growth occurs into the ferrite at an extremely slow rate (also see ref. (9)). At this time (30 min.), site D appears to be activated, having a time dependent nucleation rate that is extremely slow and approaches steady-state very slowly. This site would be difficult to detect if the overall transformation to austenite was rapid (i.e. such a slow rate would lead to so few intragranular nucleation events that detection by TEM would be very difficult). As stated earlier concerning austenite formation at ferrite grain boundaries, cementite is most likely present. Concerning site D, it is proposed that the nucleation catalyst is a cementite particle in the matrix. Two points should be made concerning this site. First, the observed rod-like morphology suggests that these austenite particles nucleated in the ferritic matrix. Second, cementite should be even more important for matrix nucleation compared to grain boundary nucleation, since there is no reduction in the interfacial free energy in the matrix without cementite. If site D is not associated with cementite, nucleation would be easier at clean grain boundaries. Although this may occur (no attempt has been made to show a one-to-one correspondence between grain boundary cementite before annealing and grain boundary austenite particles after annealing), there are still numerous clean boundaries even after 5 hours. For these reasons, it is expected that matrix nucleation of austenite occurs on cementite particles. At 740°C these cementite particles are certainly dissolving to feed the growing austenite that has nucleated elsewhere, but it is believed that complete dissolution has not occurred.

Matrix nucleated austenite has been observed by others, in particular by Judd and Paxton (2), in spheroidized, plain carbon steels, where cementite had been freed from grain boundaries. In fact, using simplifying assumptions, they determined an incubation time ($\tau$) of about 30 sec. at 750°C. More recently, Furukawa et. al. (10) mentioned austenite in the ferritic matrix and showed micrographs indicating that the austenite is rod-like in morphology, but they neglected to mention the starting structure. The steel they used had a similar composition to that used in the present study, but annealing was done at 750°C for only 2 minutes before cooling. The above results suggest that site D could be activated well before the 30 minute incubation time that we have suggested. Although determining $\tau$ should be difficult, we can surely conclude that the incubation time for sites A,B and C (at grain boundaries) is less than that for site D (i.e. $\tau$ g.b. $< \tau$ matrix). Judd and Paxton (2) provide the approximation that $\tau$g.b. is about one-fourth to one-third of $\tau$ matrix for the steels studied at 750°C. Comparison of these results can only be approximate considering the wide variation in compositions and starting structures. From the above considerations, it is suggested that in the steel studied at 740°C, austenite nucleation depends to a great extent on $\Delta G_v$ (i.e. a ready supply of carbon such as a cementite particle), while reduction in interfacial energy by nucleation at interfaces is of secondary importance. The combination of these two conditions results in an overwhelming reduction in $\Delta G^*$.

Considering second-phase morphologies, those observed agree with another study (11) of a face-centered cubic/body-centered cubic system where nuclei in the matrix were predominately rod-like and grain boundary nuclei exhibited a characteristic Widmanstätten morphology. The allotriomorphs at ferrite/pearlite interfaces can be attributed to growth along the interface which occurs because the interface is enriched in carbon which is a result of impingement and coarsening during pearlite growth.

## Acknowledgements

The authors are grateful to the Atlantic Richfield Foundation for financial support in the form of a Research Fellowship for S.W.T. The authors also wish to acknowledge Dr. G. R. Speich and Dr. M. G. Burke of U. S. Steel Research (Monroeville, PA) for helpful discussions and for the provision of specimen material.

## References

1. E. Walldow, "The Mechanism of the Solution of Cementite in Carbon Steel and the Influence of Heterogeneity," J. Iron and Steel Inst., No. 2(1930) pp. 301-341.

2. R.W. Judd and H.W. Paxton, "Kinetics of Austenite Formation from a Spheroidized Ferrite-Carbide Aggregate," Trans. TMS-AIME, 242 (1968) pp.206-215.

3. G.R. Speich and A. Szirmae, "Formation of Austenite from Ferrite and Ferrite-Carbide Aggregates," Trans. TMS-AIME, 245 (1969) pp.1063-1074.

4. K.C. Russell, "Nucleation in Solids," pp. 219-268 in Phase Transformations; ASM, Metals Park, Ohio, 1970.

5. C.I. Garcia and A.J. DeArdo, "Formation of Austenite in 1.5 Pt Mn Steels," Met. Trans. A, 12A (1981) pp.521-530

6. G. Kurdjumov and G. Sachs, "Uber den Mechanismus der Stahlhärtung," Z. Phys., 64 (1930) pp. 325-343.

7. H.W. Paxton, "The Formation of Austenite," pp.3-12 Transformation and Hardenability in Steels, Climax Molybdenum Co. Symposium (1967).

8. G.A. Roberts and R.F. Mehl, "The Mechanism and the Rate of Formation of Austenite from Ferrite-Cementite Aggregates, "Trans. of ASM, 31 (1943) pp. 613-650.

9. G.R. Speich, V.A. Demerest and R.L. Miller, "Formation of Austenite During Intercritical Annealing of Dual-Phase Steels," Met. Trans. A, 12A (1981) pp.1419-1428.

10. T. Furukawa, H. Morikawa, H. Takechi and K. Koyama, "Process Factors for Highly Ductile Dual-Phase Steels," pp.281-303 in Structure and Properties of Dual-Phase Steels, R.A. Kot and J.W. Morris, eds.; AIME, New York, N.Y., 1979.

11. G.C. Weatherby, P. Humble and D. Borland, "Precipitation in a Cu-0.55 wt.% Cr Alloy," Acta Met., 27 (1979) pp.1815-1828.

# SOME RECENT STUDIES OF FERROUS MARTENSITES

C. M. Wayman

Department of Metallurgy and Mining Engineering
University of Illinois at Urbana-Champaign
Urbana, Illinois 61801

## Abstract

This paper reports on recent work on ferrous martensites carried out at the University of Illinois. Both plate and lath martensites have been investigated in some detail, using Fe-Ni-Mn, Fe-C, and Fe-Cr-C alloys. Many of the crystallographic features associated with lath martensite have been clarified by studying Fe-Ni-Mn alloys. Accurate measurements of these are reported. Details of the lath martensite/austenite interface are now in hand and will be described, along with their crystallographic implications. It will be shown that the lath martensites in Fe-Ni-Mn alloys and carbon steels are essentially identical in nature. The kinetics of the slow isothermal growth of lath martensite below room temperature have been measured and these results are also given. Studies of plate martensite with a $\{225\}$ habit plane in an Fe-Cr-C alloy have led to two important observations: (1) macroscopic plates are actually made up of subplates, and (2) the clustering of carbon atoms in martensite at a very early stage of aging (or tempering) can be imaged and analyzed by transmission electron microscopy. Other instances where macroscopic plates of martensite appear to be comprised of subplates are pointed out, as well as previous work appearing in the literature which indicates that the planar clustering of carbon atoms in aged martensite is not peculiar to alloy steels.

## Introduction

The importance of the martensitic transformation in ferrous materials needs not be emphasized. Nevertheless, as a researcher in this field I find it interesting to ask myself from time to time how many things could not be done if hardened steels were not available. Even a partial listing here is staggering. I also ask, are we still unknowingly dealing with a diamond in the rough which could be much improved?

As one surveys our current understanding of ferrous martensites it is seen that this has occurred in a rather evolutionary way, with occasional spurts of revolution here and there. There are some interesting historical milestones associated with our understanding of the hardening of steel and it is interesting to recount some of these which have appeared during the 20th century. But before doing this we should out of respect go back into the 19th century to note that Osmond in 1895 (1) suggested using the term martensite, in honor of the German metallographer A. Martens, to designate the microconstituent found in quenched steels. Martens was quite active in microstructural characterization some 20-30 years earlier. Other interesting historical aspects dealing with the hardening of steels just before or after the turn of the century have been presented earlier by Professor Cohen. (2) In particular, he covers the many contributions made by H. M. Howe.

Two important milestones were introduced in 1924. Bain (3) proposed what is now usually referred to as the "Bain strain" as the mode of structural deformation during the austenite → martensite transformation. In the same year, French and Klopsch (4), from cooling curves, suggested a "split transformation" of austenite; the original austenite is split or divided between two different actions, one occurring at a high (Ar') temperature, the other at a low (Ar") temperature. Accordingly, it is the austenite that is unused at Ar' which transforms at Ar". We now identify Ar" as the $M_s$ temperature.

Another significant development was reported in 1938 when the now well-known Jominy bar was introduced to measure the end-quench hardenability of steels. (5) This was soon (1942) followed by the concept of calculated hardenability, where hardenability was calculated from chemical composition, by Grossman.(6)

Then, in 1947 Holloman and Jaffee (7) published the well-known book "Ferrous Metallurgical Design." At the time, it was evident that the choice of steels and heat treatments was based primarily upon experience rather than upon sound scientific principles. Holloman and Jaffee assimilated a number of scientific principles according to which the design of steels could be based upon a knowledge of the transformations that occur, their mechanical behavior, and the flow of heat during heat treatment.

One also sees in a number of instances that progress in understanding martensite in steels may have been somewhat retarded because the tradition has been to conduct research on commercially existent steels which themselves are frequently quite complex. As a case in point here, consider the typical lath martensite which forms in quenched low-carbon steels. Direct information concerning the relationship of this martensite to its parent austenite has been difficult to obtain. Essentially complete transformation of the austenite occurs because of the high $M_s$ temperature involved. Similarly our understanding of the tempering of the low carbon lath martensites has been also hampered by the high $M_s$ temperatures. Such steels actually temper during the quenching process (auto tempering) and the experimenter is in fact not dealing with virgin martensite when he examines it in the "as-quenched" condition.

Fortunately, researchers in recent times have turned to various non-commercial "laboratory" alloys, in which much progress has been made. Some examples are as follows.

Winchell and Cohen (8) used Fe-Ni-C alloys (0.02 → 0.96% C, 30.8 → 15.1% Ni) to determine the role of carbon in the strengthening of virgin (untempered) martensite. Their alloys were balanced in composition to adjust the $M_s$ temperature to -35°C in all cases, thus avoiding tempering effects. The study showed conclusively that the carbon dependent strengthening of virgin martensite is simply a solid-solution hardening effect, which significantly overrides any strengthening contributions from the intrinsic martensite substructure (microtwinning, dislocations) and carbide precipitation.

Similarly, Tamura and coworkers (9) used Fe-Ni-C alloys (Fe-33Ni → Fe-25Ni-0.9C) to determine the martensite morphology as a function of transformation temperature, in which cases with decreasing transformation temperatures the martensite morphology changed from lath → lenticular → thin plates.

Efsic and Wayman (10) used an Fe-24.5 at% Pt alloy for a study of the martensite transformation crystallography. It was found that the measured quantities, habit plane, orientation relationship, and magnitude and direction of the shape strain were essentially in perfect agreement with these same quantities predicted by the phenomenological crystallography theory.

As will be shown later, significant insights into the crystallographic characteristics of lath martensite have been reached by the study of carbonless Fe-Ni-Mn alloys.

It is impossible in a thirty minute lecture or a few typewritten pages to present a thorough coverage of the martensitic transformation in ferrous alloys. Fortunately a number of important topics in the field are being covered by other authors at this symposium and a review of the entire field has recently appeared.(11) With this in mind, I have chosen to discuss some recent studies of lath and plate martensites which have been carried out at the University of Illinois during the past few years. Most of this work involves modern transmission electron microscopy and diffraction techniques, but other methods such as optical microscopy and physical properties measurements are also involved.

## Experimental

All materials studied were prepared by vacuum melting followed by hot rolling into slabs of final thickness ranging from 6-25 mm. The nominal and actual compositions (wt.%) from chemical analysis are as follows:

| | |
|---|---|
| Fe-0.2C | Fe-0.2C-0.51Mn-0.45Si |
| Fe-0.4C | Fe-0.42C-0.52Mn-0.48Si |
| Fe-21Ni-4Mn | Fe-21.15Ni-4.07Mn-0.045C |
| Fe-20Ni-5Mn | Fe-20.2Ni-5.65Mn-0.009C |
| Fe-8Cr-1C | Fe-7.90Cr-1.11C |

After homogenization, various heat treatments to produce martensite were used. All specimens were austenitized in evacuated quartz capsules.

| | |
|---|---|
| Fe-0.2C and Fe-0.4C | austenitized for 1 hr at 1173K, quenched into iced water |
| Fe-21Ni-4Mn | austenitized for 1 hr at 1223K, quenched into iced water, further cooled to below room temperature |
| Fe-20Ni-5Mn | austenitized for 6 hrs at 1423K, water quenched |
| Fe-8Cr-1C | austenitized for 1 hr at 1500K, water quenched and further cooled to 223-233K |

For electron microscopy, 3 mm disc specimens were prepared by twin-jet electropolishing, using conventional electrolytes. Specimens for optical microscopy examination were mechanically ground and then chemically polished using an $HF-H_2O_2-H_2O$ solution. Electrical resistivity measurements were made using specimens which typically measured 1x1x50 mm. The results obtained from the study of these alloys will be grouped under headings given by the alloy composition.

## Results and Discussion

### Fe-8Cr-1C: Martensite Substructure

This alloy forms plate martensite with a $\{225\}_f^*$ habit plane. However, the macroscopic plates one sees in the optical microscope are actually composed of many individual small plates, termed subplates, which are revealed by transmission electron microscopy.(12) The subplates have the same orientation and habit plane. An example of this is shown in Fig. 1, where the coalescence plane between subplates is $(011)_b$ (which is parallel to $(111_f)$). The growth of the subplates appears to be restricted by stacking faults on $(111)_f$ in the austenite which are believed to be introduced by the martensite shape strain. Extensive deformation of the martensite occurs at the coalescence sites as shown in Fig. 2, which is mainly in the form of $(112)$ $[\bar{1}\bar{1}1]_b$ twins.

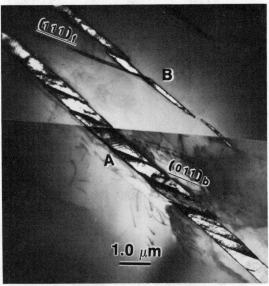

Figure 1 - Bright-field electron micrograph showing an apparently con-tinuous martensite plate (A) containing planar defects on $(011)_b$ and a group of small martensite plates (B). Fe-8Cr-1C.

On the other hand, prior to coalescence, the martensite substructure is much less complex, as shown in Fig. 3 which is a "chain" of small $(252)_f$ plates "connected" by $(111)_f$ stacking faults (or thin twins). Note that the twinning (on $(112)$ $[\bar{1}\bar{1}1]_b$) is irregular in the sense that some regions of the plates contain no twins at all. An examination of the martensite in the untwinned regions reveals the presence of dislocations, Fig. 4, with Burgers vector $\pm$ a/2 $[\bar{1}\bar{1}1]$, which is parallel to the twinning direction of the $(112)_b$ twins.

It should be noted that whether or not twins are present, the

*Subscripts f and b designate f.c.c. and b.c.t. (or b.c.c.) respectively

martensite habit plane is the same, $(252)_f$, which leads to the question as to whether the twins are transformation twins or post-transformation deformation twins caused by the constraints a plate experiences as it grows. Figure 5 shows a plate which is twinned at one side only. Yet the untwinned side has the same habit plane. Note also that the twins are "bent," as seen in the lower part of the figure.

Figure 2 - Transmission electron micrograph showing deformation structure associated with coalescence sites in a macroscopic martensite plate: Bright-field image showing coalescence planes approximately parallel to $(011)_b$ and heavy twinning on $(112)_b$. Fe-8Cr-1C.

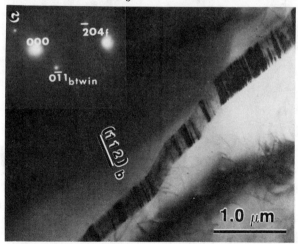

Figure 3 - Transmission electron micrograph showing a group of thin martensite plates. This is a bright-field image showing twins on $(112)_b$. Fe-8Cr-1C.

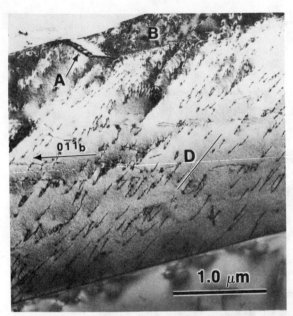

Figure 4 - Transmission electron micrograph showing the dislocation structure of the martensite. Bright-field image showing dislocations parallel to trace D, austenite area at arrow A and slightly different contrast at B. Fe-8Cr-1C.

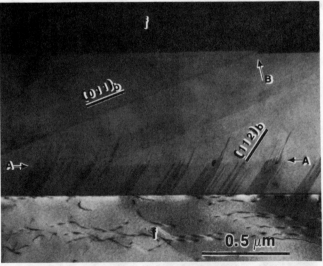

Figure 5 - Bright-field electron micrograph showing bending of $(112)_b$ twins along the line A-A in a thick martensite plate. A discontinuity (arrow B) in the martensite/austenite interface is observed in association with a planar defect on $(011)_b$. Fe-8Cr-1C.

Upon examining the martensite/parent interface at untwinned regions of the $(252)_f$ plates, many parallel screw dislocations with Burgers vector $a/2 [\bar{1}\bar{1}1]_b = a/2 [\bar{1}01]_f$ can be seen, as shown in Fig. 6. These dislocations are spaced about 13 Å apart, are difficult to image (because of carbon clustering, which will be described later) and probably accomplish the inhomogeneous shear of the crystallography theory.

The subplate characteristic of this $\{252\}_f$ alloy, as shown in Fig. 1 has also been found in other transformations. For example, Fig. 7 shows that the main $\{3\ 10\ 15\}_f$ plates appear to be made up of $\{112\}_f$ subplates in an Fe-31%Ni alloy.(13) Other examples of subplates will be shown later.

## Fe-8Cr-1C: Crystallographic Characteristics

The $\{225\}_f$ martensitic transformation has been a difficult one to understand from a theoretical point of view. The phenomenological crystallographic theory, when applied in standard form simply does not "work." The irregular distribution of internal twins in the martensite plates, Fig. 8, signals that the twins may not be fundamental to the

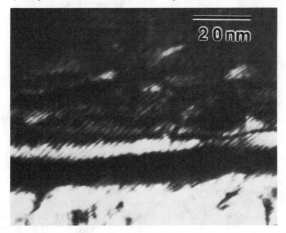

Figure 6 - Dislocation structure of the martensite/austenite interface. Weak-beam image taken using the $(110)_b$ reflection.   Fe-8Cr-1C.

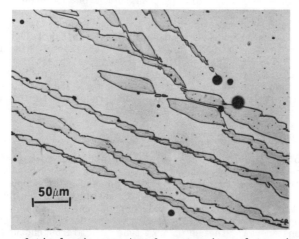

Figure 7 - Optical micrograph of martensite plates in Fe-31%Ni alloy showing that the macroscopic plates appear to be made up of subplates. Whereas the macroscopic plates have a $\{3\ 10\ 15\}_f$ habit, trace analysis of the subplates shows their habit to be near $\{112\}_f$.(13)

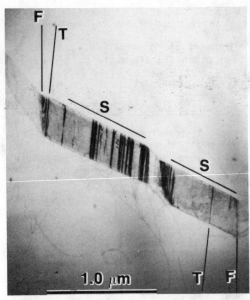

Figure 8 - Transmission electron micrograph showing thin martensite plates used for habit plane analysis. This is a bright-field micrograph showing twins in the martensite. The traces S, F and T correspond to the intersections between the martensite habit plane and the foil surface, the stacking faults and the $(112)_b$ twin plane, respectively. Fe-8Cr-1C.

transformation process, and the observation that the twins are frequently bent (Fig. 5) suggest that there may be a difference between the true transformation shape strain and that which is actually detected experimentally by measuring scratch displacements, etc. With these points of departure, we (14) have undertaken a different crystallographic approach, involving $(1\bar{3}1)[\bar{1}01]_f$ slip as the transformation inhomogeneity rather than the traditional use of $(112) [\bar{1}\bar{1}1]_b$ transformation twinning. (This is equivalent to $(2\bar{1}1)_b$ slip). Accordingly, the predicted habit plane and orientation relationship are in quite good agreement with observation. However, the predicted shape strain magnitude is substantially higher than that which has been measured by numerous investigators. This discrepancy has been provisionally rationalized on the basis that plastic accommodation of the martensite shape strain may lead to a significant difference between the true and observable shape strains. Clearly, the $\{225\}_f$ martensitic transformation is still a problem case, and requires further study and analysis.

## Fe-8Cr-1C: Carbon Clustering in Martensite

Specimens of this alloy were partially transformed to martensite ($M_s$ = 273K), aged for several days at room temperature, and then thinned and examined by transmission electron diffraction and microscopy (weak beam method).(15) After such aging diffuse scattering was observed in the diffraction patterns, a typical example of which is shown in Fig. 9, featuring "X" shaped diffraction spots. Weak beam dark field images using $\{200\}_b$ reflections reveal the gridwork type of microstructure shown in Figs. 10a and 10b. Analysis of such micrographs and their corresponding diffraction patterns shows that the grid work, due to carbon clustering, identifies four $\{012\}_b$ planes in the martensite, all of which deviate from

$(001)_b$ (the "c" axis of the Bain distortion) by about 27 degrees. The carbon clusters are spaced about 3-4 nm apart and are approximately 1-2 nm thick.

This manner of carbon clustering is not peculiar to the high Cr steel discussed here. Identical observations have been made by Nagakura an co-workers (16) for carbon steels. It remains unclear why the carbon atoms prefer to cluster in planar arrays on $\{012\}_b$ planes. However, Khachaturyan and Onismova (17) have attributed the attendant diffuse scattering to

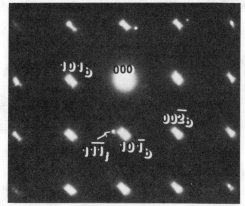

Figure 9 - Indexed diffraction pattern taken in the $[100]_b$ orientation. Diffuse scattering is associated with the fundamental martensite reflections. Fe-8Cr-1C.

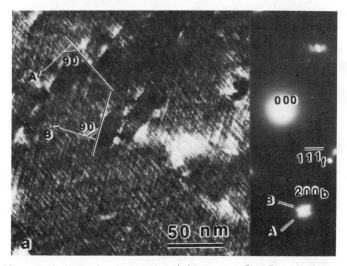

Figure 10 - Weak-beam images taken (a) in the $[010]_b$ orientation by using the $(200)_b$ reflection in the insert diffraction pattern and (b) in the $[100]_b$ orientation by using the $(020)_b$ reflection in the insert diffraction pattern. The directions A and B in (a) and C and D in (b) indicate directions of intense diffuse scattering, which are perpendicular to linear traces in the contrast images. Light regions in the weak-beam images are due to strain fields associated with dislocations. Fe-8Cr-1C.

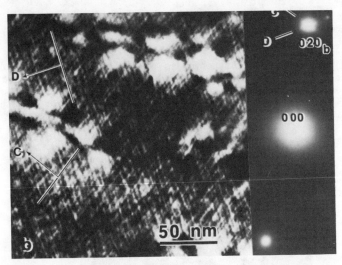

elastic distortions resulting from the short range ordering of C atoms in octahedral sites in the martensite lattice.

Further details on the Fe-8Cr-1C martensite substructure, crystallography and carbon clustering may be found in References (12), (14) and (15), respectively.

## Fe-20Ni-5Mn:  Crystallography and Substructure of Lath-Martensite

The martensite which forms isothermally below room temperature in this alloy appears to be identical to that which forms in quenched low-carbon steels, but in the Fe-Ni-Mn alloy, substantial amounts (e.g., 90%) of retained austenite can be obtained, thus allowing the investigator to avoid the experimental problems mentioned earlier in this paper.

The austenite-martensite orientation relationship was measured for some twenty isolated laths and the mean result (18) was found to be

$$(111)_f \parallel (011)_b; \quad [\bar{1}01]_f \ 3.9^0 \ \text{from} \ [\bar{1}\bar{1}1]_b .$$

This relationship was determined with an error of only a fraction of one degree by using (tilting to) a specimen orientation where the electron beam is parallel to $[111]_f$ and $[011]_b$, in which case the angle between $[\bar{1}01]_f$ and $[\bar{1}\bar{1}1]_b$ can be directly measured. Errors arising from the elongation of reciprocal lattice points are also eliminated in this orientation. Within packets of laths, adjacent laths were usually found to be of the same variant and orientation. Occasional misorientations up to 2° were found, but in this connection it should be noted that the experimentally detected orientation relationship for a single, isolated lath will vary along the length of the particular lath.

The thick layers of austenite observed between parallel laths of the same variant indicate that the laths do not form by self-accommodation. They appear to be independently nucleated, and according to an earlier report by us (19), the lath density in a typical packet is about $10^{10}/cm^3$, which implies a very high nucleation density.

The habit plane of the Fe-Ni-Mn laths is irrational and close to

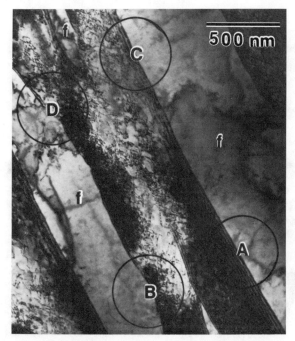

Figure 11 - Bright-field TEM micrograph taken at low magnification showing two adjacent laths which have merged. Regions designated f are austenite. Diffraction patterns taken at regions A, B, C and D show that the orientation may change along a single lath. Fe-20Ni-5Mn.

$(575)_f$ [equivalently $[1\bar{5}4)_b$]. Hence, 12 apparent habit planes are observed although 24 variants of the orientation relationship may be found. The procedure for determining the habit plane using electron microscopy and diffraction is considered to be somewhat accurate. The habit planes were determined for isolated laths in specimens which were tilted in the electron microscope such that the martensite/austenite interface was precisely ($\pm 1°$) parallel to the incident beam.

Other morphological observations show that one side (interface) of a given lath is highly planar, while the opposite side is somewhat curved and irregular, suggesting that the thickening of a lath takes place mainly in one direction, away from the initial straight interface. Austenite dislocation arrays associated with the straight and irregular lath interfaces are very different. In the vicinity of straight lath interfaces, the austenite dislocation density is low, but is high in the vicinity of the irregular interface. This again suggests that the laths start out as a planar interface, and thicken mainly in the direction of the irregular interface.

Transmission electron microscopy shows that within a given lath, screw dislocations in all four $<111>_b$ directions are found. This is typical for a mechanically deformed b.c.c. metal. However, it is also observed that one out of the four sets of $<111>_b$ dislocations is dominant in presence, leading to the suggestion that the dominant set is evidently imposed by matrix constraints opposing the martensite shape strain.

Figure 11 is a transmission electron micrograph showing two adjacent

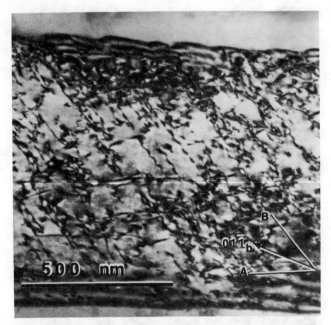

Figure 12 - Bright-field TEM image showing dislocations using the two-beam $(01\bar{1})_b$ condition. Two directions A and B are prominent. Fe-20Ni-5Mn.

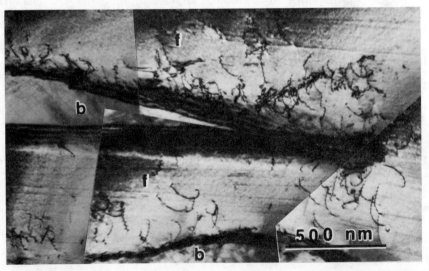

Figure 13 - Typical austenite (f) dislocation structure associated with an isolated martensite (b) lath. Fe-20Ni-5Mn.

laths which have grown together. The laths appear to completely "weld" together in some regions although they are separated by low angle boundaries at other regions. The martensite orientation relationships taken from the encircled regions A-D show that the orientation difference at A-B is 1.4° while that at C-D is zero. Figure 12 is a typical micrograph showing the internal dislocations within a lath. And finally,

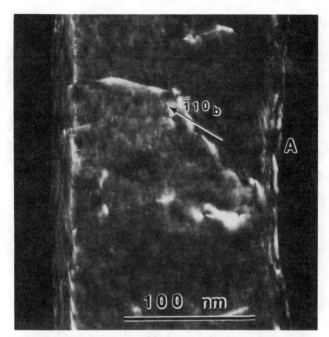

Figure 14 - Weak-beam TEM image obtained using the $(\bar{1}10)_b$ reflection, showing a single set of parallel interface dislocations on both sides of a martensite lath. Local changes in the dislocation line direction can be seen on the side of the interface designated A. Fe-20Ni-5Mn.

Fig. 13 shows the higher austenite dislocation content in the vicinity of the upper curved, irregular interface.

A more detailed accounting of the above observations may be found in Reference (18).

## Fe-20Ni-5Mn: The Martensite/Austenite Interface

Using the Fe-20Ni-5Mn alloy (20) the misfit dislocation structure at the lath martensite/austenite interface has been observed for the first time. The interface contains a single set of dislocations with Burgers vector $a/2\ [1\bar{1}1]_b = a/2\ [0\bar{1}1]_f$ and spacing of 26-63 Å, Fig. 14. These dislocations are some $10\text{-}15°$ from being in the pure screw orientation. Relative to the austenite the observed dislocation line direction is $[05\bar{7}]_f$. However, on an atomic scale, the dislocations are probably pure screw dislocations, but appear out of this orientation because of a stepped interface structure. Analysis shows that the interface dislocations are glissile and probably accomplish the lattice invariant shear (slip) of the crystallography theory. That is, they accommodate the misfit between the martensite and austenite lattices. Another example of the observed interface dislocations is shown in Fig. 15. The experimentally determined dislocation line directions are shown in Fig. 16.

## Lath Martensite Crystallography

Using the above mentioned new results from the Fe-20Ni-5Mn alloy, i.e., habit plane, orientation relationship and characteristics of the interface dislocations, some calculations were made (21) using the

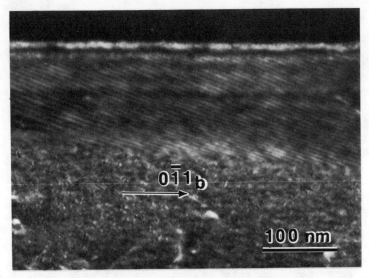

Figure 15 - Weak-beam image of interface dislocations obtained using the $(0\bar{1}1)_b$ reflection. The foil orientation is approximately $[\bar{1}\bar{1}1]_b$. Fe-20Ni-5Mn.

phenomenological crystallographic theory. The above features were reasonably well predicted using the usual Bain distortion and two lattice invariant shears (i.e., the double shear theory (22,23)), one shear on $(111)[\bar{1}01]_f$ and the other on $(100)[0\bar{1}1]_f$. (Note from Fig. 17 that the experimental dislocation lines do not lie on the initial Bain cone, indicating that the original single shear theory will not work.) However, the theoretically predicted shape deformation magnitude is quite large $m_1 = 0.96$. This is to be compared to a shape strain magnitude of 0.21, typically measured for ferrous plate martensites. Earlier work (24) indicates that the shape strain for lath martensite is indeed higher than

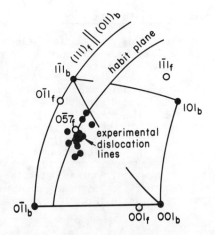

Figure 16 - Experimentally measured interface dislocation line directions for 13 different laths. The average dislocation line direction is near $[0\bar{5}7]_f$. Fe-20Ni-5Mn.

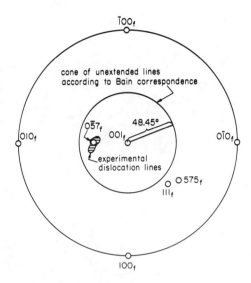

Figure 17 - Stereographic projection showing the cone of unextended lines according to the Bain correspondence and the experimentally measured dislocation line directions.

that observed for plate martensite and experimental values for the former as high as $m_1 = 0.33$ have been estimated. There is however, still quite a discrepancy between 0.33 and 0.96, and it has been suggested (21) that because of extensive accommodation deformation in the martensite (i.e., the observed high density of dislocations), the experimentally observable shape strain magnitude may be smaller than the actual magnitude. Unfortunately, because of the size of the laths, typically only 0.3 $\mu$m in width (19), it is quite difficult to measure the shape strain. By way of contrast, it is not uncommon to make shape strain measurements on martensite plates which are as wide or wider than 30 $\mu$m. (10) This is a factor of 100 compared to the laths.

According to the above theoretical analysis, one expects a stepped interface structure, as shown in Fig. 18. This is a highly coherent interface consisting of one set of steps on $(111)_f$ with associated dislocations, and one set of screw dislocations in the direction $[0\bar{1}1]_f$.

## Fe-0.2C and Fe-0.4C: Further Studies of Lath Martensite

Having obtained new results for lath martensite in Fe-20Ni-5Mn, it was decided (25) that the lath martensite in carbon steels should be reexamined in this light. Thus, Fe-0.2C and Fe-0.4C steels were studied. In view of recent work (26) attempts were made to find thin films of retained austenite between adjacent laths to use as a basis of reference. An example is such austenite films in the 0.2C steel is shown in Fig. 19, a dark field TEM micrograph taken using the $(002)_f$ reflection. Using such austenite films and both steels, the average orientation relationship was determined to be $(111)_f \parallel (011)_b$; $[\bar{1}01]_f$ 3.1$^0$ from $[\bar{1}\bar{1}1]_b$. As before (18) most of the adjacent laths within a packet are of the same orientation.

Contrary to earlier reports by various investigators, twin related laths with a packet were not found in either of the carbon steels examined

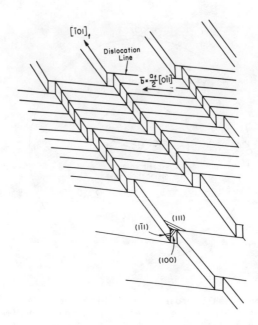

Figure 18 - Schematic representation of the interface structure in the austenite.

although twin reflections were seen. This is now considered. Figure 20a is a dark field micrograph of the 0.2%C steel taken using a twin reflection. The bright regions clearly give the impression of laths in twin relation. However, when the same specimen is properly tilted, the contrast image appears as seen in Fig. 20b. The thin bands appearing in bright contrast are $\{112\}$ twins seen edge-on which are presumably induced in the laths as the laths grow. Such twinning is even more profuse in the

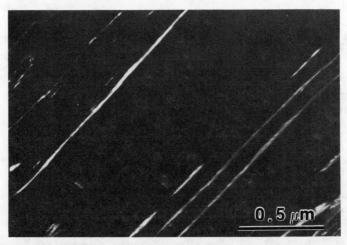

Figure 19 - Transmission electron micrograph showing continuous layers of retained austenite between martensite laths in an Fe-0.2%C alloy. Dark-field image taken using the $(002)_f$ reflection.

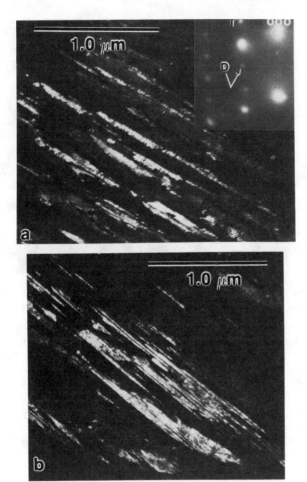

Figure 20 - Transmission electron micrographs of lath martensite in an Fe-0.2%C steel.  (a) is a dark field image taken using a twin reflection showing apparently twin related laths.  However (b) represents the same specimen tilted to a different orientation and shows thin $(112)_b$ twins which are seen edge on, rather than apparent laths.  The $<311>_b$ diffraction pattern insert in (a) shows both twin reflections (T) and double diffraction reflections (D).

harder Fe-0.4C laths.  But the main point is that overlapping thin twins (i.e., not viewed edge-on) have a more massive appearance and give the impression of a lath when imaged in dark field using a twin reflection. The heavy deformation twinnng described above was usually localized to one side of a given lath, presumably as a result of the accommodation of the shape deformation.

As argued for the Fe-Ni-Mn laths of martensite (27), the observation of layers of austenite between adjacent laths indicates that cooperative growth of adjacent laths by self-accommodation is unlikely.

Finally, the results obtained from investigating the two carbon steels

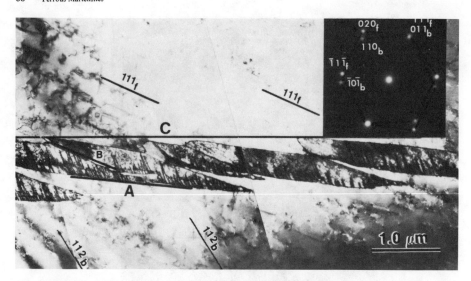

Figure 21 - Transmission electron micrograph showing a group of small martensite plates formed in a specimen transformed athermally in liquid nitrogen. See text for details. Fe-21Ni-4Mn.

were found to be in good agreement with those obtained for the Fe-20Ni-5Mn alloy, leading to the conclusion that all ferrous lath martensites have a common crystallography.

## Fe-21Ni-4Mn:  Athermal and Isothermal Martensite

After certain heat treatments plate martensite will form both athermally and isothermally in this alloy.  Figure 21 is a TEM micrograph showing the segmented nature of a plate which had formed athermally during quenching into liquid nitrogen.(28)  Several interesting features are seen in Fig. 21.  The planar segments of the inidvidual plates (marked A) are close to $(121)_f$.  The coalescence sites (marked B) are parallel to $(111)_f$ and $(011)_b$.  The macroscopic habit plane (marked C) is close to $(252)_f$.  It is also seen that the subplates are twinned at one side only.  Another example is shown in Fig. 22a.  These observations closely parallel those given earlier for the $(252)_f$ martensite formed in the Fe-8Cr-1C alloy. Figure 22b is a dark field image showing the coalescence planes and regions of retained austenite.

Figure 23 shows a very early growth stage for the isothermal plates (188°K for 2.5 min).  At this stage the low density of $(112)_b$ twins extends completely across the plates with habit plane $(121)_f$.  Figure 24a shows a later growth stage for the isothermal plates (188°K for 2.5 min).  The corresponding dark field micrograph, Fig. 24b shows that the $(112)_b$ twins in the thinner plate extend across the plate, while those in the thicker plate are concentrated to the upper side of the plate.  Numerous observations showed that the twins extend across the martensite only for comparatively thin ($<$ 0.3 $\mu$m) plates.

Figure 25 shows several martensite subplates formed during isothermal transformation at 188°K for 8 min, which represents a later growth stage than seen in Figs. 23 and 24.  A thick plate of this group was selected for a more detailed analysis.  A region (marked A-A) consisting of a high density of $(112)_b$ twins exists within the plate and defines a plane which

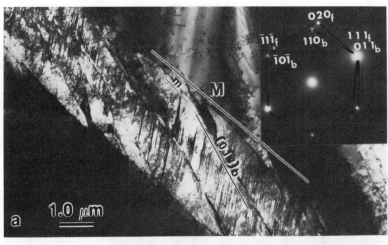

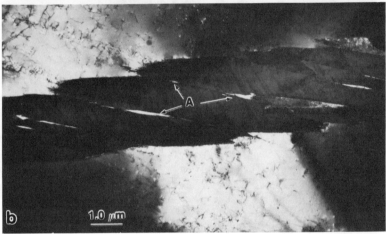

Figure 22 - (a) Transmission electron micrograph showing a macroscopic athermal martensite plate having a habit plane $(252)_f$ shown as M and containing a group of small plates with habit plane parallel to trace m. The coalescence plane of the small martensite plates is parallel to $(011)_b$. (b) dark-field image showing thin layers of retained austenite between adjacent martensite units. Fe-21Ni-4Mn.

deviates by about 19° from $(111)_f$. This region is parallel to the habit plane of a thinner plate and to a segment of the thick plate, as indicated by the traces B. A large segment of the thick plate (marked C), on the other hand, is rotated by an angle of about 7° towards $(111)_f$ and deviates from $(111)_f$ by about 12°. This habit plane "rotation" was frequently observed for somewhat thick plates in the isothermally transformed specimens. In contrast, such a habit plane rotation was not observed for plates formed athermally, perhaps indicating that there is a temperature dependent relaxation of transformation strains in the case of the isothermally formed plates. Clearly, further work is necessary to clarify the habit plane rotation.

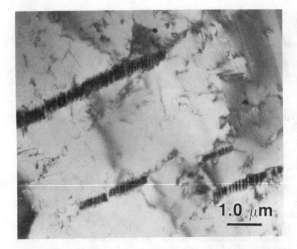

Figure 23 - Electron micrograph showing a very early growth stage of isothermal plates. A low density of $(112)_b$ twins extends across the plates. Fe-21Ni-4Mn.

Figure 24 - (a) Bright-field image showing two nearby martensite plates, one thin and the other somewhat thicker, formed after reacting at 188 K for 2.5 min. Fe-21Ni-4Mn.

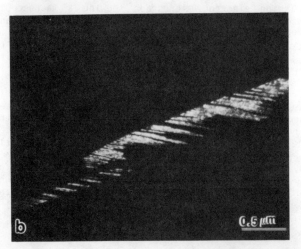

(b) dark-field image using a twin reflection showing a different distribution of twins in the thin and thicker plates.

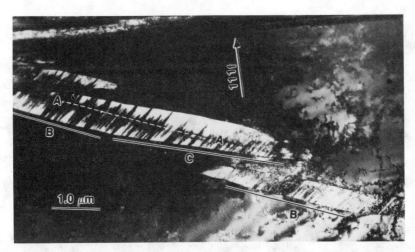

Figure 25 - Electron micrograph showing a "rotation" of the habit plane towards $(111)_f$ during the thickening of the martensite. The initial habit plane, B, apparently becomes C. The midrib of twins, A, is parallel to B. Fe-21Ni-4Mn.

## Fe-21Ni-4Mn: Lath Martensite Growth Rates

Using the Fe-21Ni-4Mn alloy, cine' films of the nucleation and growth of martensite laths at -80°C were taken, from which frame-by-frame analyses were made.(29)   The martensite laths were observed to grow very slowly, as shown in Fig. 26.   The measured lengthening rate was about 2.5 x $10^{-3}$ cm/sec.   This is $10^{-8}$ times the lengthening rate reported for plate martensite.(30)  The lath thickening rate (see Fig. 27) was found to be about 1/10th the lengthening rate.   From the growth observations, the activation energy was determined to be 14,000 cal/mole.   The slow growth observed is believed to be associated with the movement of interfacial dislocations at the martensit/austenite interface.

As mentioned earlier, a characterisic of ferrous lath martensites appears to be that one side of a lath features a planar interface, while at the other side, the interface is more irregular.   In the present case, it was observed that laths of a different habit plane variant (predominantly only a single variant) were nucleated at the irregular sides of the laths.   This is clearly seen in Fig. 28.

## Fe-21Ni-4Mn: Athermal and Isothermal Components

Under certain conditions both athermal plates and isothermal laths of martensite will form in the same specimen.(31)   The plates appear to form in a random manner, while the laths form in parallel groups or packets, as shown in Fig. 29.   These two different morphologies display different kinetics.   Figure 30 shows the variation in electrical resistance as a function of temperature during cooling (AB) and heating (BC).   The region of the curve between $M_s$ and $M_f$ corresponds to the formation of athermal plates, while the resistance decrease upon heating corresponds to the formation of isothermal laths.   The athermal martensite in this case preceded the isothermal martensite.   In this alloy the isothermal martensite can form during heating, or by holding below room temperature. The upper curve (DE) in Fig. 30 corresponds to the second cooling-heating cycle for the identical specimen.   Note that no further transformation has taken place.

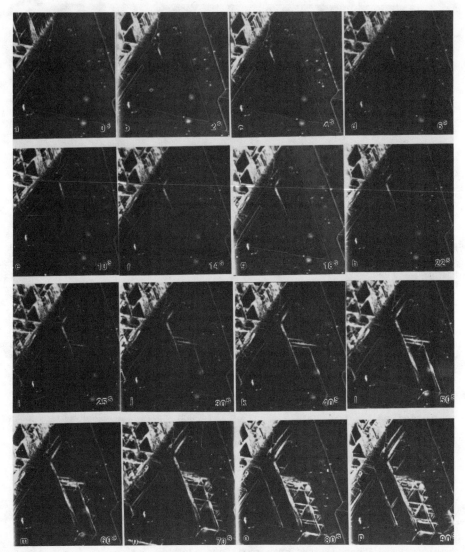

Figure 26 - Film sequence showing the formation of isothermal martensite at -80°C in an Fe-21Ni-4Mn alloy.  Observation times (in seconds) following the initial cooling to -80°C and holding there for 6.75 minutes are indicated in the lower right hand corner of each micrograph.    Elapsed frames such as these were used to determine the lengthening and thickening rates of the martensite laths.

Although in the above case, the plate martensite preceded the lath martensite, the opposite can happen in the same Fe-21Ni-4Mn alloy.(32) Figure 31 is a typical example of the surface relief due to the isothermal formation of lath martensite at -70°C.  However, upon removal of 12.1 μm of material from the initial surface by chemical polishing and etching, the overall lath appearance (i.e., equilateral triangles) still prevails (Fig. 32), but some of the laths (see arrow) appear to be undergoing fragmentation.  Closer examination at high magnification shows that laths

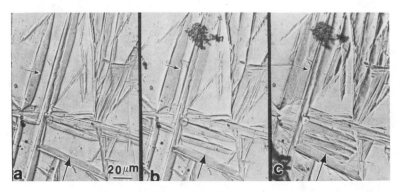

Figure 27 - Optical micrographs showing details of the thickening of martensitic laths.  The thickening process is "one-sided" and proceeds by the advancement of the irregular side of the laths and not the planar side, which apparently corresponds to the initial habit plane.

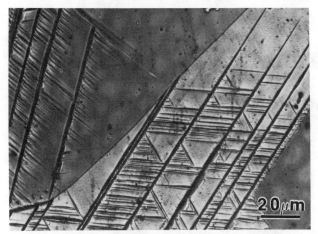

Figure 28 - Optical micrograph showing the secondary nucleation of martensite laths from the irregular side of the initially formed laths.Fe-21Ni-4Mn.

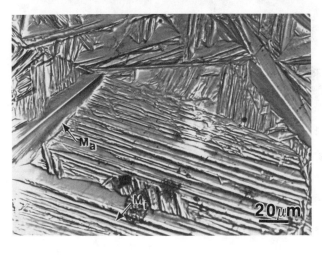

Figure 29 - Optical micrograph showing surface relief due to the formation of athermal ($M_a$) and isothermal ($M_i$) martensite plates and laths, respectively. Fe-21Ni-4Mn.

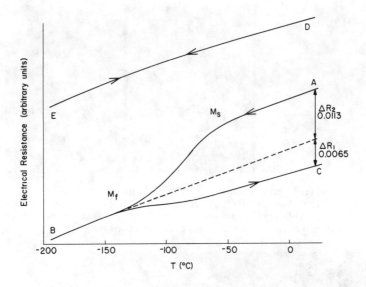

Figure 30 - Electrical resistance vs. temperature curves corresponding to athermal and isothermal martensite formation. See text for description.

Figure 31 - Optical micrograph showing surface relief from the isothermal formation of lath martensite at -70°C in an Fe-21Ni-4Mn alloy.

actually consist of aligned arrays of subplates, as seen in Fig. 33. These subplates although aligned along $(111)_f$ actually have a habit plane which is $(121)_f$ or $(252)_f$. The preference for one variant of subplate indicates that the plates were induced by the shape deformation of the lath "lying above" them. The laths were not found after some 100 μm were removed from the surface (as in Fig. 33) indicating that it is easier for the laths to form near a free surface.

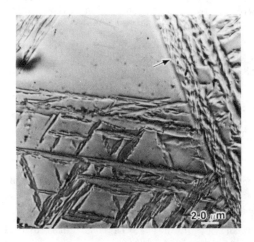

Figure 32 - Optical micrograph of the region shown in Fig. 1 after the removal of 12.1 μm of material from the initial surface.    Note the fragmented appearance of the laths at the region indicated by the arrow.

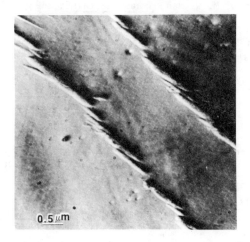

Figure 33 - Same region as shown in Figs. 31 and 32, but after removal of 102 μm of material from the initial surface.    This micrograph was taken at a higher magnification and shows that the fragmented laths are actually groupings of small plates of the same variant along $\{111\}_f$.    See text.

## Acknowledgements

The recent work on ferrous martensites described has been supported entirely by the U.S. Army Research Office, and it is a pleasure to acknowledge participation in their program of basic research. It is also a pleasure to acknowledge my past association with Kunio Wakasa, Peter Sandvik and Da-Zhi Yang, whose work formed the basis of most of this paper.

## References

1.    F. Osmond, Soc. d'Encouragement pour l'Industrie Nationale, 10 (1985) p. 480.

2.    M. Cohen, Trans. Met. Soc. AIME, 224 (1962) p. 638.

3.    E. C. Bain, Trans. AIME, 70 (1924) p. 25.

4.    H. J. French and O. Z. Klopsch, Trans. ASM, 6 (1924) p. 251.

5.    W. E. Jominy and A. L. Boeghehold, Trans. ASM, 26 (1938) p. 574.

6.    M. A. Grossman, Trans. AIME, 150 (1942) p. 226.

7.    J. H. Holloman and L. D. Jaffe, Ferrous Metallurgical Design, John Wiley and Sons, Inc. New York, 1947.

8.    P. G. Winchell and M. Cohen, Trans. ASM, 55 (1962) p. 347.

9.    T. Maki, S. Shimooka, S. Fujiwara and I. Tamura, Trans. Japan Inst. Metals, 26 (1975) p. 35.

10.   E. J. Efsic and C. M. Wayman, Trans. TMS/AIME, 239 (1967) p. 873.

11.   M. Cohen and C. M. Wayman, "Fundamentals of Martensitic Reactions," pp. 445-468 in Metallurgical Treatises, J. K. Tien and J. F. Elliott, eds., Metallurgical Society AIME, Warrendale, PA, 1982.

12.   B. P. J. Sandvik and C. M. Wayman, "The Substructure of $(252)_f$ Martensite Formed in an Fe-8Cr-1C Alloy," Met. Trans., 14A (1983) p. 2469.

13.   R. L. Patterson and C. M. Wayman, Acta Met., 24 (1966) p. 347.

14.   B. P. J. Sandvik and C. M. Wayman, "Some Crystallographic Characteristics of the $(252)_f$ Martensitic Transformation in Fe-Alloys," Met. Trans., 14A (1983) p. 2455.

15.   B. P. J. Sandvik and C. M. Wayman, "Direct Observations of Carbon Clusters in a High-Carbon Martensitic Steel," Metallography, 16 (1983) p. 429.

16.   M. Kusunoki and S. Nagakura, J. Appl. Cryst., 14 (1981) p. 329.

17.   A. G. Khachaturyan and T. A. Onisomova, Fiz. Metal. Metallov., 26 (1968) p. 973.

18.   B. P. J. Sandvik and C. M. Wayman, Met. Trans., 14A (1983) p. 809.

19.   K. Wakasa and C. M. Wayman, Acta Met., 29 (1981) p. 973.

20. B. P. J. Sandvik and C. M. Wayman, Met. Trans., 14A (1983) p. 823.

21. B. P. J. Sandvik and C. M. Wayman, Met. Trans., 14A (1983) p. 835.

22. A. F. Acton and M. Bevis, Mat. Sci. Eng., 5 (1969-70) p. 19.

23. N. H. D. Ross and A. G. Crocker, Acta Met., 18 (1970) p. 405.

24. K. Wakasa and C. M. Wayman, Acta Met., 29 (1981) p. 1013.

25. B. P. J. Sandvik and C. M. Wayman, Metallography, 16 (1983) p. 199.

26. M. Sarikawa and G. Thomas, J. de Physique, Colloque C4, Suppl. 12, 43 (1982) p. 563.

27. K. Wakasa and C. M. Wayman, Acta Met., 29 (1981) p. 991.

28. D. Z. Yang, B. P. J. Sandvik, and C. M. Wayman, "On the Substructure of Athermal and Isothermal Martensites formed in an Fe-21Ni-4Mn Alloy, Met. Trans., submitted.

29. D. Z. Yang and C. M. Wayman, "On the Slow Growth of Isothermal Lath Martensite in an Fe-21Ni-4Mn Alloy," Acta Met., in press.

30. R. F. Bunshah and R. F. Mehl, Trans. AIME, 197 (1953) p. 1251.

31. D. Z. Yang and C. M. Wayman, "Athermal Martensitic Transformation with an Isothermal Component in an Fe-21Ni-4Mn Alloy, Metallography, in press.

32. D. Z. Yang and C. M. Wayman, "Lath Martensitic Transformation with a Plate Component in an Fe-21Ni-4Mn Alloy," Scripta Met., 17 (1983) p. 1377.

# THE FORMATION OF α' AND ε' MARTENSITES IN FERROUS ALLOYS

## WITH LOW STACKING FAULT ENERGY

K. Shimizu

The Institute of Scientific and Industrial Research,
Osaka University,
8-1, Mihoga-oka, Ibaraki, Osaka 567,
Japan.

The α' and ε' martensites thermally-induced and hydrogen-induced in Fe-Mn-C alloys and an Fe-Cr-Ni steel, respectively, have been examined by optical and electron microscopy observations and electron and X-ray diffraction. The Fe-Mn-C alloys with higher manganese and lower carbon compositions exhibit both α' and ε' martensites, while those with lower manganese and higher carbon compositions mainly α' martensites. The α' martensites are observed in ε' martensite bands as well as in the γ matrix. The α' martensites in ε' martensite bands are clearly shown to form via the γ→ε'→α' transformation processes, and they are crystallographically analysed in detail. The crystallographic data are well explained by a phenomenological calculation for the hcp ε' to bcc α' transformation, using the {011}<01Ī> lattice invariant shear system found as stacking faults in α' martensites.

The α' and ε' martensites hydrogen-induced in an Fe-Cr-Ni steel are observed in the interior and near to the surface of the specimens, respectively, and at intermediate depths small α' martensite segments are observed at intersections of ε' martensite bands on two planar systems. Based on these observations, the correlation between the α' and ε' martensites in ferrous alloys or steels with a low stacking fault energy γ phase is discussed.

-------

Martensite morphology in ferrous alloys and steels is well known to have a large influence on the mechanical properties, such as the strength and toughness. Thus, many morphological studies have so far been carried out on the martensites in ferrous alloys and steels, and generally lath and lenticular shapes of α' martensites (bcc or bct structure) have been recognized to form in a wide range of ferrous alloys and steels [1]. It has also been clarified by means of electron microscopy [2] that dislocations and twin faults always exist within the lath and lenticular martensites, respectively. In addition to lath and lenticular shapes of α' martensites, butterfly and parallel-sided plate martensites have also been found to form in some ferrous alloys [3][4]. The formation of such various shapes of martensites has been attributed to a change of carbon composition, $M_S$ temperature or other factors [4], although the exact origin of such a variation has not yet been solved. On the other hand, in some ferrous alloys and steels with a low stacking fault energy $\gamma$ phase, parallel-sided plate martensite with a hcp structure ($\epsilon'$) is known to form in addition to the bcc $\alpha'$ martensite (5)(6). The $\alpha'$ martensite in these ferrous alloys and steels is sometimes formed densely in an $\epsilon'$ martensite band or in regions between two $\epsilon'$ martensite bands. Concerning such formation of $\alpha'$ martensites, three different mechanisms have so far been proposed: In the first mechanism, $\epsilon'$ martensite bands are first formed in the $\gamma$ matrix, and then $\alpha'$ martensites are successively formed in the $\epsilon'$ martensite bands (7)(8). In the second one, $\epsilon'$ martensite bands are first formed similarly, but successive formation of $\alpha'$ martensites occurs in the $\gamma$ matrix, nucleating from strained regions around boundaries between the $\epsilon'$ martensite and $\gamma$ matrix or from crossing points of two $\epsilon'$ martensite bands (9)(10). In the third one, $\alpha'$ martensites are first formed in the $\gamma$ matrix, and then $\epsilon'$ martensite bands are formed between two $\alpha'$ martensites to release the stress caused by the formation of the two $\alpha'$ martensites (11)(12). Howiver, none of the mechanisms has been verified directly, and some questions still remain as to the order of the formation of $\alpha'$ and $\epsilon'$ martensites.

Recently, attention has been paid to hydrogen embrittlement of some steels, and many studies have been carried out on the martensitic transformation induced by hydrogenation. As a result, it has been shown that $\alpha'$ and $\epsilon'$ martensites are formed by the hydrogenation in some Fe-Cr-Ni steels, but no details are known as to their distribution along the depth-wise direction and the correlation between the two kinds of martensites.

In the present paper, recent works (13)(14) by the present author on thermally-induced $\alpha'$ and $\epsilon'$ martensites in Fe-Mn-C alloys are introduced, which were carried out in order to know first the variation in external morphology and internal structure of martensites with manganese and carbon compositions, and second the order of the formation of those martensites, being analysed by a phenomenological calculation for a $\epsilon' \rightarrow \alpha'$ transformation process. Supplementary, very recent work by the present author on hydrogen-induced $\alpha'$ and $\epsilon'$ martensites in an Fe-Cr-Ni steel is briefly described, which has been done in order to understand their formation behavior along the depth-wise direction and the correlation between them (15).

## Variation of martensite morphology with manganese and carbon compositions in Fe-Mn-C alloys

### Experimental.

Four kinds of Fe-Mn-C alloys were prepared, and their chemical compositions are shown in Table 1, which are averaged ones of those analysed for the top and bottom sides of the ingots. Details of the preparation method of the ingots and subsequent heat-treatments etc. were described in the original papers (13). In preparing specimens for optical and electron microscopy, attention was paid to some surface effects, such as decarburization, on the martensitic transformations, removing the surface layer after the final heat treatment.

Table 1.   Chemical compositions of the alloys used (in wt%)

|  | Top side | | Bottom side | | Mean | |
|---|---|---|---|---|---|---|
|  | %Mn | %C | %Mn | %C | %Mn | %C |
| A | 11.9 | 0.48 | 12.1 | 0.47 | 12.0 | 0.48 |
| B | 8.6 | 0.54 | 8.7 | 0.59 | 8.7 | 0.57 |
| C | 5.2 | 0.82 | 5.2 | 1.04 | 5.2 | 0.93 |
| D | 4.1 | 1.22 | 4.0 | 1.27 | 4.1 | 1.25 |

Results.

Optical microstructures. Ms temperatures of the γ to ε' or the γ to α' transformation in Specimens A, B, C and D were measured beforehand, and they were about 243, 239, 222 and 175 K, respectively. Figures 1(a) to (d) show typical optical microstructures taken from Specimens A to D, respectively, each of which was cooled to a temperature a little below the respective Ms. Apparently various morphologies of martensites were observed from place to place in each specimen, but the characteristic morphology of Specimen A to D can be represented by Figs. 1(a) to (d), respectively. In (a), α' martensites are formed in ε' martensite bands, and never observed in the matrix γ phase. As will be verified later, the α' martensites are formed via the γ→ε'→α' transformation processes. (b), which was taken from Specimen B with a lower manganese and a higher carbon compositions than Specimen A, clearly indicates that α' martensites can form directly in the γ matrix as well as in ε' martensite bands, although the morphologies differ between the α' martensites formed in the γ matrix and ε' martensite. With decreasing manganese composition and increasing carbon composition, α' martensites are observed only in the γ matrix, and ε' martensite bands and their internal α' martensites are not observed, as seen in (c) and (d). One side of the α' martensite

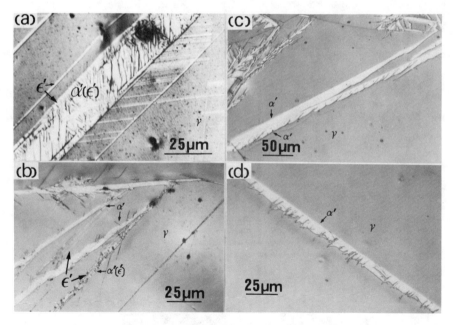

Figure 1. Optical micrographs of α' martensites in ε' martensite bands, (a), in γ matrix and ε' martensite bands, (b), and in γ matrix only, (c) and (d). (a) to (d) were taken from Specimens A to D, respectively.

plates in (c) and (d) is straight, while the other side is not straight and rather irregular, and the irregularity in (d) seems to be much greater than that in (c). Such a variety in martensite morphology with manganese and carbon compositions is consistent with the phase diagram by Schumann (5).

The straight side of α' martensite plates is very similar to that of α' martensite plates in Fe-Cr-C alloys (16), although both the sides of Fe-Cr-C martensite plates were straight. The other irregular side is similar to that of Fe-Ni martensite plates. However, even in the present Fe-Mn-C alloys, if α' martensite plates are very small corresponding to the very early stage of growth, both the sides are straight, this being verified by electron microscopy and being similar to that in Fe-Cr-C alloys (17). Therefore, the irregular boundary can be considered to be formed in a somewhat later stage of martensite growth by an interaction with some lattice defects in the adjoining γ matrix, and the other straight boundary is considered to be the nucleation side of martensite plates.

Electron microscopy. Figures 2(a) to (d) are typical electron micrographs of α' martensites in Specimens A to D, respectively. In (a), α' martensites are observed only in ε' martensite bands, which is denoted α'(ε'), and in (b) they are observed in both the γ matrix and ε' martensite bands. However, in (c) and (d), α' martensites are observed only in the γ matrix. Details of the α' martensites in ε' martensite bands will be described in the next section, but the structures of (b), (c) and (d) will be explained here.

In (b), the α' martensites formed in the γ matrix are larger in size than those formed in the ε' martensite bands, and their internal structures are also different from each other. That is, band-like structures with high contrast are seen in the former α' martensites while fine striations with faint contrast are seen in the latter ones. The band-like structures are internal twins on the {112} planes, as have so far been observed in many other ferrous martensites (2), and the fine striations are stacking faults on the {011} planes, as will be described in the next section. In this way,

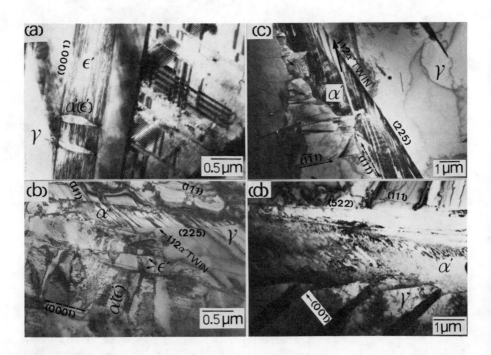

Figure 2. Electron micrographs showing α' martensites in ε' martensite band of Specimen A, (a), in both γ matrix and ε' martensite of Specimen B, (b), and only in γ matrix of Specimens C and D, (c) and (d), respectively.

the $\alpha'$ martensites formed in the $\gamma$ matrix are greatly different from those formed in $\varepsilon'$ martensite bands. Reflecting a comparatively low stacking fault energy of the $\gamma$ matrix, stacking faults or thin $\varepsilon'$ martensite bands are observed near the side boundaries of $\alpha'$ martensites, which might be formed to release stress at these boundaries.

In Fig. 2(c), which was taken from Specimen C, a twinned $\alpha'$ martensite is observed in the $\gamma$ matrix. However, one side of the plate is straight, while the other is not straight, exhibiting indentations or fang-like projections which correspond well to the irregular boundaries in the corresponding optical micrograph, Fig. 1(c). Similar structures of non-straight boundaries are also observed in Specimen D.

In Fig. 2(d), which was taken from Specimen D, a large $\alpha'$ martensite plate is observed, which is surrounded by straight and non-straight boundaries. However, the non-straight boundary is not accompanied by indentations or fang-like projections, but instead, another variant of martensite is associated with the non-straight boundary. Such an association of another variant of martensite is also observed in Specimen C, but the tendency of such an association is greater in Specimen D than in Specimen C. The association may also be attributed to stress release.

## The $\gamma \to \varepsilon' \to \alpha'$ Martensitic Transformations in an Fe-Mn-C Alloy

Experimental.

Specimen A in the preceding section was used to study the $\gamma \to \varepsilon' \to \alpha'$ transformation processes, because $\alpha'$ martensites in it were observed only in $\varepsilon'$ martensite bands and never in the $\gamma$ matrix. Austenitized specimens were air-cooled to room temperature and then further sub-zero cooled gently to various temperatures to produce $\varepsilon'$ and $\alpha'$ martensites. 5% nital solution was used as etchant to distinguish the hexagonal $\varepsilon'$ martensite from the $\gamma$ matrix, as Schumann did (5).

Results.

Verification of the $\gamma \to \varepsilon' \to \alpha'$ transformation processes. Figure 3 is a series of optical micrographs taken from a successively cooled specimen, and shows that $\alpha'$ martensites form in $\varepsilon'$ martensite bands. (a) is the micrograph of an etched surface of the $\gamma$ matrix which was retained at room temperature after air-cooling. Grain boundaries of the $\gamma$ matrix are observed, but no other significant structures are observed. (b) and (c) show unetched and etched structures, respectively, of the same place as (a) after the specimen was cooled to 238 K ($M_S$ point of this specimen is known to be about 243 K), and (d) is an enlargement of (c). The formation of $\varepsilon'$ and $\alpha'$ martensites is clearly seen in both the unetched and etched structures, $\alpha'$ martensites being in band A but no $\alpha'$ martensites in band B of $\varepsilon'$ martensite. The specimen was further cooled to 218 K. (e) is the micrograph taken from the unetched surface of the same place as (d), and there are seen surface relief effects associated with the successive formation of $\alpha'$ martensites in band B and those associated with the formation of new $\varepsilon'$ and $\alpha'$ martensites in the $\gamma$ matrix region adjacent to band B. (f) is the etched structure of the same place as (d) and (e), and $\alpha'$ martensites are clearly revealed in the $\varepsilon'$ martensite bands, although band B expands in combination with the newly formed $\varepsilon'$ and $\alpha'$ martensites in the adjacent $\gamma$ matrix region. The series of optical micrographs, thus, unambiguously indicates that $\alpha'$ martensites form in $\varepsilon'$ martensite bands via the $\gamma \to \varepsilon' \to \alpha'$ transformation processes and therefore that the mother phase of the $\alpha'$ martensites is the hexagonal $\varepsilon'$ martensite. So, the $\alpha'$ martensites must be distinguished from those directly formed in

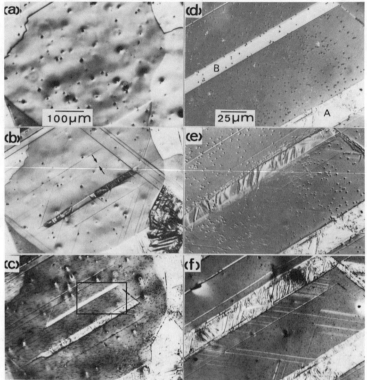

Figure 3.   Series of optical micrographs showing that α' martensites form
in ε' martensite bands via the γ→ε'→α' transformation processes.

the γ matrix in considering their formation mechanism.

   Habit planes of the α' martensites in ε' martensite bands.   Six variants
of α' martensites were observed in an ε' martensite band, making three pairs
by two variants whose surface traces deviate a little from each other (14).
Besides, three traces of α' martensite bands were sometimes observed. In such
a case, four traces of ε' martensite bands were useful in determining orientations of the γ matrix on two surfaces perpendicular to each other.   By
using the orientations thus determined (the particular ε' martensite band
containing α' martensites being set up parallel to the (111)γ plane), two
surface trace analysis has been carried out for the habit planes of six variants of α' martensites.   The habit plane normals obtained are shown in a
stereogram,   Figure 4, which is referred to the γ matrix lattice.   They make
three pairs by two  corresponding to the pairing of their surface traces as
observed.   The three pairs are close to three {11$\bar{2}$}γ poles, (1$\bar{2}$1), ($\bar{1}\bar{1}$2) and
($\bar{2}$11), respectively, which are perpendicular to the (111)γ habit plane of the
ε' martensite band containing the α' martensites.   The habit plane normals
of each pair deviate from the respective {11$\bar{2}$}γ pole in opposite ways by
about 7°, and they are not equivalent as far as the γ matrix lattice is concerned.   However, they become equivalent if they are referred to the hexagonal ε' martensite lattice, assuming the Shoji-Nishiyama orientation relationship, (111)γ // (0001)ε' and [101]γ // [1210]ε', as shown in Figure 5.
This is further evidence that the mother phase of α' martensites is not
the γ matrix but the ε' martensite.   In Fig. 5, an enlargement of the neigh-

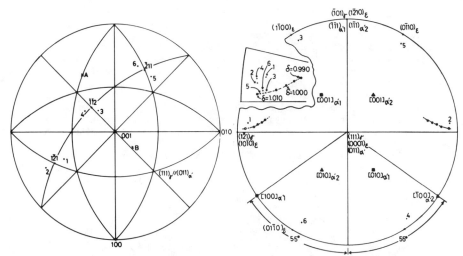

Figure 4. Stereographic representation of habit plane normals, which was referred to the γ matrix lattice. A and B represent the normals of two surfaces analysed, and 1 to 6 the habit plane normals of six variants of α' martensites in one ε' martensite band.

Figure 5. Stereographic representation of habit plane normals of the six martensites and crystallographic axes of paired two α' martensites, which was referred to the ε' martensite lattice. Calculated habit plane normals by a phenomenological theory are also plotted (see text).

borhood of the $(10\bar{1}0)_{\varepsilon}'$ or $(1\bar{2}1)_{\gamma}$ pole is inserted, and the equivalent six habit plane normals are transfered into the enlarged part. From the insertion, the experimental error or scatter is known to be within 2.5°. The paired two α' martensites are sometimes observed in contact with each other forming a junction plane, and the junction planes of three pairs are found by two surface trace analysis to coincide with the three $\{11\bar{2}\}_{\gamma}$ or $\{1\bar{1}00\}_{\varepsilon}'$ planes, respectively, which are the middles of the paired habit planes.

<u>Relation between the paired two α' martensites</u>. Figure 6(a) is an example of electron micrographs of α' martensites mainly consisting of two

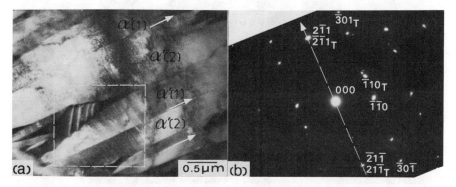

Figure 6. Electron micrograph of paired two α' martensite variants, forming a junction plane, (a), and twin-related electron diffraction pattern, (b). (b) was taken from the framed area in (a).

variants.  The martensites are paired by two  with a crystallographic junction plane ($\uparrow$), corresponding to that in optical micrographs.  (b) is an
electron diffraction pattern taken from the framed area in (a).  It consists
of two zone patterns which are twin-related to each other with respect to the
$(2\bar{1}1)_{\alpha'}$ plane, and the $2\bar{1}1_{\alpha'}$ reciprocal lattice vector is perpendicular to
the trace of junction plane in (a).  This means that the paired two $\alpha'$ martensites are twin-related to each other with respect to the $(2\bar{1}1)_{\alpha'}$ plane.  It
can thus be concluded that the junction plane between paired two $\alpha'$ martensites in the big $\epsilon'$ martensite band is nothing but the twin plane of the paired martensites.  Here, it should be noted that the habit planes of paired
martensites also have a mirror relation to each other martensite with res-
pect to the $\{2\bar{1}1\}_{\alpha'}$, $\{1\bar{1}00\}_{\epsilon'}$ or $\{11\bar{2}\}_{\gamma}$ plane.  This relation is well explained by the Shoji-Nishiyama and Kurdjumov-Sachs orientation relationships
between the $\gamma$ and $\epsilon'$ lattices and between the $\gamma$ and $\alpha'$ lattices, respectively, that is, by the following orientation relationship among the $\gamma$, $\epsilon'$ and
$\alpha'$ lattices.

$$(111)_{\gamma} \text{ // } (0001)_{\epsilon'} \text{ // } (011)_{\alpha'},$$

$$<\bar{1}01>_{\gamma} \text{ // } <1\bar{2}10>_{\epsilon'} \text{ // } [\bar{1}\bar{1}1]_{\alpha'} \text{ or } [1\bar{1}1]_{\alpha'},$$

and it is shown in Fig. 5, where crystallographic axes for only one pair of
twin-related three pairs are plotted.

Internal defects of the $\alpha'$ martensites.  Figure 7 shows a series of electron micrographs, which were taken by tilting a specimen.  In the photograph are seen a big $\epsilon'$ martensite band and internal six variants of $\alpha'$
martensites.  The six $\alpha'$ martensites are paired by two, for example, 1 and
2, 3 and 4, and 5 and 6 in (a), as have been observed.  Striations are seen
in the 1, 2 and 3 variants of (b), in the 3, 5 and 6 variants of (c), and in

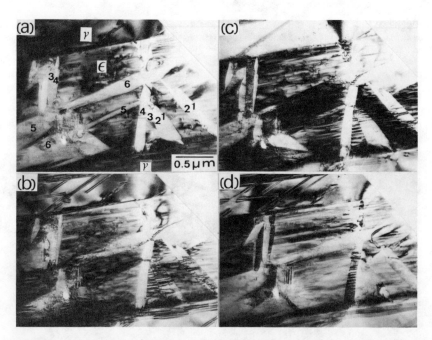

Figure 7.  Series of electron micrographs taken by tilting a specimen, exhibiting striations in the six variants of $\alpha'$ martensites along the same
direction, possibly on the $(011)_{\alpha'}$ planes.

the 4 variant of (d). All the striations are observed along the same direction, and are parallel to those in the $\varepsilon'$ martensite band. The latter striations are parallel to the $(0001)_{\varepsilon'}$ or $(111)_\gamma$ habit plane trace between the $\varepsilon'$ martensite and $\gamma$ matrix, and therefore they are considered to be due to an overlapping of $(0001)_{\varepsilon'}$ stacking faults. According to the orientation relationship between $\varepsilon'$ and $\alpha'$ martensites, the striations in all the six $\alpha'$ martensites are parallel to the $(011)_{\alpha'}$ plane. This means that some planar defects exist on the $(011)$ plane of $\alpha'$ martensites, although such planar defects have not so far been found in $\alpha'$ martensites with a bcc or bct lattice. If so, displacement vector of the planar defects can not be considered except the $[01\bar{1}]_{\alpha'}$ direction, because another possible $[\bar{1}\bar{1}1]_{\alpha'}$ shear direction on the $(011)_{\alpha'}$ plane is parallel to the Burgers vector of perfect dislocations.

### Phenomenological Consideration on the $\varepsilon'$ to $\alpha'$ transformation.

#### Lattice parameters and correspondence.

It has been shown in the previous section that the mother phase of $\alpha'$ martensites in the alloy A is the hexagonal $\varepsilon'$ martensite, and that the lattice invariant shear system of the $\varepsilon'$ to $\alpha'$ transformation is $(011)[01\bar{1}]_{\alpha'}$ or $(0001)[1\bar{1}00]_{\varepsilon'}$. Lattice parameters of the $\varepsilon'$ and $\alpha'$ martensites have been measured by X-ray diffraction as follows;

$\qquad$ a = 0.2539, c = 0.4124 nm for the $\varepsilon'$ martensite,
$\qquad$ a = 0.2883 nm for the $\alpha'$ martensite.

The X-ray diffraction study was reported elsewhere (18) as well as that of the $\gamma$ matrix. Since a tetragonality or orthorhombicity of the $\alpha'$ martensite (19) could not be detected because of some broadening of the X-ray diffraction lines, the $\alpha'$ martensite is regarded as a body-centered cubic (bcc) lattice in the present calculation.

Necessary data are now ready for the phenomenological calculation of the $\varepsilon'$ to $\alpha'$ transformation. In carrying out the calculation, a face-centered orthorhombic (fco) lattice is introduced as an intermediary one in the hexagonal close-packed (hcp) lattice of $\varepsilon'$ martensite. Atomic positions in the fco lattice do not correspond exactly to those in the hcp lattice, but this can be overcome by considering that atomic shuffles occur during the hcp to

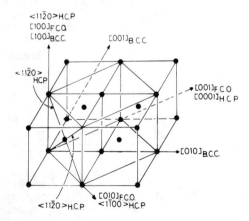

Figure 8. Lattice correspondences among the hcp $\varepsilon'$ martensite, intermediary fco, and bcc $\alpha'$ martensite lattices.

bcc transformation, as in the case of bcc to hcp transformation in a Au-47.5 at%Cd alloy (20). Thus, the calculation has first been done for the fco to bcc transformation. Lattice correspondence between the fco and bcc lattices is set up as shown in Figure 8, since the planar parallelism of $(0001)_\varepsilon$' // $(011)_\alpha$' planes has been adopted. The lattice correspondences among the hcp, fco and bcc lattices are, therefore, as follows;

$$<11\bar{2}0>_{hcp} \rightarrow [100]_{fco} \rightarrow [100]_{bcc},$$
$$<\bar{1}100>_{hcp} \rightarrow [010]_{fco} \rightarrow [01\bar{1}]_{bcc},$$
$$\text{and } [0001]_{hcp} \rightarrow [001]_{fco} \rightarrow [011]_{bcc}.$$

Phenomenological calculation.

Following the theory developed by Wechsler, Lieberman and Read (21), the phenomenological calculation has been carried out by an electronic computer. In Table 2 are shown the calculated habit plane normals, directions and magnitudes of shape strain (as well as the normal and parallel components to the habit plane), magnitudes of lattice invariant shear, and orientation relationships for two variants of $[1\bar{2}10]_\varepsilon$' // $[\bar{1}\bar{1}1]_\alpha$' (Variant I) and $[1\bar{2}10]_\varepsilon$' // $[1\bar{1}1]_\alpha$' (Variant II) of six variants with the $(0001)_\varepsilon$' // $(011)_\alpha$' planar parallelism. These numerical values are the ones for δ = 1.010 (dilation parameter) where the calculated and experimental habit plane normals are in good agreement with each other. The habit plane normals of the two variants are plotted in Fig. 5 and its insertion together with those for δ = 0.990 and 1.000. It is seen from Fig. 5 and its insertion that the calculated values for δ = 1.010 are closest to the experimental ones compared with those for other values of δ. Considering this, good agreement between the calculated and experimental habit planes would be obtained even for δ = 1.000 if lattice parameters were measured so precisely as to be able to clarify some tetragonality or orthorhombicity of α' martensite and if they were used in the calculation.

Table 2.  Crystallographic data calculated.

| | Habit plane normals | Direction of shape strain | Magnitude of shape strain | Normal component, parallel component | Magnitude of lattice invariant shear | Orientation relationship $(0001)_\varepsilon$'$(011)_\alpha$' $[1\bar{2}10]_\varepsilon$'$[\bar{1}\bar{1}1]_\alpha$' $[1\bar{2}10]_\varepsilon$'$[1\bar{1}1]_\alpha$' |
|---|---|---|---|---|---|---|
| Variant I [1$\bar{2}$10] //[$\bar{1}\bar{1}$1] | 0.71763 -0.02822 -0.68941 0.09443 | -0.14135 0.76653 -0.62519 -0.03982 | 0.21166 | 0.07265 0.19880 | 0.01895 | 0.92° 0.17° ----- |
| Variant II [1$\bar{2}$10] //[1$\bar{1}$1] | -0.68941 -0.02822 0.71763 0.09443 | 0.62519 -0.76653 0.71763 0.03982 | 0.21166 | 0.07265 0.19880 | 0.01895 | 0.92° ----- 0.17° |

Both the habit plane normals and crystallographic axes of the two variants are in a mirror relation to each other variant with respect to the $(10\bar{1}0)_\varepsilon$', $(2\bar{1}1)_\alpha$' or $(1\bar{2}1)_\gamma$ plane. According to the Table 2, directions and magnitudes of shape strain for the two variants are nearly inverse and the same, respectively, to each other variant, and therefore the mirror- or twin-related two variants are considered to be formed to release strain due to the formation of each other variant. Similar crystallographic relations

have been obtained  between other paired two variants, although details are
not described here because they are done elsewhere (14).

### Hydrogen-induced $\varepsilon'$ and $\alpha'$ martensites in an Fe-Cr-Ni steel.

#### Experimental procedures.

A commercial SUS 304 type stainless steel has been used in the present
experiment, and it was hot-forged, homogenized and hot-rolled to obtain about
0.3 mm thick sheet.  After the sheet was solution-treated at 1273 K for 30
min, specimens with suitable sizes were cut from the sheet and then hydrogen-
charged cathodically in a 1N phosphoric acid solution at a current density
of 0.1 A/cm$^2$, the charging time being 90 min..

Thin foils for electron microscopy were made by two kinds of methods.
The first was used to examine the microstructure just beneath the surface of
a hydrogenated specimen, and involved covering the surface with an insulating
paint and electro-polishing from the other side.  The second method was used
to examine the microstructure at various depths from the surface, in which
a hydrogenated specimen was first electro-polished to remove a required layer
from the surface and then subjected to the thinning procedures as in the
first method.

#### Results.

X-ray diffraction study.  Figure 9 shows a series of X-ray diffractometer traces from a hydrogenated specimen.  (a) was taken before the hydrogenation, and only exhibits a sharp 111 austenite ($\gamma$) reflection.  (b) was
taken immediately after (2 min.) hydrogenation, and shows that the 111$\gamma$ reflection becomes a little weak in intensity and two broad intensity maximum
appear at lower and higher angle sides of the 111$\gamma$ reflection, the lower
angle side maxima pushing up the tail of the 111$\gamma$ reflection.  The lower and
higher angle side maximum are identified to be reflections from a hydride
and an $\varepsilon'$ martensite, respectively, the $\varepsilon'$ martensite reflection being indexed
to be 10·1.  (c) and (d) were taken after ageing at room temperature for 1 h
and 64.5 h, respectively, and show that those reflections change with ageing
time at room temperature in intensity and position.  In (c), a broad intensity maxima corresponding to the 110 reflection of an $\alpha'$ martensite newly appear between the 111$\gamma$ and 10·1$\varepsilon'$ reflections, and the 10·1$\varepsilon'$ reflection

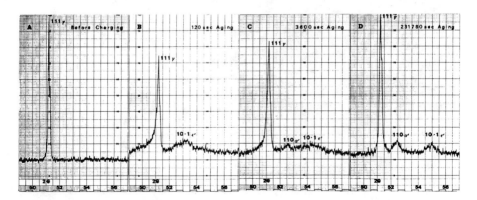

Figure 9.  X-ray diffractometer traces taken before, (a), and after, (b),
hydrogenation and after subsequent ageing at room temperature, (c), (d).

moves to the high angle side, lowering the intensity of the lower angle side hydride reflection. Such changes become more remarkable in (d), that is, the 111γ reflection returns to the original intensity and the hydride reflection almost disappears, and the 110α' reflection becomes more sharp and strong. The X-ray diffraction study clearly indicates that a hydride and an ε' martensite are induced by hydrogenation, and that the hydride disappears and a bcc α' martensite appears by dehydrogenation occuring in ageing at room temperature. The ε' martensite formed by hydrogenation has a lattice constant larger than that formed after dehydrogenation, because the former is produced in the austenite lattice expanded by hydrogenation, and therefore the 10·1ε' reflection moves to the higher angle side with increasing ageing time. The formation of α' martensites during ageing or dehydrogenation is considered to be due to some stress relaxation processes associated with the disappearance of hydride.

Electron microscopy study. In the above X-ray study, how the ε' and α' martensites formed during hydrogenation and ageing are correlated to each other is not clear. Thus the specimens aged for enough time have been observed by electron microscopy to examine their microstructures at various depths from the surface.

Figure 10 shows bright and dark field electron micrographs and an

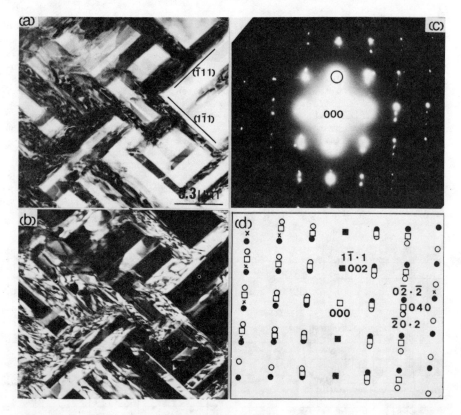

Figure 10. Bright field, (a), and dark field, (b), electron micrographs and an electron diffraction pattern, (c), and its key diagram, (d), taken from a thin foil corresponding to the surface of a specimen hydrogenated and aged at room temperature.

electron diffraction pattern taken from a thin foil corresponding to the surface layer.  In (a) and (b) are seen only $\varepsilon'$ martensite bands along two directions, which are analysed from (c) to be the {111} traces of austenite
matrix.  The dark field electron micrograph, (b), was taken by using the reflection encircled in (c), and its contrast is reversed against that in
(a).  On the other hand, microstructure at the depth of about 0.5 μm from the
surface is different from that of Fig. 10, as seen in Figure 11.  That is,
the density of $\varepsilon'$ martensite bands along two directions becomes lower, and
$\alpha'$ martensites with rod or strip shape become to be formed in the $\varepsilon'$ martensite bands or at intersections of the $\varepsilon'$ martensite bands along two directions, as clearly seen in the dark field electron micrograph, (b), taken by

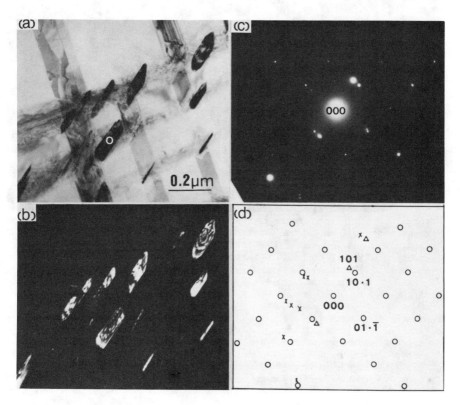

Figure 11.  Bright field electron micrograph, (a), showing $\varepsilon'$ martensite
bands and rod or strip shaped $\alpha'$ martensites, dark field electron micrograph of the $\alpha'$ martensites, (b), electron diffraction pattern taken from
the encircled area in (a), (c), and key diagram of the pattern, (d).

using a reflection from the $\alpha'$ martensite.  According to (b), the strip shape
of $\alpha'$ martensites appears to be the one grown from the rod shaped one, and
the rod shaped $\alpha'$ martensites seem to be formed in the $\varepsilon'$ martensite bands.
This appearance is consistent with a previous fact that the habit plane of
the strip shaped $\alpha'$ martensites is the same as that of $\alpha'$ martensites formed
via the $\gamma \to \varepsilon' \to \alpha'$ transformation processes.

However, at a greater  depth from the surface, larger $\alpha'$ martensites
are observed exhibiting a different habit plane index.  An example of such

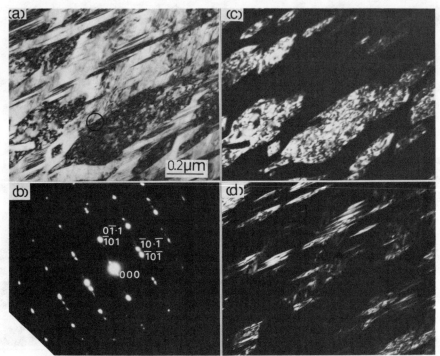

Figure 12.   Electron micrographs taken from a thin foil corresponding to a
    greater depth than 0.5 μm from the surface of a specimen hydrogenated and
    aged, (a), showing big α' martensites and more dense ε' martensite bands,
    electron diffraction pattern taken from the encircled area in (a), (b),
    and dark field images, (c) and (d), taken by using a reflection from the
    α' and ε' martensites, respectively.

electron micrographs is shown in Figure 12.   (a) is a bright field image, and
(c) and (d) are dark field images taken by using a reflection from the α' and
ε' martensites, respectively.   In these electron micrographs are seen larger
α' martensites internally dislocated and more dense ε' martensite bands than
in Fig. 10.   It may, therefore, be considered that the α' martensites with a
different habit plane are formed directly in the austenite matrix, and that
the more dense ε' martensite bands are formed by some stress caused by the
formation of α' martensites.

Discussion.

<u>Variation of martensite morphology with manganese and carbon compositions in
Fe-Mn-C alloys.</u>

It has been shown in Fig. 1 and 2 that Fe-Mn-C alloys with higher manganese and lower carbon compositions exhibit both α' and ε' martensites, while
that those with lower manganese and higher carbon compositions mainly α'
martensites.   The α' martensites are observed in ε' martensite bands as well
as in the γ matrix.   Such ε' martensite bands have been found in fcc metals
and alloys with a low stacking fault energy austenite phase, such as Co, Co-
Ni and Fe-Cr-Ni alloys, and considered to be due to a superposition of stacking faults although how the superposition was done is not unambiguously sol-

ved yet. On the other hand, α' martensites have been observed in iron alloys and steels with high stacking fault energy austenite phase, such as Fe-C and Fe-Ni alloys. Therefore, the morphological change observed in Figs. 1 and 2 can be attributed to some change of stacking fault energy accompanied with the change of manganese and carbon compositions. Actually, stacking faults or ε' martensite bands are not recognized so much in Specimens C and D in Figs. 1 and 2. Moreover, in the Specimens C and D, indentations or fang-like projections and other variants of α' martensites have been observed on the growing side of large α' martensite plates, as in Figs. 2(c) and (d). Therefore, the formation of those irregularities seems to also be related to the fact that the austenite stacking fault energy is comparatively high.

Small parallel-sided and twinned α' martensite plates might first be formed in the γ matrix, as have been observed in Fe-Cr-C alloys (17), and a stressed area might be formed in the γ matrix adjoining the small α' martensites. When such a stress is released, the martensite plates may grow into larger ones. Such a stress release may be done by a deformation in the stressed region in the γ matrix. If the stacking fault energy of the γ matrix is low, dislocations may easily be split into two Schockley partials, and then they may not easily sweep out from the slip planes. Thus, stacking faults or thin ε' martensite bands may be formed as a deformation mode to release the stress. However, a new stress field may be built up by the formation of stacking faults or ε' martensite bands. The new stress field may act as a driving force for further growth of the initial small α' martensite plates, and thus larger and parallel-sided α' martensite plates may be formed because they can grow inheriting the stacking faults or thin ε' martensite bands inside themselves, as actually observed (3). On the other hand, if the stacking fault energy of the γ matrix is higher, dislocations may not so easily be split, and therefore they may sweep out from the slip planes. Thus, the surrounding γ matrix may be perfectly released from the stress caused by the formation of initial small α' martensites, and no driving force may be available for further growth. This seems to be an origin of the formation of indentations or fang-like projections on the growing side of larger α' martensites. In fact, slip traces have frequently been observed in the adjoining γ matrix originating from the indentations or fang-like projections. Therefore, the higher the stacking fault energy of the γ matrix the greater the irregularity, being consistent with the experimental observations, Figs. 1 (c) and (d).

Morphology change with depth from the specimen surface in a hydrogenated Fe-Cr-Ni steel.

It was described in the above section that the martensite morphology in Fe-Mn-C alloys varied with manganese and carbon compositions, that is, with stacking fault energy of the γ matrix. On the other hand, Figs. 10 and 12 indicate that martensite morphology can change with the depth from specimen surface in a hydrogenated Fe-Cr-Ni steel, even though the stacking fault energy is supposed to be low and constant. This means that martensite morphology varies with not only stacking fault energy of the matrix but also another factor. In the present case of a hydrogenated Fe-Cr-Ni steel, the other factor is considered to be a stress relaxation process which occurs during dehydrogenation accompanied with ageing at room temperature and varies with the depth from specimen surface.

Immediately after hydrogenation, only ε' martensites are produced without regard to the depth from specimen surface, Fig. 9. In this stage, the γ matrix lattice may be uniformly expanded by the formation of a hydride, and its stacking fault energy may be as low as that of the original γ matrix, since ε' martensites are observed. However, Fig. 9 also indicates that α' martensites begin to appear when the hydride disappears by dehydrogenation occuring in ageing at room temperature. The formation of the α' martensites is observed at parts deeper than about 0.5μm depth from the specimen surface, and may be attributed to an increase of stacking fault energy or a stress relaxation process during the disappearance of hydride. Since ε' martensite bands are observed near the specimen surface, Fig. 10, the α' martensites are considered to be formed in a stress relaxation process but not by an increase of stacking fault energy. Details are not clear yet as to the mechanism of how the formation of α' martensites is related to the stress relaxation process accompanied by the disappearance of hydride.

By the way, rod shaped segments of α' martensites have been observed along the intersections, <110>γ, of ε' martensite bands on two {111}γ planes in the hydrogenated Fe-Cr-Ni steel, Fig. 11. Venables also observed (10) similar α' martensites near the intersections of two ε' martensite bands, but the α' martensites were so big that whether they nucleated in the ε' martensite bands or in the adjacent γ matrix could not be definitely distinguished. However, the length of the present α' martensite segments is almost the same as the projected width of ε' martensite bands. This seems to indicate that the α' martensites probably nucleate in the ε' martensite bands, and that they can grow into the adjacent γ matrix exhibiting a strip shape. On the other hand, α' martensites are observed to nucleate directly in the γ matrix, as mentioned above, and ε' martensite bands are cooperatively formed arround the α' martensites, Fig. 12, so as to release stress or strain caused by the formation of those α' martensites.

As mentioned in the beginning, three mechanisms have been proposed as to the cooperative formation of α' and ε' martensites. According to the present observations, however, all the three mechanisms seem to be operative depending on the specimen conditions. Which of the three mechanisms is operative may be determined by a delicate correlation between stacking fault energy and stress or strain field.

## Possible model for the (011)[01$\bar{1}$]α' planar defects in Fe-Mn-C martensites.

The α' martensites in ε' martensite bands of Fe-Mn-C alloys have been clearly demonstrated to be formed via the γ→ε'→α' transformation precesses, and (011)[01$\bar{1}$]α' planar faults have been found in the α' martensites. Considering the planar faults as lattice invariant shears, a phenomenological calculation has been carried out as to the γ→ε'→α' transformation processes. The calculated habit plane was in good agreement with the experimentally analysed one. On the other hand, (011)[$\bar{1}\bar{1}$1]α' or (0001)[1$\bar{2}$10]ε' system is also possible as a lattice invariant shear system on the same (011)α' or (0001)ε' plane. However, if this shear system were operated in the ε' to α' transformation, dislocation lines would be observed in the α' martensites, as in other Fe-Mn-C and Fe-Cr-Ni alloys (22)(23), since the shear direction is parallel to the Burgers vector of perfect dislocations. The fact that no dislocations were observed and instead planar faults were observed in the present α' martensites rules out the possibility of (011)[$\bar{1}\bar{1}$1]α' or (0001) [1$\bar{2}$10]ε' shear.

However, the (011)[01$\bar{1}$]α' planar faults have never been observed in α' martensites with a bcc or bct lattice. So, a possible model for the occurrence of the faults should be discussed. Figure 13(a) and (b) are projections of atomic arrangement of the hcp ε' and bcc α' martensite lattices onto the (0001)ε' and (011)α' planes, respectively. In the figures, A and B atoms are in the first and second layers, respectively. Comparing (b) with (a), the hcp to bcc transformation can be thought to be performed by a lattice distortion of those layers and a shuffle of the second layer to the first layer, the amount of the shuffle being 1/12·a along the [1$\bar{1}$00]ε' direction. However, the stacking position of the second layers in the bcc lattice is not

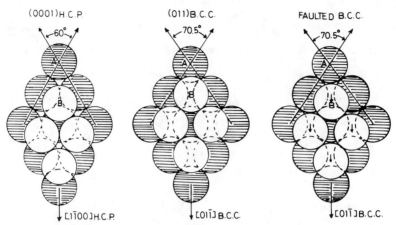

Figure 13.  Projections of atomic arrangements of A and B layers of the hcp,
bcc and faulted bcc lattices onto the $(0001)_{ε'}$ and $(011)_{α'}$ planes, respectively.

so stable compared  with that in the hcp lattice.  Therefore, the shuffling
may not always be carried out on all the second layers.  If so, some of the
second layers may be stacked at the position shown by arrows in (c), inheriting the stacking position in the hcp lattice, and $(011)_{α'}$ planar faults may
be introduced in the bcc α' martensite.  According to Table 2, magnitudes of
lattice invariant shear, g, are 0.019 for both the Variant I and II, and
this is equivalent to that $(011)[01\bar{1}]_{α'}$ planar faults are inserted every
fifteenth layers of the $(011)_{α'}$ planes.  The occurrence of the $(011)_{α'}$ planar
faults may depend on the degree of stacking fault energy of the ε' martensite
or γ matrix.  If stacking fault energy were a little higher, dislocation
lines would be observed in α' martensites, corresponding to the occurrence of
the other $(011)[\bar{1}\bar{1}1]_{α'}$ lattice invariant shear as in other Fe-Mn-C and Fe-Cr-
Ni alloys.

## References

(1)   Z. Nishiyama; Martensitic Transformations, Fundamental Part, (1971),
Maruzen, Tokyo; (1978), Academic Press, Inc., New York.
(2)   K. Shimizu and Z. Nishiyama; Metall. Trans., 3 (1972), 1055.
(3)   J. A. Klostermann and W. G. Burgers; Acta Metall., 12 (1964), 355.
(4)   I. Tamura, T. Maki and H. Hato; Trans. JSIJ, 10 (1970), 163.
(5)   H. Schumann; Arch. Eisenhuttenw., 38 (1967), 647; 40 (1969), 1027.
(6)   B. Cina; J. Iron 7 Steel Inst., 177 (1954), 406.
(7)   R. Lagneborg; Acta Metall., 12 (1964), 823.
(8)   L. Mangonon and G. Thomas; Metall. Trans., 1 (1970), 1577.
(9)   J. F. Breedis and W. D. Robertson; Acta Metall., 11 (1963), 547.
(10)  J. A. Venables; Phil. Mag., 7 (1962), 35.
(11)  J. Dash and H. M. Otte; Acta Metall., 11 (1963), 1169.
(12)  A. J. Goldman, W. D. Robertson and D. A. Koss; Trans. Metall. Soc. AIME,
230 (1964), 240.
(13)  Y. Tanaka and K. Shimizu; Trans. JIM, 21 (1980), 34.
(14)  K. Shimizu and Y. Tanaka; Trans. JIM, 19 (1979), 685.
(15)  H. Kubo, N. Kato and K. Shimizu; in preparation, (1983).
(16)  K. Shimizu, M. Oka and C. M. Wayman; Acta Metall., 19 (1971), 1.
(17)  K. Shimizu, M. Oka and C. M. Wayman; Acta Metall., 18 (1970), 1005.

(18) Y. Tanaka and K. Shimizu; Trans. JIM, 21 (1980), 42.
(19) G. V. Kurdjumov; Phys. Met. and Metall., 42 (1976), No. 3, 66.
(20) D. S. Lieberman, M. S. Wechsler and T. A. Read; J. Appl. Phys., 26 (1955), 473.
(21) M. S. Wechsler, D. S. Lieberman and T. A. Read; Trans. AIME, 197 (1953), 1503.
(22) Z. Nishiyama, K. Shimizu and S. Morikawa; Mem. Inst. Sci. Ind. Res., Osaka Univ., 21 (1964), 41.
(23) K. Suemune and K. Ooka; J. JIM, 30 (1966), 428 (in Japanese).

# A STATISTICAL APPROACH TO MARTENSITIC TRANSFORMATIONS IN STEELS

A. J. Pedraza[*] and D. Fainstein Pedraza[**]

Department of Metallurgy and Institute of Materials Science
University of Connecticut, Storrs, CT    06268

The average atomic displacements from the parent to the product lattice during a martensitic transformation may be taken as the actual atomic displacements or as an integral of all the possible different displacements that the atoms may undergo.  The latter approach has been adopted in a statistical theory based on a generalization of the concept of lattice correspondence.  The function of a fictitious lattice (the D-lattice) that is defined in that approach is outlined.  An analysis of published experimental data on the {225} transformation in steels and on the $Fe_3Pt$ martensite provides with the basic crystallographic conditions for developing a quantitative theory of the crystallography of martensitic transformations.  The atomic mechanism of the transformation is briefly described, as it can be inferred from the crystallography and from the associated strain that yields on a free surface a characteristic relief.  Martensitic growth is thus fully characterized by a registering stage where segments of one set of compact lines of the austenite phase register in the vicinity of the interface with the martensitic lattice.  This process is followed by a homogeneous motion that locates the atoms of the registered layer in their sites of the product phase.  This theory accounts for all the crystallographic features of plate-like martensites in steels and in other ferrous alloys.  An application to the Fe-8%Cr-1%C alloy serves to illustrate the power of the present approach.

------

In martensitic growth, the experimental evidence of an invariant plane strain associated with each plate is a clear indication of the existence of some type of space correlation between the atomic positions of the parent and product phases.  The problem is then how to relate this spatial (static) correlation with the actual one that involves time as well.  In this connection, the transformation (dynamic) interface plays a fundamental role because its motion brings into evidence the sequential character of the transformation process.  Considering this aspect, it may be generally stated that the interface topography can change along the process, particularly if one

---

* Present Address:  Dept. of Chemical, Metallurgical and Polymer Engineering,
                    University of Tennessee, Knoxville, Tennessee 37916
**Present Address:  Metals and Ceramics Division, Oak Ridge National
                    Laboratory, Oak Ridge, Tennessee 37830

considers that the two phases it connects are generally incommensurate. That statement is further supported by experimental results that show that compact planes of the two structures do not meet edge to edge along the habit plane that is invariant. A dynamic interface can be defined such that its motion proceeds by a conservative process, the volume change occurring normal to it. In this picture the interface topography consists of a patchwork of distorted austenitic planes in a layer adjacent to it. Those planes are distorted at different levels in different directions, forming a true "devil staircase", i.e., unpredictable. The distorsions are induced by the presence of the interface in its character of bi-dimensional defect and are essentially confined to the unstable phase.

The preceding description of the interface helps to understand the meaning of the idea of a certain independence of the atomic movements at the interface level in martensitic transformations. We have used the latter concept several years ago (1) in order to generalize the lattice correspondence and hence formulate an approach that departs from the concept of relating the two real lattices by means of a unique virtual process. Although some kind of "fitting" is produced at each interface position as it moves along, our assumption is that the actual process can and does change in time. The observation of the macroscopic deformation a posteriori of the transformation must then be related to the way the atoms in the vicinity of the interface interact and to the kind of simpler possible movements that permit an atom to attain the proper position in the final lattice.

The statistical approach we have developed is to be understood as a limit where the multiplicity of actual processes as the interface moves, is very high. In this framework we define the atomic displacements at the interface level solely on the average. The constraints we recognize here are: 1) The atoms may occupy upon transformation only immediate neighboring positions in the product phase. 2) The only final change that can be described as a result of the entire transformation process is the experimentally observed invariant plane strain.

The first requirement can be fulfilled by selecting a probability distribution $P_{I_k F_k}$ ($I_k$ and $F_k$ denote respectively initial and final lattice positions) that permits to define a fictitious lattice D whose sites are given by

$$_o\vec{k} = \sum_{I_k} P_{I_k F_k} \; {}^I\vec{k} \tag{1}$$

I.e., the D lattice sites are weighted averages of the initial I lattice positions the atoms at which have a certain probability of occupying upon transformation the product F lattice site ${}^F\vec{k}$. The second constraint acknowledged above can be fulfilled if the product lattice is obtained from the D lattice by an invariant plane strain which is that experimentally measured, i.e.,

$$^F\vec{k} = S_E \; _o\vec{k} \tag{2}$$

In the previous crystallographic approaches, the shape deformation does not generate the martensite lattice. It does in the present theory because the shape deformation operates over the D lattice rather than over the I one. That is, in the Bowles Mackenzie theory for instance

$$^F\vec{k} = \frac{1}{\delta} \; S_E \; P_2 \; {}^I\vec{k}$$

where $P_2$ is the complementary strain. Here, however, the D lattice sites are not obtained from the parent lattice by means of any deformation. I.e.,

$$_o\vec{k} \neq \frac{1}{\delta} \; P_2 \; ^I\vec{k}$$

At variance with a static view, the D lattice has no real existence. It may be physically conceived as an average lattice that the interface "confronts". Moreover, it is a useful vehicle for analysing experimental results through the parameters of the probability distribution that link it to the initial phase lattice. It should be stressed that the multiplicity of actual processes taking place as the interface moves does not refer to one given interface position. It is a consequence of the interface motion which allows for the process to change along. This approach should not be considered equivalent to a multiple shear theory.

## The Statistical Approach and its Use for Analysing Experimental Information

An inverse procedure can be used to obtain the D lattice out of the product phase lattice, as it is clear from equation (2). In order to derive $S_E$ from experimental measurements it is necessary to establish the relationship of the initial (I) lattice and the D one with regard to their shape and their volume per atom. It can be shown (ref. 2, Part I) that there is no shape difference between a macroscopic volume of I and the corresponding volume of D, the reason being that the two lattices are not related by any sort of deformation procedure. Furthermore, the two lattices have the same volume per atom.

The D lattice has a statistical value. Its relationship with the I lattice reveals the kind of geometrical features that should be maintained all along the transformation process. Hence, it becomes the proper instrument for describing a virtual process, i.e., time independent, for the transformation.

The two most important conclusions that have been drawn from the analysis of experimental results are: 1) One of the functions of the invariant plane strain is to reduce the atomic distorsions at the interface level necessary to allow the transformation to proceed; 2) Independently of the orientation relationship of parent and product and of the particular alloy there is a unique orientation relationship that links the I and D lattices whereby one compact plane of each is parallel to the other and one compact direction in each of those planes are parallel to each other. The D compact plane and direction in question are related to an F compact plane and direction, respectively.

## The Transformation Mechanism

The relevant crystallographic relationships are established between a set of compact planes and lines of the I lattice and the D one. The latter thus reveals what are the requisites to be fulfilled for the growth front to advance. Indeed, the atoms in the austenite layer that is adjacent to the interface should be set in an unstable condition whereby a strong interaction arises all along the front pushing up the transforming region so as to get the atoms in that layer into their correct positions in the growing phase. We propose that such a condition is imposed by the product phase that forces the atoms of the matrix to register with the potential of the other phase. In fundamental terms, we should formulate a problem consisting

in having a layer of a given phase (the austenite) under an external potential (that of the martensite) that has a strength gradient (away from the
interface).  We then want to find what are the positions the atoms will
occupy on the average under the influence of the two potentials - their own
interaction and the external one.

Fortunately, in the case of martensitic transformations there is an
experimental evidence that allows us to infer what the atomic positions for
registering are, without solving the equations that express such a complex
problem.  The existence of an invariant plane strain singles out those
positions.  This is so because that collective and unique motion of all the
atoms in the transforming layer must be preceded by another process.  The
registration process must also be effected by all the atoms in the layer,
but cannot be related to a homogeneous atomic displacement.  This process
makes the involved layer highly unstable as a whole and hence the homogeneous deformation motion that produces the shape change must follow.  We
thus conclude that the registering positions are D lattice sites.  It should
be kept in mind that the register occurs in the vicinity of the interface
and that is why the D-lattice is fictitious and not real.  Its significance
is thus seen to go beyond the particular choice of a probability distribution that relates it to the matrix phase (I) lattice.

The registering event highlights the strong interaction of the austenitic layer involved with the whole martensitic phase through the continuity
of atom rows of the latter into this layer.  The register is not random,
indeed.  All the previous experimental information emphasizes that segments
of compact austenite lines register into D compact lines.  However, there
can occur changes of the register pattern for a number of reasons (ref. 2,
part II) and if such changes are frequent the process appears as having a
random component that can be properly quantified by means of a statistical
approach.  We thus regain in the analysis of the transformation mechanism
the use of the D lattice and its related probability distribution as a tool
for the analysis of experimental results.

The register pattern can be described as consisting of a patchwork of
flat subregions from the austenite layer the atoms of each subregion being
uniformly registered with the martensite phase.  Being the register a collective process that involves segments of compact lines, those subregions
must contain these compact directions.  The manifested anisotropy of the
phenomenon allows to reason that only one compact direction can be involved
in the register process and that all the subregions must be portions from
the same crystallographic austenite planes.  The pair selected for ferrous
martensites (fcc $\rightarrow$ bcc/bct) is $[1\bar{1}0]_I$ $(110)_I$.

The above defined patchwork thus consists of portions of austenite
planes that conveniently distorted register into D lattice sites.  The patchwork itself is therefore defined as made up of portions of D planes of the
same type.  These planes must have the same density of sites as the corresponding I planes, and hence the same spacing, since the transformation process is conservative.  However, the I and D planes are not identical.

## The D Lattice Calculation

The analysis outlined in the preceding section allowed us to deduce the
characteristics of the D lattice as follows:

i)    There is a unique orientation relationship between the D and the I
      lattices whereby one compact plane of one is parallel to one compact
      plane of the other.  Also, one compact line in each related plane is

parallel to the other.

ii) The patchwork plane is defined as a D lattice plane that is parallel to and has the same spacing as the related (aar) plane in the austenite lattice.

As a corollary of i) and ii), the spacing of the D compact line is seen to equal that of the related I line in the respective compact planes.

The D lattice points are obtained by applying the reciprocal of the shape strain matrix to the F lattice vectors. Taking into account the above mentioned characteristics of the D lattice, a system of two coupled equations has been deduced (ref. 2, Part II) that relates the (two) parameters that yield the habit plane with those (two) parameters that specify the orientation relationship. Two conditions limit the possible solutions of those two equations, viz, the magnitude of the associated shape strain that is restricted by the constraining matrix (typically m < ∿ 0.23 for steels) and the least and maximum distance between neighbor registering sites.

A family of possible orientation relationships and related habit planes is thus obtained that is a function of the lattice parameters of the parent and product phases. For any particular solution of the family the parameters of the shape deformation are also obtained. The present theory is thus seen to be fully quantitative with regards to the crystallographic features of martensitic transformations.

## Results

A detailed application of the theory abstracted here to plate like growth is presented in ref. 2, Part III. That application has been made to three model alloys selected on the basis that they exhibit a parent/product orientation relationship representative of the various ranges observed in experiment. These alloys are: 1) Fe-8%Cr-1%C with a near Kurdjumov-Sachs orientation, 2) $Fe_3Pt$ with a near Nishiyama-Wassermann orientation and 3) Fe-31%Ni with an intermediate one. The symmetry of the final lattice and the magnitude of the parent to product volume change are shown to have a distinct effect on the possible orientation relationships. In particular, the K-S case is shown to be most unlikely for the $Fe_3Pt$ alloy.

The whole range of experimental measurements on the crystallography of ferrous martensites with a plate-like morphology is explained without adding any particular hypothesis that should apply to one case but not to others. The presence of internal twinning or/and the scatter of strain vectors and habit planes associated with one plate are consistent with the theory for the pertinent cases.

## References

1.  A. J. Pedraza and D. Fainstein Pedraza, Acta Met. 25, 87 (1977).

2.  A. J. Pedraza and D. Fainstein Pedraza, "The Martensitic Transformation in Ferrous Alloys" Part I. Analysis of Experimental Results, Part II. Theory and Part III. Applications, to be published.

TEMPERING AND STRUCTURAL CHANGE IN FERROUS MARTENSITIC STRUCTURES

George Krauss

AMAX Foundation Professor
Colorado School of Mines
Department of Metallurgical Engineering
Golden, Colorado 80401

This paper reviews information, largely generated in the last decade,
regarding the tempering of martensite in steels. Major new findings
include carbon atom clustering and other aging reactions prior to
transition carbide formation, identification of the transition carbide as
the orthorhombic eta-carbide ($\eta$-carbide), replacement of the transition
carbide by chi-carbide ($\chi$-carbide) instead of cementite ($\theta$-carbide) in
high carbon steels, the microsyntactic intergrowth of chi-carbide,
cementite, and other related carbides, eventual transformation of the
chi-carbide to cementite, and the characterization of segregation and
co-segregation of impurity and alloying elements during tempering. New
instrumentation allows determination of the microdistribution of alloying
elements as well as the characterization of the structural components in
tempered steels. A classification system which designates aging stages of
tempering as A1, A2, A3, .... and the carbide-forming stages of tempering
as T1, T2, T3, ... is used and suggested for the incorporation of known
tempering reactions as well as those which will be identified and
characterized in the future.

## Introduction

Almost all steels hardened by the formation of a martensitic
microstructure are also tempered. The major purpose of tempering is to
introduce the degree of ductility and toughness necessary for a given
application while preserving as much as possible the strength and hardness
of the as-quenched martensitic microstructure. Tempering may be performed
at any temperature between room temperature and the lower critical ($Ac_1$)
to produce a wide variety of structures and properties in any given steel.

The mechanisms associated with the structural changes produced by
tempering are best grouped into stages, many of which overlap and develop
concurrently. Several extensive reviews of tempering and the various
tempering stages associated with the structural changes were presented in
the early and middle 1970's [1-3]. More recent developments were presented
in the Peter G. Winchell Symposium on the Tempering of Steel [4]. The
present paper reviews the newer information and understanding of the
structural changes produced by tempering of martensitic microstructures in
iron-carbon alloys and low-alloy carbon steels.

The approach taken here is to preserve the classification of the stages of tempering largely developed by Cohen and his colleagues in the 1950's [5-8]. Those stages of tempering emphasize the development of various carbides in the tempering temperature ranges used in commerical practice. Recent work, however, has shown that considerable structural rearrangement of carbon atoms occurs prior to the development of the first identifiable carbide. Following Olson and Cohen [9], these precarbide precipitation reactions may be considered to be aging reactions. Thus a classification system where the symbols A1, A2, ... identify various aging reactions associated with tempering, and T1, T2, ... identify stages of tempering associated with carbide formation, represents a system consistent with our present stage of knowledge. Such a classification system would also permit the addition of new aging and tempering stages as they are discovered and confirmed in the future without introducing the confusion of continually renumbering the classical stages of tempering widely used for the last 25 years.

This paper follows the structural changes produced by tempering as they develop with increasing temperature (and time). Mechanical property changes produced by tempering are mentioned here only in association with embrittlement phenomena. However, the mechanical property toughness changes ultimately determine alloy design, the specification of tempering heat treatments, and the safe limits of operation for a given application.

## Aging Reactions

The very fine scale structural changes which develop on heating as-quenched martensite from room temperature or below to temperatures near 100°C have received considerable attention in the last decade. Prior to 1970, it was known that considerable carbon atom arrangement occurred in Fe-Ni-C martensite even well below room temperature [10, 11]. In plain carbon steels containing less the 0.2 pct carbon, at room temperature almost all of the carbon is segregated to the dislocation and/or subboundary structure of the lath martensite [1, 12]. Cooling from the high $M_s$ temperature of the low carbon steels to room temperature allows ample time for the transfer of the carbon from octahedral sites in freshly formed martensite to the substructural features [12]. Even in the austenite, there is evidence that carbon atom rearrangement may take place. The latter evidence, developed by Ansell and his colleagues by means of rapid quenching techniques, shows that $M_s$ temperature is sensitive to the cooling rate of the austenite, and therefore, to the time-dependent rearrangement of carbon in austenite prior to martensite formation [13-15].

Recent x-ray diffraction [16, 17] electrical resistivity [18], transmission electron microscopy (TEM) [19], and atom probe field ion microprobe [20] studies now indicate that as many as three unique structural arrangements of carbon may develop during the aging of martensite prior to any carbide precipitation. Most of those studies [16-18, 20] were performed on Fe-Ni-C alloys with subzero $M_s$ temperatures, but the TEM work by Nagakura and his colleagues was performed on high carbon Fe-C alloys. Generally, the effects of dislocation substructures have not been considered in the various analyses, but recently differential thermal analysis measurements of the enthalpy changes which occur during tempering indicate that segregation of carbon atoms to lattice defects as well as carbon atom clustering as described below may occur during aging [21, 22].

Two studies have shown systematically the progress of tempering, especially that part associated with the aging and early carbide formation. One, by Winchell and colleagues [16, 17] followed the changes in the shapes, positions, and integrated intensities of martensitic (200), (020), and (002) x-ray diffraction peaks as martensite in an Fe-18Ni-1C alloy was tempered. Figure 1 shows the dramatic changes which develop in the (002) peak of the Fe-Ni-C martensite with increasing tempering temperature.

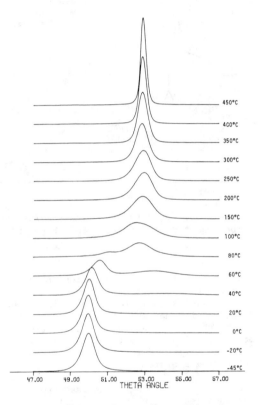

Figure 1 – (002) x-ray diffraction peak profiles for martensite in an 18Ni-1C-Fe alloy at temperatures shown for times between 1 and 2.5 hours. From Chen and Winchell [17].

The other study by Sherman, et.al. [18], measured electrical resistivity of specimens of Fe-Ni-C alloys containing 18 to 24 Ni and 0.003 to 0.62 C, first quenched in liquid nitrogen to form virgin martensite and then up-quenched to temperatures between 80 and 350°C for tempering. Figure 2 shows a schematic electrical resistivity curve with the changes typically produced by the tempering of the Fe-Ni-C martensites. The slight, almost immediate decrease in resistivity is attributed to a small amount of martensite formation, but the major feature of Figure 2 is a peak attributed to rearrangement of carbon atoms prior to carbide formation. Carbide formation is assumed to develop at the change in slope (as noted in Figure 2) in the electrical resistivity curve after the peak.

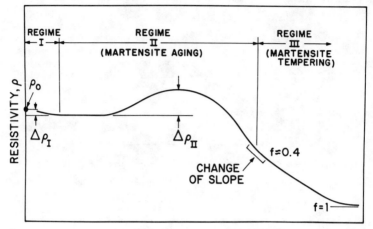

TIME OR TEMPERATURE

Figure 2 - Schematic curve of resistivity vs. aging time/temperature for the early aging and tempering stages in martensite. From Sherman, Eldis, and Cohen [18].

The temperature dependence of the rate at which the aging resistivity peak develops is characterized by an activation energy which increases gradually from 75 k joule to 100 k joule from the beginning of the resistivity increase, through the maximum, to the decline of the resistivity [18]. This range of activation energies agrees with that for the diffusion of carbon in ferrite and martensite [23]. Thus, carbon diffusion in the bcc or bct lattice of ferrous martensites is clearly identified with the aging stages of tempering.

The carbon atom rearrangement during aging leads to several structural modifications of the as-quenched martensite. There is good agreement from a variety of experimental techniques that the first aging stage (A1) consists of carbon atom clustering. In as-quenched martensite, carbon atoms are trapped in randomly distributed and separated c-axis octahedral sites. Mössbauer spectroscopy has provided evidence for such clustering in as-quenched martensite [24], although some investigators interpret Mössbauer spectra of fresh martensite as showing some fraction of carbon in tetrahedral sites [25]. The A1 stage thus develops as carbon atoms cluster to adjacent octahedral sites in the martensite. By means of the analysis of integrated intensity changes of (002) x-ray peaks during this aging stage, Winchell and his colleagues estimated that two to four carbon atoms were associated in the clusters [16]. Atom probe analysis also provides evidence of the development of carbon clusters as well as of the depletion of the carbon in the martensitic matrix [20]. The maximum carbon content of the carbon-rich areas was about 10 at pct, and the authors suggest that many of the clusters form on dislocations or twins in the matrix martensite.

Evidence for the carbon atom clustering has been presented by Nagakura, et.al. [19, 26] by means of TEM.  In that work, diffuse intensity spikes around the fundamental electron diffraction spots were attributed to carbon atom clustering on c-oriented octahedral sites.  Darkfield images of the clusters with the contrast due primarily to iron atom displacements, showed that the clusters were about 1 nm in size, in good agreement with conclusions based on the x-ray diffraction studies [16].

Beyond the A1 stage, indirect x-ray diffraction and electrical resistivity measurements are not able to differentiate further structural changes due to aging, but high resolution TEM by Nagakura and his associates [19] has identified two other structures, a modulated structure (A2) and a long period superlattice (A3).  The x-ray measurements [17], however, do show peak shifts (Figure 1), and therefore, a gradual decrease in tetragonality of the martensite in the tempering temperature range where these other aging structures develop.

The A2 structure consists of modulation of the martensite structure on (102) planes due to clusters of carbon atoms about 1 nm in size spaced at 1 to 2 nm on (102) planes [19, 27].  The planes on which the clusters concentrate are spaced about 1 nm apart with the intervening regions depleted of carbon.  Figure 3(a) shows fringes parallel to (102) planes and Figure 3(b) shows white patches associated with the modulated structure. Satellite spots around the fundamental spots in the electron diffraction patterns are associated with the modulated structure.

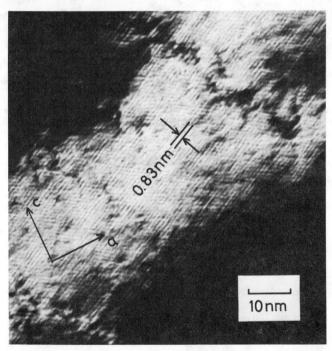

Figure 3 - Modulated A2 structure in martensite of an Fe-1.31C alloy tempered at 70°C for 1 hour.  Courtesy of Professor S. Nagakura, et.al. [19].  (a) darkfield electron micrograph taken with 002 fundamental spot and its satellites.

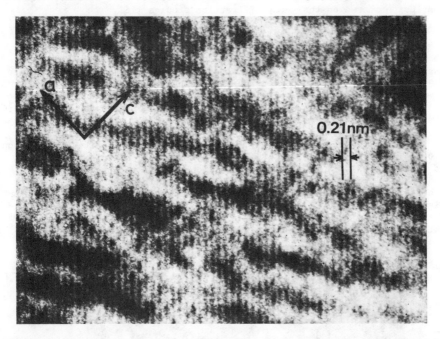

Figure 3(b) – High resolution electron micrograph taken with beam parallel
to [010]. The fringes are (101) lattice fringes and the white patches are
due to the modulated structure.


The A3 structure is identified by superstructure spots which develop
in electron diffraction patterns of martensite tempered at the upper end of
the aging temperature range, i.e., 60 to 80°C [19, 28]. The electron
diffraction patterns are interpreted as representing a long-period ordered
phase with an orthorhombic structure characterized by lattice parameters a'
= b' = 2a and c' = 12c, where a and c are the lattice parameters of the
martensite. The A3 structure, according to Nagakura, et.al., has the
composition $Fe_4C$; this composition has also been frequently associated by
other investigators with later stages of aging in martensite [19].

Olson and Cohen [19] have reviewed the various types of experimental
evidence for the A1 and A2 aging stages of tempering, and suggest further
work be performed on Fe-15Ni-1C and Fe-25Ni-0.4C alloys for which
considerable observations already exist. Olson and Cohen do not consider
the A3 stage for which strong evidence has been developed [19]. The
success of Nagakura and his colleagues in characterizing the A1, A2, and A3
structures by TEM of Fe-C alloys show that, with care in electropolishing
and with sufficiently high carbon content, future work on Fe-C alloys and
steels is justified to characterize the aging stages of tempering which are
the precursors to the practically important carbide-forming stages of
tempering (T1, T2, ...) in commercial steels.

## Tempering Stages

Early work has established at least four tempering stages associated with carbide formation in martensite [1-3]. These stages may be listed as follows:

T1: The formation of a transition carbide and the lowering of the carbon content of the matrix martensite.

T2: The transformation of retained austenite to ferrite and cementite.

T3: The replacement of transition carbide and low-carbon martensite by cementite and ferrite.

T4: The development of alloy carbides or secondary hardening in alloy steels.

The following sections describe some of the more recent findings regarding these stages of tempering.

### T1: Transition Carbide Formation

The transition carbide which forms in T1 was first identified by Jack [29] as epsilon carbide ($\varepsilon$-carbide) with a hexagonal structure and a composition of about $Fe_{2.4}C$. Early TEM of specimens tempered in the T1 range, roughly 100 to 200°C, indicated that the epsilon carbide was precipitated as thin plates, about 10 nm in width and up to 50 nm or more in length [30].

The above characterization of the structure of martensite tempered in T1 has been substantially changed in the last decade. Hirotsu and Nagakura [31-33] identified in medium and high carbon steels the transition carbide as eta-carbide ($\eta$-carbide) with an orthorhombic structure and a composition corresponding to $Fe_2C$. The carbon atoms occupy a sublattice within the eta-carbide lattice. Further, the actual transition carbide particles are quite fine, about 2.0 nm in diameter and arranged in rows within the matrix martensite as shown in Figure 4 [34]. It is now apparent that the larger plate-like features formerly [30] associated with TEM images of the transition carbide are a result of the strain fields associated with the carbide precipitation.

Figure 4 - Darkfield electron micrograph of fine transition (eta-carbide) carbides in martensite of an Fe-1.22C alloy tempered at 150°C [34].

In addition to the precipitation of the transition carbide, evidence is accumulating that significant plastic flow and substructural rearrangement occur during T1.  Figure 5 shows a bright field TEM micrograph of the substructure in a plate of martensite in an Fe-1.22C alloy tempered at 150°C for 1 hour.  The white areas are essentially dislocation-free and are bounded by regions of dark strain contrast.  It is in the latter regions that fine transition carbides are revealed by dark-field illumination, Fig. 4.  The cell-like substructure or martensite tempered in T1 has formed as a result of considerable rearrangement of the substructure in as-quenched high carbon martensite [35] and this rearrangement is associated with significant plastic flow within individual martensite plates [36].  Figure 6 shows the surface relief or plastic flow generated in martensite plates by tempering an Fe-1.2%C martensite at 200°C.  The specimen surface was polished flat prior to tempering.  The plastic deformation apparently accommodates lattice strains which accompany transition carbide precipitation and the associated volume changes in the martensitic matrix.  Chen and Winchell [17], on the basis of observing changes in the population of some martensite plate orientations, suggest twinning or detwinning as a plastic deformation mode to accommodate lattice strains produced during carbide precipitation.

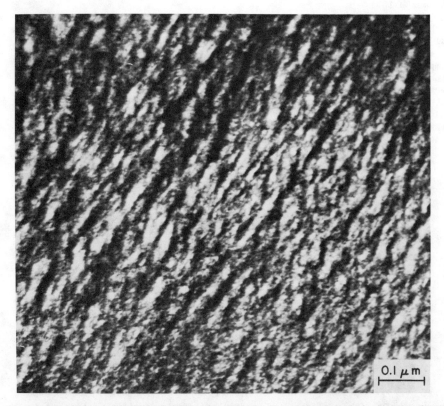

Figure 5 - Bright field electron micrograph of substructure in martensite of an Fe 1.22C alloy tempered at 150°C.

Figure 6 - Surface relief produced by tempering martensite of an Fe-1.22C alloy at 200°C for 30 minutes. The specimen surface was polished prior to tempering [36].

In regard to the effects of transition carbide precipitation on the structure of martensite, Chen and Winchell [17] have shown that a diffuse (002) peak develops at an angle that corresponds to a negative tetragonality of the martensite matrix at tempering temperatures below 100°C. They attributed this negative tetragonality to coherently diffracting comains with a lattice parameter smaller than that of ferrite because of coherency strains induced by transition carbide precipitation.

The practical importance of volume changes due to carbide precipitation was noted in a study of stress relaxation during the tempering of hardened AISI 52100 steel and a series of plain carbon steels [37]. In that study stress relaxation was attributed to the plastic flow which accompanied local volume changes produced by transition carbide and cementite precipitation.

## T2:  Retained Austenite Transformation

As the carbon content of steel increases, the $M_s$ temperature decreases and the amount of austenite retained at room temperature increases [38, 39]. The retained austenite transforms to mixtures of ferrite and cementite during tempering between 200 and 300°C. The activation energy for this process is about 115 k joules per mole [5, 36], consistent with the activation energy for the diffusion of carbon in austenite [40].

The transformation of austenite during tempering is referred to as the second stage of tempering. (T2). The insights developed in the last decade regarding T2 are related primarily to low and medium carbon steels. By means of high-resolution TEM and precision dark-field illimination, Thomas [41] has shown that retained austenite is present as thin, continuous layers between martensite laths in steels in which the lath martensite morphology develops. This retained austenite survives the aging and T1 tempering stages and then decomposes to large, relatively continuous interlath carbides.  Figure 7 shows quantitatively the effect of tempering on austenite content in several medium carbon steels tempered at the temperatures shown for 1 hour [42].  The interlath carbides which have formed in the 4340 steel tempered at 350°C are shown in Fig. 8 [43].  These interlath carbide arrays are detrimental to toughness and are associated with the transgranular mode of tempered martensite embrittlement which develops in medium carbon steels tempered between 200 and 400°C [41, 43–45].

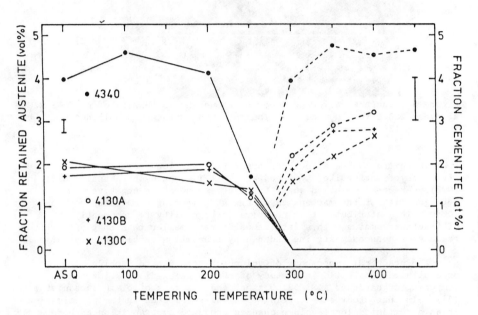

Figure 7 – Retained austenite and cementite as a function of tempering temperature for several medium–carbon steels [42].

## T3:  Formation of Ferrite and Cementite

Concurrent with the transformation of retained austenite, the transition carbide within the martensite units is eventually replaced by cementite and the carbon content of the tempered martensite falls to that of ferrite.  These processes are referred to as the third stage of tempering (T2).  Recent evidence shows that the carbide which forms in high carbon martensite in T3 is often chi–carbide (x-carbide) rather than cementite.  Nevertheless, with prolonged tempering in the T3 range, the chi–carbide is replaced by cementite [19, 46].

In low and medium carbon steels, cementite nucleates within the laths of martensite, as shown in Figure 9 [47].  The cementite is initially in the form of fine platelets, which form on $(110)_m$ habit planes [44].  As cementite forms in low C martensite, the martensite lath boundary area per unit volume drops very rapidly as a result of elimination of low angle lath boundaries [48].  The remaining large angle lath boundaries are pinned by the carbides formed in both T2 and the early T3 stages, and the lath

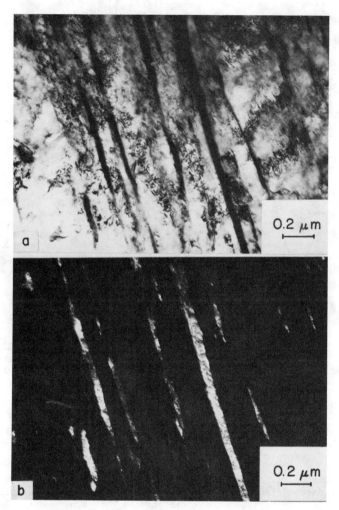

Figure 8 – Interlath cementite formed in martensitic structure of 4340 steel tempered at 350°C [43]. (a) Bright field electron micrograph (b) Darkfield electron micrograph taken with (210) cementite diffracted beam.

morphology of the as-quenched martensite, although coarsened, is stabilized to very high tempering temperatures [48]. Eventually, with prolonged tempering, for example 12 hours at 700°C, a coarse dispersion of spheroidized carbides in a matrix of equiaxed ferrite develops. The equiaxed ferrite appears to be result of grain growth, not recrystallization, because:

1.  the driving force (strain energy) for recrystallization is reduced by recovery and polygonization of the martensite substructure during the intermediate T3 period of stability of the cementite-ferrite lath structure; and

Figure 9 - Bright field electron micrograph of martensite in an Fe-0.2C alloy tempered at 400°C for 1 minute.

2.   the cementite particles have coarsened to the point where they no
     longer effectively pin the residual, high angle, lath boundaries [48,
     49].  Therefore, the boundaries rearrange themselves into an equiaxed
     network in order to lower interfacial energy.

     As noted above, evidence is accumulating that the first carbide formed
in T3 in high carbon steels is chi-carbide [3, 19, 46, 50].  Chi-carbide
has a monoclinic structure and the stoichiometry $M_5C_2$ compared to the
orthorhombic structure and the $Fe_3C$ stoichiometry of cementite [51].
Figure 10 shows carbide formed in the martensitic microstructure of an
Fe-1.22 pct C alloy tempered at 350°C.  These carbides have been identified
as chi-carbides by electron diffraction and Mössbauer spectroscopy [46].
Two morphologies of the chi-carbide are shown:  an intragranular morphology
and a morphology associated with formation at martensite plate interfaces.
Another morphology, not shown, forms on the transformation twins in
martensite plates [46, 50].

     Nagakura, et.al. obtained TEM images of lattice planes within carbide
particles formed during T3.  Within a single carbide particle lattice
fringe spacings corresponding to chi-carbide and cementite as well as
higher order carbides of general composition $Fe_{2n+1}C_m$ were found [19
52].  An example of these findings is shown in Figure 11.  The intergrowth
of the various carbides was referred to as microsynthectic growth.  As

tempering in T3 proceeds, cementite eventually replaces the chi-carbide in tempered high carbon martensite by short range displacement of iron atoms and carbon diffusion [19, 52].

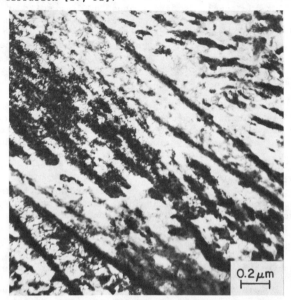

Figure 10 - Chi-carbide formed in martensitic structure of an Fe-1.22C alloy tempered at 350°C.

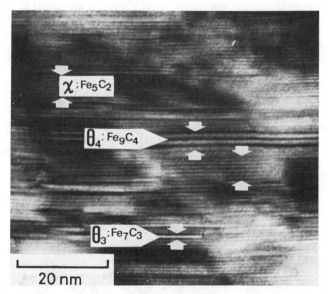

Figure 11 - Lattice image of microsyntactic intergrowth of various carbides in martensite of an Fe-1.3C alloy tempered at 270°C for 1 hour.   Courtesy of Professor S. Nagakura et.al. [19].

## T4:   Effects of Composition and Alloy Carbide Formation

Steels containing carbide-forming elements, such as Cr, Mo, W, V, and Nb, can form alloy carbides when tempered at 500°C or above [1, 2].  For example, in a 2.25Cr-1Mo steel, tempering above 500°C results in the replacement of cementite successively by $Fe_3C + Mo_2C$, $Fe_3C + Mo_2C + Cr_7C_3$, $M_{23}C_6 + Cr_7C_3$, and $M_{23}C_6 + M_6C$ carbide dispersions if time and temperature of tempering (or high temperature service) is sufficient.  The alloy carbides may form directly from cementite particles or may nucleate independently of the cementite on the substructure of the tempered martensite [54].

The alloying elements in steels, especially if they form alloy carbides, either significantly retard softening or produce secondary hardening during tempering.  Grange, et.al. published the results of a systematic study on the effect of various elements on the retardation of softening on a wide range of martensitic steels tempered between 200 and 700°C (400 and 1300°F) [72].  Examples of the results for tempering temperatures of 260 and 540°C are shown in Fig. 12.  At 260°C, the carbide-forming elements have little effect on retarding softening, in contrast to their strong effect on maintaining hardness by alloy carbide precipitation during tempering at 540°C.  Silicon, a non-carbide forming element, however, significantly retards softening in specimens tempered at 240°C.

New information regarding the reasons for the behavior of Si-containing steels during tempering in the overlapping T2 and T3 stages has recently been obtained by the use of the atom probe and Scanning TEM x-ray microanalysis [55].  Two conflicting explanations for the strong retarding effect of Si on cementite formation have been proposed.  The one explanation is based on the observation of the enrichment of Si in the transition carbide relative to cementite which has very low solubility for silicon [56].  The other explanation assumes that Si is rejected from the first-formed cementite.  The resulting high Si layer surrounding the cementite particles severely limits growth of the cementite because Si must diffuse away from the cementite interface for growth to continue [57].  Also, the high matrix interface content of Si raises the activity of carbon, decreases the flux of carbon atoms at the cementite interface and thus decreases the cementite growth rate.  The latter effect of Si has been demonstrated in the growth of carbides after the quenching of ferrite [1] and the growth of proeutectoid cementite grain boundary allotriomorphs in high carbon steels [58].

The atom probe study showed that the transition carbides are not enriched in silicon, the solubility of silicon in cementite is extremely low, and that a Si-rich layer does indeed form in the tempered martensite matrix adjacent to the cementite-ferrite interface.  Thus the silicon rejection explanation of the role silicon plays in tempering is directly confirmed [55].

Figure 12 shows that phosphorus has a strong retarding effect on softening during tempering at 260°C.  The mechanism of this effect has not been studied, but it has been demonstrated elsewhere that phosphorus initially stimulates cementite grain boundary allotriomorph formation [54].  Once formed, the rate of cementite growth decreases significantly.  The latter very sluggish growth has been attributed to silicon rejection from the cementite [58], and to chromium partitioning during later growth stages in Fe-Cr-C alloys and chromium-containing steels [60-61].  Phosphorus appears to concentrate at cementite-matrix interfaces in later growth stages [59], and thereby may retard growth in a manner analogous to that of silicon.  More work is required to understand the role of phosphorus during tempering.

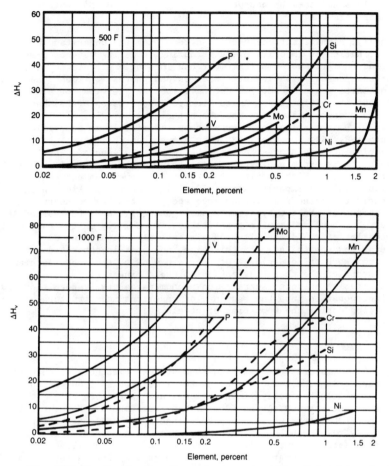

Figure 12 - Effect of alloying elements on retention of hardness during tempering at 260°C (500°F) and 540°C (1000°F). ΔHV (Vickers hardness) is the hardness difference between an Fe-C alloy and a steel containing given amounts of the elements indicated. From Grange et.al [72].

Within the latter stages of tempering, information regarding alloy carbide composition has been developed as the result of the application of new instrumentation. Analytical electron microscopy has been used to identify carbides in highly alloyed Cr-Mo-V-Nb steels [62]. Also with the use of electron probe microanalysis (EPMA) together with selected are diffraction, Pilling and Ridley [63] have shown that chemical changes within complex carbides can develop with increasing tempering time in low carbon 2.25 pct Cr-1 pct Mo steel [63]. The chromium-based carbides, $M_{23}C_6$ and $M_7C_3$, showed little change in composition with increasing tempering time at 700°C, but the molybdenum-based carbides, $M_2C$ and $M_6C$, showed an increase in molybdenum content as tempering time increased. This supports the explanation of the variable effect of molybdenum on suppressing temper embrittlement by phosphorus in alloy steels [64-66]. Initially molybdenum suppresses temper embrittlement by the formation of $Mo_3P$, but with increasing tempering or service time, molybdenum forms molybdenum-based carbides, such as $M_2C$. As a result the $Mo_3P$ dissolves and the phosphorus is free to segregate and contribute to temper embrittlement.

Finally, significant advances in determining and understanding the microdistribution of impurity elements during tempering have been made in the last decade. Temper embrittlement has long been associated with the segregation of metalloid elements, such as phosphorus, antimony, and tin to prior austenite grain boundaries during tempering, service, or slow cooling between 350°C to 550°C [67]. Auger electron spectroscopy (AES) has verified such segregation, and has shown that the severity of embrittlement increases with increasing concentration of impurity atoms at critical interfaces [68, 69]. However, AES also shows that the substitutional alloying elements may segregate with the impurity elements. The latter co-segregation is explained by a regular solution thermodynamic model which incorporates terms which specify the interaction energies between transition metal and impurity atoms [70, 71]. The interaction energies in turn contribute to the segregation free energies and the boundary concentrations of the two types of atoms. Recently, with the use of AES, the theory has been made quite quantitative, especially in the case of the co-segregation of P and the alloying elements Ni, Mn, Cr, and Mo [71]. The co-segregation effects have, to this time, not been incorporated into the overall framework of tempering. However, in view of the importance and advanced state of the characterization of these fine-scale, non-precipitation effects, segregation phenomena deserve recognition and formal incorporation into the sequence of reactions and structural changes which develop during the tempering of steels.

Summary

Table I summarizes the reactions and stages of structural changes which occur during the tempering of martensitic steels, iron-carbonalloys, and iron-nickel-carbon alloys. The temperature ranges are only approximate, and are usually based on 1-hour treatments. However, alloy carbide formation, ferrite boundary rearrangements, and segregation phenomena may take much longer. The list has grown from similar lists published in earlier reviews [1, 2]. Major additions have been made at both the very low temperature and the very high temperature regions of tempering. The former regime includes structural changes produced by aging of martensite prior to carbide formation, while the latter regime includes the segregation and co-segregation of impurity and alloying elements atoms within a context of changing alloy carbide chemistry and distribution.

Even within the more firmly established central stages of tempering, T1, T2, and T3, new information has been developed in the last decade regarding carbide crystal structure, stoichiometry, and distribution . In particular, eta-carbide in T1, chi-carbide in the early part of T3, and the interlath decomposition of austenite to cementite in T2 in low and medium carbon steels have been identified and characterized.

Much of the new information generated, as in the period prior to 1970, is related to the characterization of carbides formed during various stages of tempering. Relatively little attention has been paid to changes in dislocation substructure and martensite interface structures. However, it is the dynamic dislocation interactions with matrix dislocation substructure and carbide distributions which determine the flow stresses (or strain hardening) and the fracture resistance of tempered steels. Thus, continued effort in the characterization of the ferritic substructure

TABLE I - Tempering Reactions in Steel

| TEMPERATURE RANGE (°C) | REACTION AND SYMBOL (if designated) | COMMENTS |
|---|---|---|
| -40 to 100 | Clustering of 2 to 4 carbon atoms on octahedral sites of martensite segregation of carbon atoms to dislocations boundaries | Clustering is associated with diffus spikes around fundamental electron diffraction spots of martensite |
| 20 to 100 | Modulated clusters of carbon atoms on (102) martensite planes (A2) | Identified by satellite spots around electron martensite |
| 60 to 80 | Long period ordered phase with ordered carbon atoms arranged (A3) | Identified by superstructure spots in electron diffraction patterns |
| 100 to 200 | Precipitation of transition carbide as aligned 2 nm diameter particles (T1) | Recent work identifies carbides as eta (orthorhombic, $Fe_2C$); Earlier studies identified the carbides as epsilon (hexagonal, $Fe_{2.4}C$) |
| 200 to 350 | Transformation of retained austenite to ferrite and cementite (T2) | Associated with tempered martensite embrittlement in low and medium carbon steels |
| 250 to 700 | Formation of ferrite and cementite; eventual development of well spheroidized carbides in a matrix of equiaxed ferrite grains (T3) | This stage now appears to be initiated by chi-carbide formation in high carbon Fe-C alloys |
| 500 to 700 | Formation of alloy carbides in Cr, Mo, V, and W containing steels. The mix and composition of the carbides may change significantly with time (T4) | The alloy carbides produce secondary hardening and pronounced retardation of softening during tempering or long time service exposure around 500°C |
| 350 to 550 | Segregation and co-segregation of imurity and substitutional alloying elements | Responsible for temper embrittlement |

as well as the carbides in tempered specimens would be justified in the years ahead.

Finally, the recent characterization of the structural changes and the mechanisms which contribute to these changes during tempering are due, in large part, to the development of instrumentation with greater lateral, depth, and chemical resolution. In particular, high resolution, analytical TEM, scanning transmission electron microscopy, AES, and the atom probe were either not available or in limited use in 1970. The continued systematic application of the new instrumentation and that yet to be developed should significantly increase future understanding of the structural changes produced by tempering and the effect of these changes on the mechanical properties and service performance of tempered steels.

## Acknowledgements

The support of the Army Research Office and the AMAX Foundation for studies in tempering of medium and high carbon steels at the Colorado School of Mines is deeply appreciated.

## References

1.    G.R. Speich and W.C. Leslie, "Tempering of Steel", Metallurgical Transactions, 3 (1972), pp. 1043-1054.

2.    G.R. Speich, "Tempered Ferrous Martensitic Structures" in Metallography Structures and Phase Diagrams, vol. 8, Metals Handbook, Eighth Edition, American Society for Metals, 1973.

3.    Y. Imai, "Phases in Quenched and Tempered Steels", Transactions Japan Institute of Metals, 16 (1975), pp. 721-734.

4.    The Peter G. Winchell Symposium on the Tempering of Steel, Metallurgical Transactions A, 14A (1983), pp. 991-1145. (Also available as a separately bound collection of paper, TMS-AIME, Warrendale, Pennsylvania.)

5.    C.S. Roberts, B.L. Averbach, and M. Cohen, "The Mechanism and Kinetics of the First Stage of Tempering", Transactions ASM, 45 (1953), pp. 576-604.

6.    B.S. Lement, B.L. Averbach, and M. Cohen, "Microstructural Changes on Tempering Iron-Carbon Alloys", Transactions ASM, 46 (1954), pp. 851-881.

7.    F.E. Werner, B.L. Averbach, and M. Cohen, "The Tempering of Iron-Carbon Martensite Crystals", Transactions ASM, 49 (1957), pp. 823-841.

8.    B.S. Lement, B.L. Averbach, and M.Cohen, "Further Study of Microstructural Changes on Tempering Iron-Carbon Alloys", Transactions ASM, 47 (1955), pp. 291-319.

9.    G.B. Olson and M. Cohen, "Early Stages of Aging and Tempering of Ferrous Martensite", Metallurgical Transactions A, 14A (1983), pp. 1057-1065.

10. P.G. Winchell and M. Cohen, "Solid Solution Strengthening of Martensite by Carbon" in Electron Microscopy and Strength of Crystals, Interscience Publishers, (1963), pp. 995-1007.

11. P.G. Winchell and M. Cohen, "The Strength of Martensite", Transactions ASM, 55 (1962), pp. 347-361.

12. G.R. Speich, "Tempering of Low-Carbon Martensite", Transactions TMS-AIME, 245 (1969), pp. 2553-2564.

13. S.J. Donachie and G.S. Ansell, "The Effect of Quench Rate on the Properties and Morphology of Ferrous Martensite", Metallurgical Transactions A, 6A (1975), pp. 1863-1875.

14. G.S. Ansell, S.J. Donachie, and R.W. Messler, Jr., "The Effect of Quench Rate on the Martensitic Transformation in Fe-C Alloys", Metallurgical Transactions, 2 (1971), pp. 2443-2449.

15. D.G. Hennessy, V. Sharma, and G. Ansell, "Tempering Kinetics of Alloy Steels as a Function of Quench Rate and Ms Temperatures", Metallurgical Transactions A, 14A (1983), pp. 1013-1019.

16. P.C. Chen, B.O. Hall, and P.G. Winchell, "Atomic Displacements Due to C in Fe-Ni-C Martensite", Metallurgical Transactions A, 11A (1980), pp. 1323-1331.

17. P.C. Chen and P.G. Winchell, "Martensite Lattice Changes During Tempering", Metallurgical Transactions A, 11A (1980) pp. 1333-1339.

18. A.M. Sherman, G.T. Eldis, and M. Cohen, "The Aging and Tempering of Iron-Nickel-Carbon Martensite", Metallurgical Transactions A, 14A (1983), pp. 995-1005.

19. S. Nagakura, Y. Hirotsu, M. Kusunoki, T. Suzuki, and Y. Nakamura, Metallurgical Transactions A, 14A (1983), pp. 1025-1031.

20. M.K. Miller, P.A. Beaver, S.S. Brenner, and G.D.W. Smith, "An Atom Probe Study of the Aging of Iron-Nickel-Carbon Martensite", Metallurgical Transactions A, 14A (1983), pp. 1021-1024.

21. M. van Rooyen and E.J. Mittemeijer, "Differential Thermal Analysis of Iron-Carbon Martensites", Scripta Metallurgica, 16 (1982), pp. 1255-1260.

22. E.J. Mittemeijer and F.C. van Doorn, "Heat Effects of Precipitation Stages on Tempering of Carbon Martensites", Metallurgical Transactions A, 14A (1983), pp. 976-977.

23. M. Hillert, "The Kinetics of the First Stage of Tempering", Acta Metallurgica, 7 (1959), pp. 653-658.

24. W.K. Choo and R. Kaplow, "Mössbauer Measurements on the Aging of Iron-Carbon Martensite", Acta Metallurgica, 21 (1973), pp. 725-732.

25. R. Kaplow, M. Ron, and N. DeCristofaro, "Mössbauer Effect Studies of Tempered Martensite", Metallurgical Transactions A, 14A (1983), pp. 1135-1145.

26. S. Nagakura, K. Shiraishi, and M. Toyoshima, "High Resolution Electron Microscopy Observation of Iron-Carbon Martensite", Proceedings of the First JIM International Symposium on New Aspects of Martensitic Transformation, (1976), The Japan Institute of Metals, Sendai, Japan, pp. 299-304.

27. M. Kusunoki and S. Nagakura, "Modulated Structure of Iron-Carbon Martensite Studied by Electron Microscopy and Diffraction", Journal of Applied Crystallography, 14 (1981), pp. 329-336.

28. S. Nagakura and M. Toyoshima, "Crystal Structure and Morphology of the Ordered Phase in Iron-Carbon Martensite", Transaction of the JIM, 20 (1979), pp. 110-110.

29. K.H. Jack, "Structural Transformations in the Tempering of High Carbon Martensitic Steel", Journal of the Iron and Steel Institute, 169 (1951), pp. 26-36.

30. M.G.H. Wells, "An Electron Transmission Study of the Tempering of Martensite in a Fe-Ni-C Alloys", Acta Metallurgica, 12 (1964), pp. 389-399.

31. Y. Hirotsu and S. Nagakura, "Crystal Structure and Morphology of the Carbide Precipitated from Martensitic High Carbon Steel During the First Stage of Tempering", Acta Metallurgica, 20 (1972), pp. 645-655.

32. Y. Hirotsu and S. Nagakura, "Electron Microscopy and Diffraction Study of the Carbide Precipitated at the First Stage of Tempering of Martensitic Medium Carbon Steel", Transaction JIM, 15 (1974), pp. 129-134.

33. Y. Hirotsu, Y. Itakura, K-C. Su, and S. Nagakura, "Electron Microscopy and Diffraction Study of the Carbide Precipitated from Martensitic Low and High Nickel Steels at the First Stage of Tempering", Transactions JIM, 17 (1976), pp. 503-513.

34. D.L. Williamson, K. Nakazawa, and G. Krauss, "A Study of the Early Stages of Tempering in an Fe-1.2%C Alloy", Metallurgical Transactions A, 10A (1979), pp. 1351-1363.

35. G. Krauss and A.R. Marder, "The Morphology of Martensite in Iron Alloys", Metallurgical Transactions, 2 (1971), pp. 2343-2357.

36. T.A. Balliett and G. Krauss, "The Effects of the First and Second Stages of Tempering on Microcracking in Martensite of an Fe-1.22% Alloy", Metallurgical Transactions A, 7A (1976), pp. 81-86.

37. R.L. Brown, H.J. Rack, and M. Cohen, "Stress Relaxation During the Tempering of Hardened Steel", Materials Science and Engineering, 21 (1975), pp. 25-34.

38. C.S. Roberts, "Effect of Carbon on the Volume Fractions and Lattice Parameters of Retained Austenite and Martensite", Transactions AIME, 197 (1953), pp. 203-204.

39. A.R. Marder and G. Krauss, "The Morphology of Martensite in Iron-Carbon Alloys", Transactions ASM, 60 (1967), pp. 651-660.

40. C. Wells, W. Batz, and R.F. Mehl, "Diffusion Coefficient of Carbon in Austenite", Transactions AIME, 188 (1950), pp. 553-560.

41. G. Thomas, "Retained Austenite and Tempered Martensite Embrittlement", Metallurgical Transactions A, 9A (1978), pp. 439-450.

42. D.L. Williamson, R.G. Schupmann, J.P. Materkowski, and G. Krauss, "Determination of Small Amounts of Austenite and Carbide in Hardened Medium Carbon Steel by Mössbauer Spectroscopy", Metallurgical Transactions A, 10A (1979), pp. 379-382.

43. J.P. Materkowski and G. Krauss, "Tempered Martensite Embrittlement in SAE 4340 Steel", Metallurgical Transactions A, 10A (1970), pp. 1643-1651.

44. M. Sarikaya, A.K. Jhingan, and G. Thomas, "Retained Austenite and Tempered Martensite Embrittlement in Medium Carbon Steels", Metallurgical Transactions A, 14A (1983), pp. 1121-1133.

45. F. Zia-Ebrahimi and G. Krauss, "The Evaluation of Tempered Martensite Embrittlement in 4130 Steel by Instrumented Charpy V-notch Testing", Metallurgical Transactions A, 14A (1983), pp. 1109-1119.

46. C-B. Ma, T. Ando, D.L. Williamson, and G. Krauss, "Chi-Carbide in Tempered High Carbon Martensite", Metallurgical Transactions A, 14A (1983), pp. 1033-1045.

47. T. Swarr and G. Krauss, "The Effect of Structure on the Deformation of As-Quenched and Tempered Martensite in an Fe-0.2%C Alloy", Metallurgical Transactions A, 7A (1976), pp. 41-48.

48. R.W. Caron and G. Krauss, "The Tempering of Fe-C Lath Martensite", Metallurgical Transactions, 3 (1972), pp. 2381-2389.

49. R.M. Hobbs, G.W. Lorimer, and N. Ridley, "Effect of Silicon on the Microstructure of Quenched and Tempered Medium-Carbon Steels", Journal of the Iron and Steel Institute, 210 (1972), pp. 757-764.

50. Y. Ohmori, "χ-Carbide Formation and Its Transformation into Cementite During the Tempering of Martensite", Transactions JIM, 13 (1972), pp. 119-127.

51. K.H. Jack and S. Wild, "Nature of X-Carbide and Its Possible Occurrence in Steels", Nature, 212 (1966), pp. 248-250.

52. S. Nagakura, T. Suzuki, and M. Kusunoki, "Structure of the Precipitated Particles at the Third Stage of Tempering of Martensitic Iron-Carbon Steel Studied by High Resolution Electron Microscopy", Transactions of JIM, 22 (1981), pp. 699-709.

53. R.G. Baker and J. Nutting, "The Tempering of 2 1/4Cr-1Mo Steel After Quenching and Normalizing", Journal of the Iron and Steel Institute, 192 (1959), pp. 257-268.

54. R.W.K. Honeycombe, Steels, Microstructure and Properties, American Society for Metals, 1981, pp. 152-165.

55. S.J. Barnard, G.D.W. Smith, A.J. Garratt-Reed, and J. Vander Sande, "Atom Probe Studies: (1) The Role of Silicon in the Tempering of Steel and (2) Low Temperature Chromium Diffusivity in Bainite", in Solid Phas Transformations, H.I. Aaronson, et.al. editors, TMS-AIME, (1982).

56. B.G. Reisdorf, "The Tempering Characteristics of Some 0.4 pct Carbon Ultra-High-Strength Steels", Transactions TMS-AIME, 227 (1963), pp. 1334-1341.

57. W.S. Owen, "The Effect of Silicon on the Kinetics of Tempering, Transactions ASM, 46 (1954), pp. 812-829.

58. R.W. Heckel and H.W. Paxton, "Rates of Growth of Cementite in Hypereutectoid Steels", Transactions TMS-AIME, 218 (1960), pp. 799-806.

59. T. Ando and G. Krauss, "The Effect of Phosphorus Content on Grain Boundary Cementite Formation in AISI 52100 Steel", Metallurgical Transactions A, 12A (1981), pp. 1283-1290.

60. T. Ando and G. Krauss, "The Isothermal Thickening of Cementite Allotriomorphs in a 1.5Cr-1C Steel", Acta Metallurgica, 29 (1981), pp. 351-363.

61. T. Ando and G. Krauss, "Development and Application of Growth Models for Grain Boundary Allotriomorphs of a Stoichiometric Compound in Ternary Steels", Metallurgical Transactions A, 14A (1983), pp. 1047-1055.

63. J. Pilling and N. Ridley, "Tempering of 2.25Cr-1Mo Low Carbon Steels", Metallurgical Transactions A, 13A (1982), pp. 557-563.

64. C.J. McMahon, Jr., A.K. Cianelli, and H.C. Feng, "The Influence of Mo and P-induced Temper Embrittlement in Ni-Cr Steel", Metallurgical Transactions A, 8A (1977), pp. 1055-1057.

65. J. Yu and C.J. McMahon, Jr., "The Effects of Composition and Carbide Precipitation on Temper Embrittlement of 2.25Cr-1Mo Steel, Metallurgical Transactions A, 11A (1980), pp. 287-289 and pp. 291-300.

66. Z. Qu, and C.J. McMahon, Jr., "The Effects of Tempering Reactions on Temper Embrittlement of Alloy Steels", Metallurgical Transactions A, 14A (1983), pp. 1101-1108.

67. C.J. McMahon, Jr., "Temper Brittleness--An Interpretive Review", in Temper Embrittlement in Steel, ASTM STP No. 407, 1968, pp. 127-167.

68. H. Ohtani, H.C. Feng, C.J. McMahon, Jr., and R.A. Mulford, "Temper Embrittlement of Ni-Cr Steel by Antimony 1. Embrittlement at Low Carbon Concentration", Metallurgical Transactions A, 7A (1976), pp. 97-101.

69. S. Takayama, T. Ogura, S.C. Fu, and C.J. McMahon, Jr., "The Calculation of Transition Temperature Changes Due to Temper Embrittlement", Metallurgical Transactions A, 11A (1976), pp. 1513-1530.

70.  M. Guttman, "The Link Between Equilibrium Segregation and
     Precipitation in Ternary Solution Exhibiting Temper
     Embrittlement" <u>Metal Science</u>, 10 (1976), pp. 337-341.

71.  M. Guttman, Ph. Dumonlin, and M. Wayman, "The Thermodynamics of
     Interactive Co-Segregation of Phosphorus and Alloying Elements in
     Iron and Temper-Brittle Steels", <u>Metallurgical Transactions A</u>,
     13A (1982), pp. 1693-1711.

72.  R.A. Grange, C.R. Hibral, and L.F. Porter, "Hardness of Tempered
     Martensite in Carbon and Low Alloy Steels", <u>Metallurgical
     Transactions A</u>, 8A (1977), pp. 1775-1785.

# PREDICTION OF MICROSTRUCTURE AND HARDENABILITY IN LOW ALLOY STEELS

J.S. Kirkaldy and D. Venugopalan

Department of Metallurgy and Materials Science
McMaster University
Hamilton, Ontario, Canada L8S 4L7

A multicomponent theoretical structure based on rigorous thermodynamics and approximate isothermal phase transformation kinetic equations for ferrite, pearlite and bainite has been fitted to transformation start curves from the U.S.S. Atlas of Isothermal Transformation Diagrams, thus providing a universal predictor for the IT diagrams of most hardenability (H) steels. It is demonstrated that valid CCT curves for arbitrary cooling regimes can be derived therefrom using the additivity rule, and that the microstructural mixes thus predicted, when combined with empirical hardness formulas, yield fair to good approximations to observed Jominy curves. Thus it is confirmed that the main alloy interactions or synergisms can be understood on the basis of multicomponent thermodynamics and kinetics. With further refinement and calibration of the formulas to microstructural and hardness data an accurate predictor of CCT and Jominy curves is attainable.

## Introduction

During the first three quarters of this century it has generally been accepted that the ability of low alloy steels to be hardened to depth, the dictionary definition of hardenability, is predicated upon the nucleation and growth rates of equilibrium ferrite and pearlite and their metastable forms such as martensite and bainite. Microalloying, whether by artifact in steel-making, or by design has undermined this simplistic viewpoint. Oxygen, nitrogen, boron, vanadium and niobium have proven hardening or softening effects which do not fit into the conventional pattern. We reluctantly conclude that such effects must be included within a comprehensive analysis and synthesis. Yet the line must be drawn somewhere if descriptive economy is to be achieved. Accordingly, we shall regard these specific effects as marginal but quantifiable artifacts, recognizing that micro-alloyed steels are designed for a different thermomechanical regime than considered here.

The aim of a science of hardenability is to predict for a given geometry the microstructures of an arbitrary low alloy steel according to the relevant cooling regime, and to semi-empirically derive the significant mechanical properties therefrom. In principle, such a calculation involves as a very minimum the prediction of stable and metastable multicomponent phase diagrams, the corresponding isothermal transformation (IT) diagrams, the continuous cooling transformation (CCT) diagrams which correspond to each cooling regime and semi-empirical or empirical correlations of structure versus hardness. Since there does not exist a unique theoretical method of proceeding, nor indeed a unique set of data whereby to test a procedure, the developer is faced simultaneously with an optimization and selection problem. We know, however, that pure empiricism has served reasonably well in specific applications so our selection of the optimal procedure must not ignore the lessons of the past.

## Empirical Methods for Predicting Microstructure and Hardenability

Because hardenability depends on microstructure, the latter being derived via kinetics combined with discrete equilibrium diagrams, it is in principle discontinuous in the compositions. Notwithstanding, kinetic processes have a smoothing influence on the usual measures of hardenability so that their observational variations also tend to be reasonably smooth. Thus a Taylor series representation of hardness observations of the form

$$H = H_0 + \sum a_i C_i + \frac{1}{2} \sum_{ik} a_{ik} C_k C_i + \ldots \tag{1}$$

may appear to be justified for limited ranges of the composition $C_i$ and other variables. Typically such a formula is applied to selected depths in a Jominy bar. Since the data set is usually insufficient for quadratic regression, regression analysis is used to determine the linear coefficients only (1), and the higher order terms are neglected. Brown and James have suggested that this is an adequate procedure only over the band of a single grade (2). Such formulas must never be used out of context, and are thus unreliable for alloy development.

The conventional Grossmann method (3) is a near-relative of the linear regression method and refers to a depth of hardening index such as ideal critical diameter (50% martensite at center of bar quenched at infinite severity), often correlated to the Jominy point. The product form for the most dilute alloys

$$D_I = D_I^0(\%C, \text{ grain size}) \cdot \prod_i (1 + a_i C_i) = D_I^0 \cdot \prod_i f_i \tag{2}$$

(where $D_I^0$ is the plain carbon limit) possesses all positive terms (a posynomial). Grossmann thought that this might be sufficiently general, but, it has turned out not to be the case. Hence, even with quadratic $a_i$, a myriad of inconsistent sets of multiplying factors $f_i$ have been reported (c.f. Fig. 1).

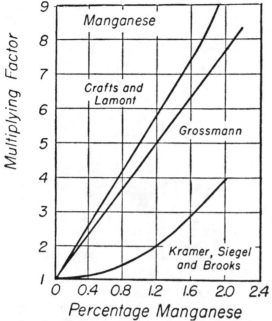

Figure 1 - Variability of Mn hardenability factors to be used in the Grossmann formula (4).

Users have found that the formula performs adequately in a given alloy range provided that the appropriate set of coefficients is chosen. Note that the Grossmann formula is less general than a quadratic regression formula, for a discrete set of its second order coefficients, like the first order coefficients from which they derive, are positive. Charts based on empirical data are available for generating complete Jominy curves from the carbon content and $D_I$ values, and these have been incorporated in a number of slide rules (Bethlehem Steel, U.S. Steel, Climax Molybdenum). Such predictors, being occasionally in error by a factor of 2, can be regarded as only rough guides in a general context. The tabular  multiplication factors published annually by ASM are equally defective (5).

Just at Volkswagenwerk has reported a number of non-linear equations (6,7) based separately on SAE bands, the U.S. Steel Atlas and Max-Planck Institute collections covering the usual composition ranges of AISI-SAE H-steels, e.g., 0.08-0.64C, 0.15  - 1.95 Si, 0.20-1.88 Mn, 0-5 Ni, 0-1.55 Cr, 0-0.52 Mo and 0-0.2 V, ASTM grain sizes G from 2 to 11 and depths from 4/16 to 25/16($J_{4-25}$) in a Jominy bar. A separate formula is given for $J_1$ ($R_c$), viz.,

$$J_1 = 52(\%C) + 1.9(\%Mn) + 1.4(\%Cr) + 33 \qquad (3)$$

The general shape formula fitting the mid-line of the SAE bands is

$$J_{4-25} = 98\sqrt{C} - 0.025 \; E^2 \; \sqrt{C} - 24 \; \sqrt{E} + 2.86E - 7$$

$$+ \; 19(\%Mn) + 6.4(\%Ni) + 20(\%Cr) + 34(\%Mo) + 28(\%V) \tag{4}$$

where E is the Jominy depth in sixteenths of an inch. A similar formula based on 37 steels from the U.S. Steel Atlas suggests that by normalization to 0.3% $S_i$ and G=8 for more generality, the constant -7 can be replaced by

$$(-7) \equiv 5(\%Si) - 0.82 \; G - 2 \tag{5}$$

This formula fits the middle of the bands at all depths to within ±2.4 RC r.m.s.; the regression fits at a 0.96 significance level. It is remarkable that this level of correlation can be obtained without taking account of the so-called synergistic, or product effects, so often attributed to Grossmann type formulas. Just nonetheless cautions that "we should not expect the formulas to predict a steel's hardenability precisely". They are to "assist the designer in determining the steel to select, and to help the metallurgist in correcting the melt". While Just's formulas may not be better than a particular Grossmann formula in the latter's favoured context (8), they should be a better overall predictor for the entire range of H-steels. As can be seen, they are in a very convenient form for use by engineers in a programmable hand calculator.

The foregoing procedures are concerned with the correlation of composition and grain size to hardness without reference to kinetics or thermal configurations. Because the available data sets are generally inadequate and often inaccurate, an insufficient number of regression coefficients can be accurately determined. For example, if we were to attempt an accurate evaluation of all the coefficients in Eqn 1 for predicting the Jominy curves of the H-steels with typically 10 input variables, there are 111 coefficients to be evaluated. To do this adequately we would require a matrix of about 25 experimental Jominy curves for each of $10^5$ variants (alloys and grain sizes), for a total of 2.5 million uniformly distributed curves. Grossmann attempted to reduce the necessary data set by assuming a linear-posynomial representation, while Just sought a rationalization in terms of the absence of certain functional dependencies between composition, E and C variables and a law of diminishing returns for the carbon effect. In both cases there is an implicit assumption concerning the complex underlying kinetic effects, and an inherent inaccuracy deriving from the tenuousness of the global hypotheses. The obvious next step is to seek an explicit representation of the kinetic effects. Following this strategy, it is possible to proceed in a purely theoretical, semi-empirical or purely empirical way.

The Creusot-Loire system follows the latter course (9,10). The hardness of a given alloy is assumed to be determined by a cooling rate weighted law of mixtures for the microstructural constituents:  martensite (as a function of carbon content), ferrite-pearlite and bainite. The phase mixture is determined by interpolation from a set of formulas based on a method of linear regression on critical cooling rates for a selected set of 341 CCT diagrams from various compendia. This method has the advantage of giving an estimate of the microstructural mix as well as the hardness distribution in an arbitrary sample for which the mean thermal decrement per unit time is locally specified in the transformation temperature range 800°-500°C. The formulas are sufficiently compact that Jominy curves have been generated by a hand-held programmable calculator (11).

In the analysis of the CCT curves by linear regression it is assumed that the logarithm of the critical cooling rate for onset, specific percent transformed or completion is linear in the composition and a separately defined grain size variable. This hypothesis, together with multiple averaging and interpolation is presumed to obviate the necessity for second order terms as in the Taylor expansion of Eqn 1. The number of necessary coefficients is thus reduced substantially. Since CCT curves and the coefficients in the law of mixtures are difficult (and expensive) to establish, and are often inaccurate due to the imprecise methodology (e.g., dilatometry and phase analysis) the formulas cannot be expected to perform appreciably better than those of Just in a hardenability application unless calibrated to actual Jominy data. At the same time, the microstructure predictions have other important applications, and so the rather onerous and expensive methodology is justified overall. The system is described in detail in references 9 and 10.

Hildenwall and Ericsson (12) at Linköping and Sakamoto et al. at Kawasaki Steel (13) have proceeded on a slightly less empirical basis by fitting Johnson-Mehl-Cahn (JMC) type formulas (14,15) to known IT diagrams. The Japanese workers have developed expressions for the coefficients in these formulas which are optimized to the U.S. Steel Atlas data set. Thus IT curves predicted via these formulas can be converted to CCT curves and microstructure via the additivity rule (see below) and appropriate cooling curves, and hence to hardness distributions via empirical formulas.

The Minitech Hardenability Predictor developed by one of us (JSK), recognizes that the amount and precision of available data is generally insufficient for a purely empirical evaluation of quadratic and higher order synergistic effects and seeks a reduction in the empirical requirements by drawing extensively on the theory of phase constitution and austenite decomposition in multicomponent alloys (16-19). The basic theoretical idea for the strong alloy interactions is a simple one, argued in early publications (20,16,18). The main additions in hardenability steels can be generally classified as either raising or lowering the $Ae_3$ line of the $Fe-Fe_3C$ phase diagram (ferrite or austenite stabilizers, respectively). Carbon, manganese, copper and nickel fall into the latter class while silicon molybdenum and vanadium fall into the former class. Chromium is an austenite stabilizer at low concentrations (<8%) and a ferrite stabilizer at high concentrations. The austenite stabilizers decrease the supersaturation for a given transformation temperature so on this basis alone would be expected to decrease reaction rates and increase hardenability. Ferrite stabilizers, on the other hand, would be expected to decrease the hardenability. However, there is the further kinetic effect that all substitutional elements tend to inhibit the austenite decomposition reactions through partitioning by phase-boundary diffusion, and in particular the austenite $\rightarrow$ ferrite + pearlite reaction. Molybdenum and chromium are notable in this respect. Because the rates decrease exponentially with temperature the best combination for an overall slow reaction rate will be a combination of strong $Ae_3$ depressors with strong rate depressors. It is no coincidence that the manganese-molybdenum combination is one of the best on a percentage added basis. These concepts are incorporated rationally into Zener-Hillert (ZH) IT ferrite and pearlite growth formulas (21,22) to provide a semiempirical prediction of IT start curves. From this a CCT start curve is derived via the additivity rule (23-25) and the combined calculations together with the cooling curve relations for a Jominy bar are used to predict the Jominy inflection point depth. This is then combined with a set of carbon-dependent empirical Jominy curve shapes to identify a particular Jominy curve. In a blind test carried out by Republic Steel Laboratories this predictor outperformed four different Grossmann type or regression predictors 5 out of 6 on alloy grades and 8 out of 8 on boron grades. A number of commercial users regard the predictions, based on accurate chemistry, as more reliable in assessing performance than routine Jominy testing.

While the IT formula in the predictor was initially calibrated to an empirical IT set, the coefficients in the linear sub-sections of the theory have now been varied within a hardness optimization over one thousand actual Jominy curves provided by the AISI and over 30 international corporations. As well, a number of quadratic and higher order carbon-alloy and alloy-alloy interactions have been introduced to produce accurate predictions at the high C (0.65%) and high alloy (Cr and Mo) end of the formula range. These effects are related to bainite and complex carbide formation and carbide retention which are not incorporated in the elementary IT formulas. There are approximately fifty independent coefficients in the full formula, comparable to the number in the Creusot-Loire formula, and these are optimized to actual Jominy data covering more than two hundred experimental and commercial grades. As well as fitting the raw data set (c.f. Figs. 2 and 3) the outputs of the program are in detailed agreement with all of the SAE H-bands for alloy and boron grades and for EX grades which replace them.

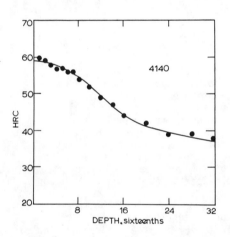 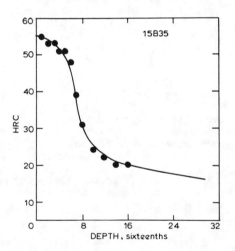

Figure 2 - Comparison of observed and predicted Jominy curve for a 4140 steel (17).

Figure 3 - Comparison of observed and predicted Jominy curve for a 15B35 steel (17).

Needless to say, such great generality and detailed accuracy in given instances does not guarantee an ability to predict a given experimental Jominy curve. Hardenability varies markedly with plant practice, inclusion content and unreported residuals such as nitrogen and oxygen. Furthermore, grain size estimates are notoriously unreliable, routine Jominy testing is subject to grave inaccuracies and fraud in reporting test results is not unknown.    Minitech currently takes the view that for optimum accuracy for on-line predictions in steelmaking the formulas must be calibrated to each grade and company practice. Of course, rigorous adherence to standard test procedures for monitoring the calculations must be maintained. It is only in this way that predictions reliable to within $\sigma = 1.5$ $R_c$ over an entire curve may be maintained. Such accuracy is essential in support of alloy substitution and steelmaking procedures which use on-line additions calculations to achieve Jominy bands which are approximately 1/3 the normal SAE width or which aim for alloy cost minimization within normal bands. German and Japanese producers have claimed such capabilities in recent years.

This article, by extending the foregoing, lays the foundations for a predictor which follows the same general principles but which is capable of generating microstructural volume fractions as well as hardness distributions for arbitrary shapes and thermal distributions.

## Structure of a Possible Complete Calculation

The general outline of a fundamental calculation could have been constructed prior to 1940 when Scheil and Avrami suggested how the description of isothermal phase transformations could be extended to continuous cooling situations (23-25). The isothermal case, through the researches of Scheil, Mehl and coworkers were reasonably well understood before 1940 (14,23) and perfected by Zener (21), Hillert (22), Cahn (26,27) and others before 1960. The elementary structure proposed by one of us (JSK) in 1973 for pearlite (16) assumes that from thermodynamics we are given an expression for the effective eutectoid temperature as a function of composition; that from nucleation theory we are given an expression for the incubation time (IT start curves) as a function of undercooling and composition; that we are given a procedure for converting an IT curve to a CCT curve; that we are given the grain size and a realistic model for a site-saturated distribution of pearlite nuclei at the end of the incubation time; and that we are given an expression for the pearlite growth velocity as a function of composition and temperature. The corresponding algorithm is given in reference 16. While of limited scope, and not particularly successful in a quantitative way, the foregoing procedure served to identify the limitations and pitfalls of a purely theoretical approach and led us to a rather compact framework for a more comprehensive and critical semi-empirical approach which incorporates the JMC global structures, the ZH local kinetic expressions and the Scheil-Avrami additivity rule.

## The Master Formulas for IT and CCT Curves (28)

Consider an empirical CCT diagram consisting of several C-curves for a set of non-overlapping cooling curves of the form

$$T = T(x,t) = T(T_{CCT},t) \tag{6}$$

where $x(T_{CCT})$ is a unique, smoothly varying location index for each cooling curve in the original test sample. We thus assume that the coupling of cooling regimes due to latent heat evolution and (or) stress is insignificant. Let X be the fraction of a unit volume transformed and assume that aging in the transformed material has no effect on the untransformed material. We will describe the overall process as a chemical reaction of arbitrary order expressed in terms of free energy evolution and prior grain size G, viz.,

$$\frac{d(\Delta F)}{dt} = -k(G,T,\dot{T},X,\dot{X})(\Delta F)^n \tag{7}$$

where the effective reaction order may be as general and complex as

$$n = n(G,T,\dot{T},X,\dot{X}) \tag{8}$$

Now without loss in generality we can substitute the free energy difference

$$\Delta F = (1-X) \; f(G,T,\dot{T},X,\dot{X}) \qquad (9)$$

and noting that f depends strongly on T and $\dot{T}$ and and $\dot{X}$ approach constants whence $(d\dot{T}/dt)(\partial f/\partial x)$ and $(\partial \dot{X}/\partial t)(\partial f/\partial \dot{X}) \sim 0$, transform Eqn (7) via Eqn (6) to

$$\frac{dX}{dt} = g(G,X,T,x) = h(G,t,x) \qquad (10)$$

verifiable via Eqn (6) and the integral over g(G,X,t,x),

$$X = X(G,t,x) \qquad (11)$$

for the total amount transformed subject to X = 0 at t = 0 along any cooling curve indexed by fixed x.  Changes in phase mixture must lie implicitly within the construction of the rate constant k.  This abstract formulation immediately suggests an alternative to the conventional Johnson-Mehl formalism for describing isothermal transformations based on the extended volume concept (14) by operating upon equation (7).  For the isothermal case and an approximately surface-localized reaction (e.g., pearlite) we can take, for example, n ≈ 1 and write Eqn 9 as

$$\Delta F \approx (1-X) \; \frac{\Delta H \Delta T}{T_0} \qquad (12)$$

where $\Delta H$ ($\sim$ constant) is the mean enthalpy change of reaction per unit volume and $\Delta T = T_0 - T$ is the undercooling.  Hence, with $\dot{T} = 0$, Eqn (7) becomes

$$\frac{dX}{dt} \approx k(G,T,X)(1-X) \qquad (13)$$

For a general reaction with point or localized nucleation and termination one can factor g in Eqn (10) such that

$$\frac{dX}{dt} = \mathcal{D}(G,T,X)X^m(1-X)^p \qquad (14)$$

The IT curve is then given by

$$\tau(X,T) = \int_0^X \frac{dX}{\mathcal{D}(G,T,X)X^m(1-X)^p} \qquad (15)$$

where $\mathcal{D}$ is to be interpreted as an effective non-zero rate coefficient and m < 1, p < 1 to assure convergence.  This can be matched for early times to JMC type formulas and is free of the ambiguities which attend the extended or phantom volume concept for long times and (or) second product origination and evolution.  A wide range of nucleation mechanisms, and mixtures of ferrite with pearlitic, bainitic and martensitic structures can be subsumed by Eqn 15 or generalizations involving rational or non-rational powers of X and 1-X. Note that Eqn (15) through (9) is inclusive of diffusion impingement effects.

We recognize next that both IT and CCT curves are solutions, $T_f(t,x)$, to the integral identity

$$\int_{T_i}^{T_f} dX \equiv X \qquad (16)$$

where X is a fixed fraction transformed (e.g., 0.01), $T_i$ is the initiation temperature for transformation (e.g., the $Ae_3$ temperature for an alloy steel) and $T_f$ is the final or limiting temperature for the integral corresponding to X. We can thus write

$$\int_{T_i}^{T_f} dX = \int \frac{dX}{dt}\, dt = \int h\, dt = \int_{T_i}^{T_f} \overline{h}\, dt = X \tag{17}$$

the line integral being independent of the path defined by Eqn (6). Thus for an infinitely fast quench followed by an isothermal hold at temperature T up to a fixed value of X we may infer from Eqns (10) and (11) the time average transformation rate as

$$\overline{h} = \frac{X}{\tau} \tag{18}$$

where $\tau(T)$ designates the IT curve. Substituting in Eqn (17) and cancelling the prescribed value of X on the left and right side we obtain with Avrami (25)

$$\int_{T_i}^{T_f} \frac{dt}{\tau} = 1 \tag{19}$$

independent of the path defined by Eqn 16. $\tau$ is therefore established as the integrating factor for the non-systemic or inexact differential, dt. This is a rigorous if rather subtle inference from Eqn 10 first recognized by Cahn (26), and represents the most general expression of the additivity rule. It is inclusive of equations (7) or (10) which are exceptionally general. In particular, the proposed $\overline{T}$ and T dependence of both k and n are capable of subsuming an arbitrary phase boundary extent (due to rate-induced instabilities), associated auto-catalytic nucleation effects and undercooling dependent nucleation mechanisms within untransformed material. It is our thesis that it is this generality which assures the success of Eqn (19) in its widely successful application.

## Kinetic Equations for Explicit Products and Mixtures

One of the problems unsolved to date pertains to finding a rational approach to mixed and competing product reactions, e.g., pearlite and ferrite as suggested by the schematic IT curve of Fig. 4. The treatment of pearlite and bainite alone is rather straightforward since site-saturated models with subsequent steady state growth are adequate representations. Ferrite transformation, on the other hand, consists of parabolic and steady state (Widmanstätten) components. Since our attempts to deal with this and the multi-product problem rigorously have been thus far unsuccessful we have taken recourse to an artificial average steady state growth model, recognizing that geometrical effects are more influential than local kinetics in determining the form of the overall reaction rate. We thus characterize the equivalent steady state ferrite reaction as a phantom discontinuous (pearlite-like) reaction in which the lamellar product is equilibrium ferrite and austenite according to the rigorous multicomponent phase diagram. If the temperature is such that the phantom austenite reaches the eutectoid composition then a subsequent or simultaneous pearlite reaction is presumed to occur in the austenite regions of the phantom austenite, its ZH type start curve being calibrated empirically against measured IT curves. To the phantom product we also assign a ZH type steady state velocity formula (21,22)

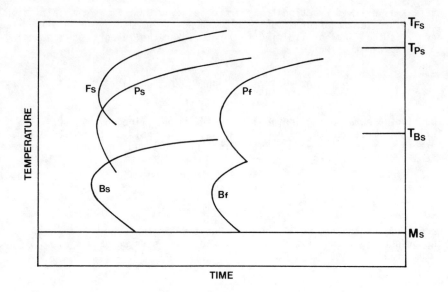

Figure 4 - Schematic isothermal transformation curve for a low
alloy steel.

$$\mathcal{D} \backsim v \backsim D\Delta T^q \tag{20}$$

where D is an effective diffusion coefficient, the exponent of the undercooling q is an empirical exponent determined by the effective diffusion
mechanism (for pearlite, q = 2 for volume and q = 3 for boundary diffusion
(20)) and the proportionality constant is to be determined empirically from
measured IT curves.

The general form of IT curve chosen in accord with the analysis described by Eqns (6) through (15) and ZH type formulas is

$$\tau(X,T) = \frac{1}{\alpha(G)D\Delta T^q} \int_0^X \frac{dX}{X^{2(1-X)/3}.(1-X)^{2X/3}} \tag{21}$$

This has the property for a single product like pearlite that it agrees with
Cahn's site saturation and local steady state models for nucleated spheres
as X→0 (15) with

$$\alpha = \beta.2^{(G-1)/2} \tag{22}$$

where G is the ASTM grain size, and β an empirical coefficient characteristic
of the particular reaction product, and it terminates the reaction by local
steady state closure of hypothetical spheres.  The implicit and very strong
assumption here is that the number of final closures is approximately equal
to the number of original nuclei.  Furthermore, the form of the exponents
in the denominator of the integrand has been chosen as the simplest possible
interpolation consistent with spherical nuclei and closures.  The ultimate
justification  lies in our ability to find coefficients β which adequately

represent all of the reported start and finish curves for ferrite, pearlite and bainite.

Experience and multi-component analysis has indicated that the diffusion resistance in reactions based on Fe-C can be represented to the first order and approximation by series resistance relations like (16,18)

$$\frac{1}{D} = \frac{1}{D_C} + \sum_{i=2}^{n} \frac{k_i C_i}{D_i} \tag{23}$$

provided the mechanism for all elements is the same. If the mechanism is not the same then an appropriate power of the undercooling appears in the denominator of some of the terms. There is considerable evidence that the phase-boundary model applies to all solutes in Fe-C pearlite at significant undercoolings. We can also conjecture, without threat of contradiction, that the hypothetical discontinuous ferrite reaction is similarly disposed. The bainite reaction, which involves no alloy partitioning ($k_i=0$) does not require expansion (23). However, it is apparent from the experimental IT curves that C, Cr and Mo all have strong retarding effects on the $B_{s,f}$ curves. In the absence of a quantitative theory we enter these effects empirically as described below. Our preliminary versions of the coefficients of the integral I in Eqn (21) optimized simultaneously over nose temperatures and times are represented for ferrite, pearlite and bainite, respectively, in terms of concentration weight %, viz

$$\tau_F = \frac{59.6(\%Mn)+1.45(\%N_i)+ 67.7\,(\%Cr)+ 244\,(\%M_o)}{2^{(G-1)/2} \cdot (\Delta T)^3 \cdot \exp(-23,500/RT)} \cdot I \tag{24}$$

$$\tau_p = \frac{1.79+5.42(\%Cr+\%Mo+4\%Mo \cdot \%N_i)}{2^{(G-1)/2} \cdot (\Delta T)^3 \cdot D} \cdot I \tag{25}$$

where D is evaluated according to eqn 23 as

$$\frac{1}{D} = \frac{1}{\exp(-27500/RT)} + \frac{0.01\%Cr+0.52\%Mo}{\exp(-37000/RT)} \tag{26}$$

and

$$\tau_B = \frac{(2.34+10.1\%C+3.8\%Cr + 19\%M_o) \cdot 10^{-4}}{2^{(G-1)/2} \cdot (\Delta T)^2 \cdot \exp(-27,500/RT)} \cdot I' \tag{27}$$

In Eqn (27) the integral I has been modified to I' to account for the very sluggish termination of the bainite reaction (see below). Note how very strong non-linearities have appeared through 1/D and $\Delta T^q$. The not-unreasonable values of the activation energies were determined by the optimal fit.

When one is concerned with an alloy and temperature for which the final structure is ferrite-austenite or ferrite-pearlite particular attention must be paid to the precise meaning of X in Eqn (21). For the discontinuous model for ferrite precipitation it is presumed that the phantom reaction goes to completion. However, the actual ferrite fraction for any phantom X(t) is $X_F= X \cdot X_{FE}$ where $X_{FE}$ is the attainable equilibrium fraction of ferrite. Thus if $\tau$ in Eqn (21) is to be expressed as an explicit function of $X_F$ we must everywhere in Eqn (21) set

$$X = X_F/X_{FE}$$    (28)

If austenite is the only second product then its volume fraction is $1-X_F$. If pearlite forms its fraction varies between 0 and $1-X_{FE}$ and we can express $\tau_P$ as an explicit function of $X_P$ by writing

$$X = X_P/(1-X_{FE})$$    (29)

We do not attempt to deal with the case where three reaction products appear. We simply assume that no bainite forms for all temperatures above the intersection of $P_S$ with $B_S$ and that no pearlite forms below the inter-section. If the intersection of $F_S$ with $B_S$ occurs below the intersection of $P_S$ with $B_S$ then we assume that below the $F_S$-$B_S$ intersection only bainite forms and that between $F_S$-$B_S$ and $P_S$-$B_S$ intersections a mix of ferrite and bainite forms. For alloys for which a segment of $P_S$ preceeds both $F_S$ and $B_S$ (Fig. 4) we assume that within that segment pearlite goes to 100% completion as some sort of degenerate structure. These assumptions have little effect on the prediction of CCT product quantities or hardnesses.

There remains the problem of how to evaluate the undercooling in Eqn (21) for ferrite, pearlite and bainite together with the $\tau = \infty$ asymptote of the IT curve ($\Delta T=0$, c.f., Equation 21). It has now been clearly established that the ferrite asymptote is the $Ae_3$ temperature (18,29,30). Fig. 5 shows our prediction of $Ae_3$ based on rigorous thermodynamic argument and multi-component data versus the asymptotes established by the U.S. Steel Atlas (31). The mean deviation is about 5°C, an insignificant error which can easily be

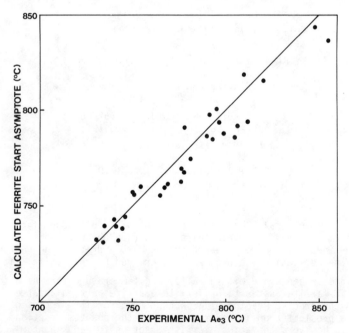

Figure 5 - Relation between the calculated ($Ae_3$) ferrite start asymptote and experimental $Ae_3$ temperature (31).

explained by experimental artifact. This procedure has been used in our foregoing numerical analysis. There still do not exist theories to define $\Delta T=0$ for the metastable phases bainite and martensite so we have accordingly evaluated our own linear regression formula for bainite from the data in the U.S.S. Atlas and Steven and Haynes (32). This is

$$T_{B_S} (°C) = 656 - 58\%C - 35\%Mn - 75\%Si - 15\%Ni - 34\%Cr - 41\%Mo \qquad (30)$$

We have adopted the $M_s$ formula of Steven and Haynes (32) and this is

$$M_s(°C) = 561 - 474\%C - 33\%Mn - 17\%Ni - 17\%Cr - 21\%Mo \qquad (31)$$

The pearlite reaction presents a special problem for in some cases the IT asymptote is determined by the multicomponent equilibrium limit of the austenite field and in other cases by a corresponding paraequilibrium limit. For example, Cr tends to partition strongly in pearlite at high temperatures so the asymptote for Cr containing steels tends towards the equilibrium value. Manganese and nickel, on the other hand, resist partitioning so Mn or Ni containing steels have apparent asymptotes which approach the para-equilibrium ideal. While we know how to calculate the stable and para-equilibrium phase diagrams in the special cases we do not yet know how to predict "a priori" the appropriate values for an arbitrary mix of thermo-dynamics and alloying elements. We proceed therefore with the following interpolation procedure for the pearlite asymptote,

$$T_{P_S} = Ap_1 + (Ae_1 - Ap_1) \frac{Cr}{(Cr+Mo+Ni)} \qquad (32)$$

Figure 6 shows the comparison between $T_{P_S}$ and the experimental pearlite asymptote from the U.S.S. Atlas (31) where $Ae_1$ and $Ap_1$ have been calculated via the rigorous thermodynamic formulas of Hashiguchi et al. (30).

## The Closure with Independent Experimentation

The empirical inputs to the foregoing concern equilibrium and metastable thermodynamics and optimization of theoretical formulas to isothermal nucleation and growth kinetics. A preliminary test of the elementary structure will now be presented based on the additivity rule (Eqn 19) for conversion of IT to CCT curves and subsequent conversion via predicted volume fractions and empirical hardness relations and cooling curves to hardness distributions or Jominy hardenability curves.

In our calculation of CCT curves for a Jominy bar, we have used the set of empirically developed Jominy cooling curves (18). The temperature along the Jominy bar as measured from the quenched end is given by

$$T(°K) = 297 + (T_a - 297)erf \frac{x}{2\sqrt{\eta t}} \qquad (33)$$

where $T_a$ is the austenitization temperature, x is the depth in cm, t is the time and the effective thermal diffusivity is the function of depth

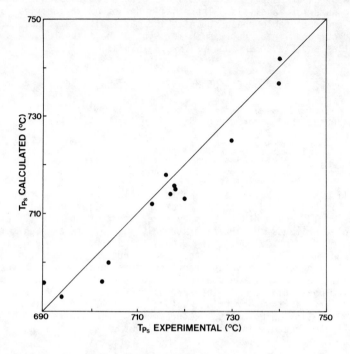

Figure 6 - Comparison of the calculated pearlite asymptote with experimental data from the USS Atlas.  Grades include 1340, 4027, 4037, 4068, 5140, 4140, 4317, 4340, 4640, 8630, 8660, 8745 and 9420. Also included is a 1%Ni-0.6%C steel.

$$\eta = \eta_2 + (\eta_1 - \eta_2)(T_a - 1116)/83.3 \ cm^2/sec \tag{34}$$

where

$$\eta_1 = (3.226 + 55.31x)/1000; \ x < 0.572$$

$$\eta_1 = (23.23 + 18.90x)/1000; \ x \geqq 0.572 \tag{35}$$

$$\eta_2 = (3.226 + 39.87x)/1000; \ x < 0.826$$

$$\eta_2 = (23.23 + 14.99x)/1000; \ x \geqq 0.826 \tag{36}$$

From eqn 33 we obtain

$$\dot{T} = -(T_a - 297) \ \frac{x}{2\sqrt{\pi\eta} \ t^{3/2}} \ exp(-x^2/4\eta t)$$

$$= -(T_a - 297) \frac{4\eta}{\sqrt{\pi}\, x^2} \text{inverf}^3\phi \, \exp(-\text{inverf}^2\phi) \qquad (37)$$

where

$$\frac{x}{2\sqrt{\eta t}} = \text{inverf}\phi = \text{inverf}\left(\frac{T-297}{T_a-297}\right)$$

$$\approx \frac{\sqrt{\pi}}{2}\phi + 0.4406\,\phi^{3.725} \qquad (38)$$

The Diamond Pyramid Hardnesses (HV) are calculated via the relations of Maynier et al. (10). Their formulas in terms of weight% are

$$HV_{(martensite)} = 127+949C+27Si+11Mn+8Ni+16Cr+21\log_{10}V_r \qquad (39)$$

$$HV_{(bainite)} = -323+185C+330Si+153Mn+65Ni+144Cr+191Mo+$$

$$\log_{10}V_r(89+53C-55Si-22Mn-10Ni-20Cr-33Mo) \qquad (40)$$

$$HV_{(pearlite-ferrite)} = 42+223C+53Si+30Mn+12.6Ni+7Cr+19Mo$$

$$+\log_{10}V_r(10-19Si+4Ni+8Cr+130V) \qquad (41)$$

where $V_r$ is the cooling rate at 700°C given in °C/hr. For mixed structures the hardness is calculated according to a weighted volume mean and converted to $R_c$ via standard correlations.

We begin with the comparison of formula-derived IT diagrams with experimental ones. Figures 7-11 show the comparative IT diagrams for AISI grades 1050, 5140, 4037, 4140 and 8630. The calculations based on our model are very good for the unalloyed case and salutary for the alloyed cases. The significant departure lies in the transformation finish times for the highly alloyed cases due to defects in the formulation of the transformation integral I. The deviations in the (F+P) regime are probably due to different nucleation mechanisms dominating at higher and lower temperatures in contradistinction to our assumption of a single nucleation model (point nucleation) throughout the entire range of temperatures. In the bainite regime the basic model predicted the 50% transformation times quite accurately as indicated in the diagrams. However, the predicted finish times tended to be less than the observed finish times. A reason for this discrepancy could lie in the unvarying steady-state growth assumed in the model. In the actual case the reaction may be slowed down at long times due to soft impingement resulting from the interaction of carbon diffusion fields in austenite and the subsequent change of the transformation mode so as to involve fractional alloy diffusion. This is expected to occur at long times where a high volume fraction has transformed. There may also be an effect of the "shear" component of the lower bainite reaction [33].

We have partially corrected for this effect empirically by inserting the term $\exp[X^2(1.9C+2.5Mn+0.9Ni+1.7Cr+4Mo-2.6)]$ in the numerator of the integrand of I in Eqn (21), transforming I to I' in Eqn (27). This is equivalent to increasing the activation energy as a measure of increase of bainite retardation. The coefficients were determined by matching the experimental and calculated $B_f$ lines at 400°C. Note that the correction term is to be deleted for all negative values of the exponent. As can be seen from Figs. 7 to 11, the adjustment has not been completely successful. The reason may lie in our failure to make a distinction between the upper and lower bainite regimes and our use of a single nucleation and closure model for all products and temperatures. Further improvement in this area awaits advance in the systematics of the bainite reactions.

In Figs. 12 to 16 we have compared the predicted CCT's for H-steel 1541 (constant cooling rate), the French grade 35NC6 (Newton's Law of cooling) and the German grades 36 Cr6, 34 CrMo4 and 16 MnCr5 (exponential cooling; $T=T_0 \exp[-b(t-t_0)]$) with experimental CCT's (34,35,36). In Fig. 13 our calculated CCT is for a constant cooling rate evaluated at the average value of the experimental cooling rate between 800°C and 500°C.

The German data provides experimental values of the % transformed (100-% martensite) and final hardnesses for each cooling curve. We have accordingly calculated these values and entered them for comparison with the experimental values in Figs. 14 to 16. If one considers that the reproducibility of CCT critical cooling times between different laboratories is not better than a factor of 2 (9), our predictions may be deemed very satisfactory.

The limitations of the model discussed earlier with reference to IT diagrams are not likely to seriously affect predictions of Jominy hardenability. The inflection point on the Jominy curve is influenced primarily by the initial softening (i.e., the position of the ferrite or bainite start nose on the IT diagram). Since the error in the present calculations is larger in the bainite finish times than in the ferrite+pearlite finish times, we expect a final microstructure with a higher volume fraction of bainite in the completely transformed cases with mixed structure. Thus near the inflection point we expect to predict higher than actual hardness. In Figures 17 and 18 we have compared predicted Jominy hardness curves according to the empirical cooling regime of Eqn (33) ff. with two measured curves in the USS Atlas (31). The predictions are reasonaly accurate. Clearly, the base IT formulas could be calibrated simultaneously to CCT and (or) Jominy data sets rather than to IT curves alone thereby becoming even more accurate predictors of the latter quantities. In further assessing the success of the procedure it should be clear to the reader that the empirical part pertains almost entirely to isothermal nucleation and diffusion rates and their temperature dependencies at the initiation of transformation. The fractions transformed, the CCT curve and the room temperature microstructures are all predicted.

## Discussion

This research has demonstrated that the predominant alloy synergisms which determine low alloy (H-steel) microstructure can be understood within the multicomponent theory of alloy phases and transformations and that the theoretical formulas when calibrated against observed IT start curves can be used to accurately predict the course of isothermal and non-isothermal transformation and the volume fraction of the ambient microstructures which result. The advantage of this procedure over a purely empirical method such as the Creusot-Loire system lies not only in pedagogy, but in transparency and adaptability in dealing with unusual alloy effects or with artifacts at the margin. For example, boron is thought to primarily retard the nucleation of ferrite. Thus one can see that a simple reduction in

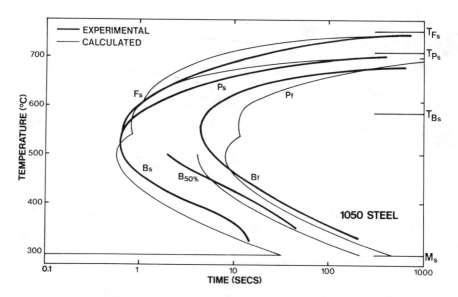

Figure 7 - Comparison between predicted IT and experimental IT diagrams for a 1050 steel. C 0.5; Mn 0.91; Si 0.13; G.S. 7-8 (31).

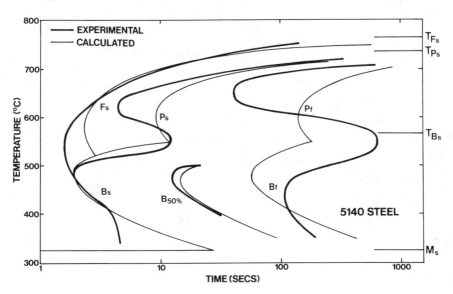

Figure 8 - Comparison between predicted IT and experimental IT diagrams for a 5140 steel. C 0.42; Mn 0.68; Si 0.16; Cr 0.93; G.S. 6-7 (31).

the overall coefficient of $\tau_F$ in (24), with a lesser effect on $\tau_p$, would generate the deeper and steeper Jominy curves commonly observed for boron steels (Figure 3).

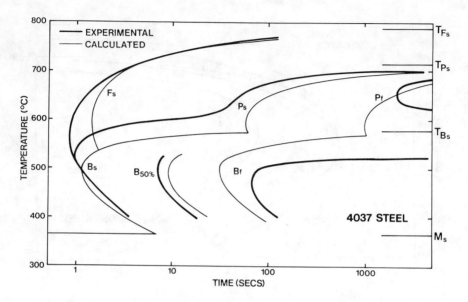

Figure 9 - Comparison between predicted IT and experimental IT diagrams for a 4037 steel.  C 0.35; Mn 0.8; Si 0.24; Mo 0.25; G.S. 7 (31).

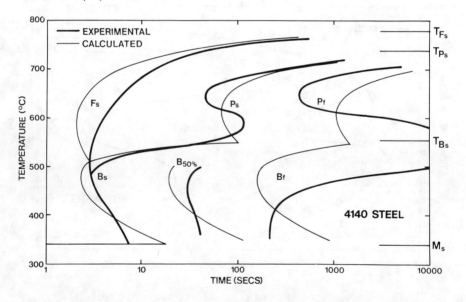

Figure 10 - Comparison between predicted IT and experimental IT diagrams for a 4140 steel.  C 0.37; Mn 0.77; Si 0.15; Ni 0.04; Cr 0.98; Mo 0.21; G.S. 7-8 (31).

Many of the other applications of comprehensive predictors of this type have been demonstrated by Creusot-Loire, Minitech Ltd. and Kawasaki Steel.  They are used in dozens of international corporations for research,

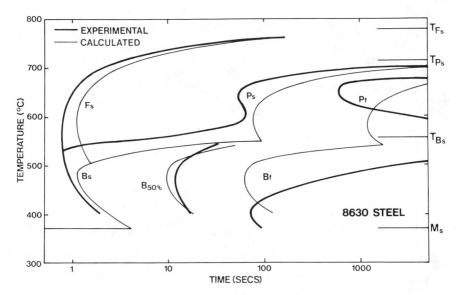

Figure 11 - Comparison between predicted IT and experimental IT diagrams for a 8630 steel. C 0.3; Mn 0.8; Si 0.29; Ni 0.54; Cr 0.55; Mo 0.21; G.S. 9 (31).

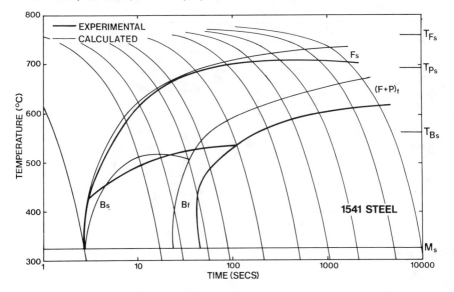

Figure 12 - Comparison of the predicted CCT for a 1541 steel with the experimental CCT diagram. C 0.39; Mn 1.56; Si 0.21; G.S. 8 (34).

quality control and on-line control of steel properties (9,10,13,19).

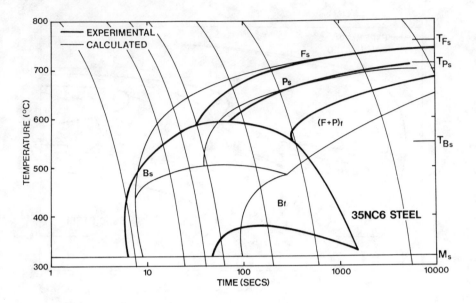

Figure 13 - Comparison of the predicted CCT for a 35NC6 steel with the experimental CCT diagram. C 0.41; Mn 0.55; Si 0.24; Ni 0.93; Cr 0.8; Mo 0.06; Cu 0.1; G.S. 11 (35). Note that $B_s$(exp) lies <u>above</u> $T_{Bs}$.

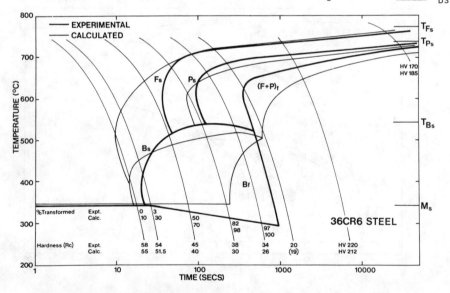

Figure 14 - Comparison of the predicted CCT for a 36 Cr6 steel with the experimental CCT diagram. C 0.36; Mn 0.49; Si 0.25; Ni 0.21; Cr 1.54; Mo 0.03; Cu 0.16; G.S. 7.5 (36).

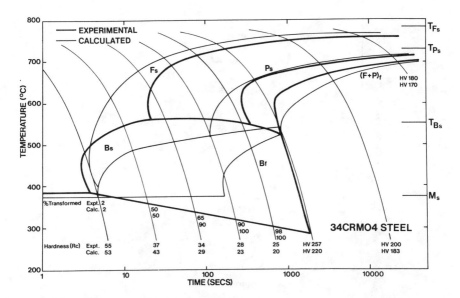

Figure 15 - Comparison of the predicted CCT for a 34Cr Mo4 steel with the experimental CCT diagram. C 0.3; Mn 0.64; Si 0.22; Ni 0.11; Cr 1.01; Mo 0.24; Cu 0.19; G.S. 9 (36).

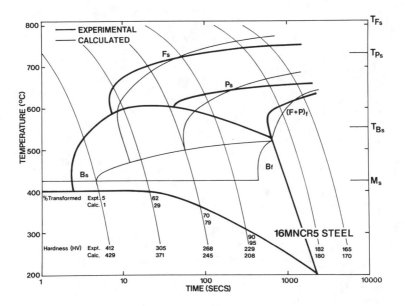

Figure 16 - Comparison of the predicted CCT for a 16Mn Cr5 steel with the experimental CCT diagram. C 0.16; Mn 1.12; Si 0.22; Ni 0.12; Cr 0.99; G.S. 6 (36).

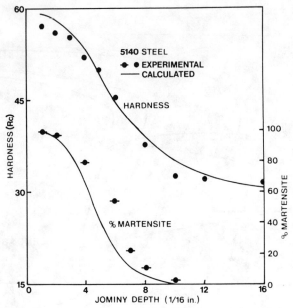

Figure 17 - Comparison of predicted % martensite and Jominy hardness curves with the observations for a 5140 steel (31).

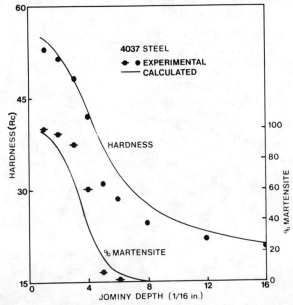

Figure 18 - Comparison of predicted % martensite and Jominy hardness curves with observations for a 4037 steel (31).

## Acknowledgements

The authors are grateful to Koichi Hashiguchi and Professor G.R. Purdy for many useful discussions.

## References

1. R.A. Grange, "Estimating the Hardenability of Carbon Steels," Met. Trans. 4 (2231) (1973).
2. G.T. Brown and B.A. James, "The Accurate Measurement, Calculation and Control of Steel Hardenability," Met. Trans., 4 (2245) (1973).
3. M.A. Grossmann, Elements of Hardenability, ASM, Cleveland, 1952.
4. Metals Handbook, American Society for Metals, Cleveland, 1948, p. 498.
5. Metal Progress Databook, Metals Park OH, June 1975, p. 141.
6. E. Just, "Formeln der Härtbarkeit," Härterei Technical Mitteilungen, 23 (85) (1968).
7. E. Just, "New Formulas for Calculating Hardenability Curves," Metal Progress, November 1969, p. 87.
8. D.V. Doane, "A Critical Review of Hardenability Predictors," p. 351 in Hardenability Concepts with Applications to Steel, D.V. Doane and J.S. Kirkaldy, ed.; AIME, Warrendale PA, 1978.
9. Ph. Maynier, J. Dollet and P. Bastien, "Prediction of Microstructure via Empirical Formulae Based on CCT Diagrams," p. 163 in Hardenability Concepts with Applications to Steel, D.V. Doane and J.S. Kirkaldy, ed.; AIME, Warrendale PA, 1978.
10. Ph. Maynier, B. Jungmann and J. Dollet, "Creusot-Loire System for the Prediction of the Mechanical Properties of Low Alloy Steel Products," p. 518 in Hardenability Concepts with Applications to Steel, D.V. Doane and J.S. Kirkaldy, ed.; AIME, Warrendale PA, 1978.
11. Private communication, Ph. Maynier.
12. B. Hildenwall and T. Ericsson, "Prediction of Residual Stresses in Case-Hardening Steels," p. 579 in Hardenability Concepts with Applications to Steel, D.V. Doane and J.S. Kirkaldy, ed.; AIME, Warrendale PA, 1978.
13. Y. Sakamoto, M. Saeki, M. Nishida, T. Tamaka and Y. Ito, "Mathematical Model Simulating Phase Transformation in Low Alloy Hot-Rolled Sheet Steel," Kawasaki Steel Co. Report, August 1981.
14. W.A. Johnson and R.F. Mehl, "Reaction Kinetics in Processes of Nucleation and Growth," Trans. AIME, 135 (416) (1939).
15. J.W. Cahn, "The Kinetics of Grain Boundary Nucleated Reactions," Acta Met., 4 (449) (1956).
16. J.S. Kirkaldy, "Prediction of Alloy Hardenability from Thermodynamic and Kinetic Data," Met. Trans., 4 (2327) (1973).
17. J.S. Kirkaldy, G.O. Pazionis and S.E. Feldman, "An Accurate Predictor of the Jominy Hardenability for Low Alloy Hypoeutectoid Steels," Proceedings of the 16th International Heat Treatment Conference, The Metals Society, London, 1976, p. 169.
18. J.S. Kirkaldy, B.A. Thomson and E.A. Baganis, "Prediction of Multicomponent Equilibrium and Transformation Diagrams for Low Alloy Steels," p. 82 in Hardenability Concepts with Applications to Steel, D.V. Doane and J.S. Kirkaldy, ed.; AIME, Warrendale PA, 1978.
19. S.E. Feldman, "The Minitech Computerized Alloy Steel Information System," p. 546 in Hardenability Concepts with Applications to Steel, D.V. Doane and J.S. Kirkaldy, ed.; AIME, Warrendale PA, 1978.
20. M.P. Puls and J.S. Kirkaldy, "The Pearlite Reaction," Met. Trans., 3 (2777) (1972).
21. C. Zener, "Kinetics of Decomposition of Austenite," Trans. AIME, 167 (550) (1946).
22. M. Hillert, "The Role of Interfacial Energy During Solid State Transformations," Jernkont. Ann., 141 (758) (1957).
23. E. Scheil, "Anlaufzeit der Austenitumwandlung," Archiv f. Eisenhüttenw., 8 (565) (1935).
24. M. Avrami, "Kinetics of Phase Change I," J. Chem. Phys., 7 (1103) (1939).
25. M. Avrami, "Kinetics of Phase Change II," J. Chem. Phys., 8 (212) (1940).
26. J.W. Cahn, "Transformation Kinetics During Continuous Cooling,", Acta Met., 4 (572) (1956).

27.  J.W. Cahn, "The Kinetics of Cellular Segregation Reactions," Acta Met., 7 (18) (1959).

28.  J.S. Kirkaldy and R.C. Sharma, "A New Phenomenology for Steel IT and CCT Curves," Scripta Met., 16 (1193) (1982).

29.  J.S. Kirkaldy and E. Baganis, "Thermodynamic Prediction of the $Ae_3$ Temperature of Steel with Additions of Mn, Si, Ni, Cr, Mo, Cu," Met. Trans., 9A (495) (1978).

30.  K. Hashiguchi, T. Fukuzumi, V. Pavaskar and J.S. Kirkaldy, "Prediction of the Equilibrium, Paraequilibrium and No-partition Local Equilibrium Phase Diagrams for Multicomponent Fe-C Base Alloys", CALPHAD, in press (1984).

31.  U.S.S. Atlas of Isothermal Transformation Diagrams, U.S. Steel Co., Pittsburgh, 1963.

32.  W. Steven and A.G. Haynes, "The Temperature of Formation of Martensite and Bainite in Low-Alloy Steels," JISI, 183 (349) (1956).

33.  H.K.D.H. Bhadeshia, "Bainite:  The Incomplete Reaction Phenomenon and the Approach to Equilibrium", Proceedings of the International Conference on Solid-Solid Phase Transformations, AIME, 1982.

34.  Metal Progress Data Book, Metals Park OH, p. 147, June 1975.

35.  Atlas Des Courbes de Transformation, IRSID, France.

36.  Atlas zur Wärmbehandlung der Stähle, Band 1, Max-Planck Institute für Eisenforschung, 1961.

# HARDENABILITY OF LOW AND MEDIUM-CARBON Mn-Cr-Ni-Mo STEELS

T.M. Scoonover* and F.B. Fletcher

Climax Molybdenum Company of Michigan
Division of AMAX, Inc.
Ann Arbor, Michigan

In a statistically designed hardenability study, 85 laboratory melted steels were tested with the standard Jominy end-quench test. The steels in the study included 0.11%C Mn-Cr-Ni-Mo quenched constructional plate steels, 0.22%C Mn-Cr-Mo low-hardenability carburizing steels, 0.22%C Cr-Ni-Mo high-hardenability carburizing steels, and 0.44%C Ni-Mo strategic (chromium-free) through-hardening steels. In three of the four steel groups molybdenum was found to be about two times more effective than chromium and manganese and four to eight time more effective than nickel in increasing $D_I$. A previously reported two-way alloy interaction between molybdenum and nickel was confirmed by the results from two groups of steels. No strong two-way interaction between manganese and chromium was detected in any of the results.

## Introduction

The hardenability of a steel is controlled by metallurgical factors, such as its austenite grain size, and by its chemical composition. Experience has shown that alloying elements used in combination usually produce a greater increase in hardenability than expected from their individual effects.

Alloy interaction effects involving molybdenum have been found in medium-carbon low alloy steels where the presence of nickel or chromium has been reported to enhance the molybdenum hardenability effect[1]. As little as 0.3%Cr or 0.7%Ni has been found to increase the effect of molybdenum in suppressing the start of the bainite transformation[2]. Increasing manganese from 0.8 to 1.4% and silicon from 0.3 to 1.5% also increases the effectiveness of molybdenum in delaying the bainite start. A strong Mo-Ni alloy interaction has also been reported in low-carbon carburizing steels when the nickel content exceeded 0.75%[3].

Regression analysis is one method which may be used to relate hardenability to chemical composition and metallurgical factors. The regression model should include all the element effects and all the interactions which an analysis of variance reveals to be significant[4].

*Now at David Taylor Naval Ship Research and Development Center, Annapolis MD.

The objective of the present investigation was to detect and assess individual alloy and alloy interaction effects in four groups of low and medium-carbon steels containing combinations of manganese, chromium, nickel and molybdenum. The Jominy end-quench test was used to measure hardenability. A group of 0.11%C Mn-Cr-Ni-Mo quenched constructional plate steels, designated Group A, and two groups of 0.22%C steels, Groups B and C, were studied because of their relevance to low and high hardenability carburizing steels; a group of 0.44%C Ni-Mo strategic (chromium-free) steels, Group D, was studied to confirm the Mo-Ni alloy interaction at medium-carbon levels. Because of limited space, it is impossible in this paper to fully describe the results of the hardenability study. A complete copy of the laboratory report detailing this study(5) may be obtained from the Climax Molybdenum Company.

## Laboratory Procedures

Twenty-two 34 kg (75 lb) heats of steel were induction melted under argon in a laboratory induction furnace. Ferromolybdenum additions were made between ingot pours to obtain a total of 85 compositions for study according to the listing shown in Table 1. The ingots were hot forged to 32 mm (1-1/4 in.) diameter bars. The 0.11%C and 0.22%C steels were normalized from 927 C (1700 F) and the 0.44%C steels from 871 C (1600 F).

Table I.  Nominal Compositional Levels of Experimental Steels

|         | C    | Mn   | Cr  | Ni  | Mo   |
|---------|------|------|-----|-----|------|
| Group A | 0.11 | 0.8  | 0   | 0   | 0    |
|         |      | 1.4  | 0.8 | 0.8 | 0.08 |
|         |      |      |     |     | 0.15 |
|         |      |      |     |     | 0.30 |
| Group B | 0.22 | 0.8  | 0   | 0   | 0    |
|         | 1.15 | 0.5  |     |     | 0.08 |
|         |      | 1.0  |     |     | 0.15 |
|         |      |      |     |     | 0.30 |
| Group C | 0.22 | 0.8  | 1.0 | 0   | 0    |
|         |      | 1.5  | 1.0 | 0.15 |
|         |      |      | 2.0 | 0.30 |
|         |      |      |     |     | 0.50 |
| Group D | 0.44 | 0.8  | 0   | 0   | 0    |
|         |      |      |     | 0.8 | 0.15 |
|         |      |      |     | 1.6 | 0.30 |

The hardenability of each steel was determined by austenitizing a Jominy bar for 20 minutes at temperature and end-quenching in accordance with standard practice. Austenitizing temperatures were 955 C (1750 F) for the 0.11%C steels, 927 C (1700 F) for the 0.22%C steels, and 843 C (1550 F) for the 0.44%C steels. Austenite grain sizes were determined for each steel.

The measure of hardenability chosen to study was Grossmann's ideal critical diameter ($D_I$). The Jominy position at which the hardness corresponds to that of 50% martensite for the carbon content of the steel--according to the correlation of Hodge and Orehoski(6)--was converted to $D_I$ using

Carney's(7) data, which relates the cooling rates at the centers of cylinders of various diameters subjected to an ideal quench to positions of equivalent cooling rate on the Jominy bar. The $D_I$ values were corrected to a common austenite grain size for each group (the average value) using Grossmann's correction factors(7). Common grain sizes were ASTM No. 7.1 for Group A, ASTM No. 8.4 for Group B, ASTM No. 8.3 for Group C, and ASTM No. 6.8 for Group D.

For Groups A, B and C, the data were subjected to analysis of variance to determine significant alloy and alloy-interaction effects. All effects with confidence level >75% were included in the initial regression model.

## Results and Discussion

### Group A

Molybdenum exhibited the strongest effect on the hardenability of the 0.11%C steels of Group A. The simple regression equation calculated by including only the effects of the individual alloying elements in the regression model is given by:

$$D_I \text{ (mm)} = 28.6Mn + 30.3Cr + 12.2Ni + 59.6Mo - 17.5 \tag{1}$$

For equation (1) the confidence level of the coefficients is >99.4%. However, the multiple correlation coefficient is 0.889 and the standard error of estimate is 9.7 mm.

The final regression equation obtained by including alloy interaction effects is given by:

$$D_I \text{ (mm)} = 11.4Mn + 21.1Cr + 34.2Mo + 4.9MnNi + 33.3CrNi \text{ (Mn-1)} \tag{2}$$
$$+ 149.6CrMo \text{ (Mn + 0.6MnNi - 1)} + 7.4$$

For equation (2) the confidence level of the coefficients is >98.5% and the multiple correlation coefficient is 0.988. The standard error of estimate is only 3.6 mm.

The simple regression equation indicates that for the composition range of Group A steels, molybdenum is, on average, about twice as effective as manganese and chromium and about four times as effective as nickel in increasing the Jominy hardenability. However, the complex regression equation shows that, like manganese and chromium, only a part of the molybdenum effectiveness was from the individual element effect with the rest coming from complex two-, three-, and four-way alloy interactions. Molybdenum shows a much larger hardenability effect in a Group A steel if it also contains chromium in combination with rather large amounts of manganese or manganese plus nickel. Not only were the hardenability effects of individual elements weak relative to alloy interactions in these low carbon steels but also the Mo-Ni two-way alloy interaction, reported in higher carbon steels(1,3) was absent.

### Group B

The simple regression equation describing the effects of the individual alloying elements in the regression model for the 0.22%C Mn-Cr-Mo low-hardenability carburizing steels of Group B is:

$$D_I \text{ (mm)} = 36.8Mn + 36.2Cr + 77.6Mo - 20.0 \tag{3}$$

For equation (3) the confidence levels of the coefficients are all 100.0%, the multiple correlation coefficient is 0.973, and the standard error of estimate is 4.8 mm.

As with Group A, the simple equation for Group B indicates that molybdenum is, on average, a little more than two times more effective than either manganese or chromium in increasing the Jominy hardenability.

Analysis of variance for Group B showed that the Cr-Mo two-way interaction and the three-way Mn-Cr-Mo alloy interaction effects were significant. The regression equation calculated for Group B is:

$$D_I \text{ (mm)} = 25.3\text{Mn} + 28.6\text{Cr} + 46.3\text{Mo} + 107.7\text{CrMo} \ (1.6\text{Mn}-1) - 4.8 \qquad (4)$$

For equation (4) the confidence level of the coefficients is >98.1%, the multiple correlation coefficient is 0.991, and the standard error of estimate is only 2.9 mm.

Comparison of this equation with the complex equation for Group A indicates that only about 0.6%Mn is required for the three-way alloy interaction involving molybdenum to contribute to hardenability at 0.22%C while 1.0%Mn was required at 0.11%C. Chromium is required at both carbon levels. Earlier CCT experiments have shown that chromium, manganese, and nickel increase the effect of molybdenum in suppressing the start of the bainite transformation in medium-carbon steels(2). The results for Groups A and B indicate that a similar interaction occurs in low-carbon steels.

Group C

The simple regression equation calculated by including only the effects of the individual alloying elements in the regression model for the 0.22%C Cr-Ni-Mo high hardenability carburizing steels in Group C is given by:

$$D_I \text{ (mm)} = 125.0\text{Cr} + 58.3\text{Ni} + 142.9\text{Mo} - 99.9 \qquad (5)$$

For equation (5) the confidence levels of the coefficients are all 100.0%, the multiple correlation coefficient is 0.982, and the standard error of estimate is 10.3 mm.

The simple regression equation for Group C indicates that molybdenum is more than twice as effective as nickel but only slightly more effective than chromium in increasing the Jominy hardenability of these 0.22%C steels. For the 0.22%C steels of Group B, molybdenum was more than twice as effective as chromium. The different effectiveness of molybdenum relative to chromium results from the higher level of chromium present in Group C than in Group B; a rapid increase in $D_I$ occurred when chromium content was increased above 1%.

The final complex regression equation obtained by backward elimination of two-way and three-way alloy interactions in Group C is given:

$$D_I \text{ (mm)} = 124.9\text{Cr} + 42.6\text{Ni} + 107.8\text{Mo} + 66.5\text{NiMo} - 91.5 \qquad (6)$$

For equation (6) the confidence level of the coefficients is >99.5%, the multiple correlation coefficient is 0.992 and the standard error of estimate is 7.5 mm. The complex regression equation for Group C confirms the Mo-Ni alloy interaction previously reported for low-carbon carburizing steels with nickel content greater than 0.75%(3).

Group D

The regression equation calculated by including only the effects of the individual alloying elements in the regression model for the 0.44%C chromium-free steels of Group D is given by:

$$D_I \text{ (mm)} = 16.0\text{Ni} + 125.9\text{Mo} + 17.7 \qquad (7)$$

For equation (7) the confidence level of the coefficients is >97.6%, the multiple correlation coefficient is 0.915, and the standard error of estimate is 10.4 mm. The equation indicates that molybdenum is, about eight times more effective than nickel in increasing the Jominy hardenability.

The multiple regression equation obtained by backward elimination of two-way and three-way alloy interactions is given by:

$$D_I \text{ (mm)} = 49.0\text{Mo} + 96.4\text{NiMo} + 30.5 \qquad (8)$$

For equation (8), the confidence level of the coefficients is >97.9%, the multiple correlation coefficient is 0.982, and the standard error of estimate is only 4.9 mm. This regression equation for Group D shows that the effect of nickel alone is insignificant and confirms the Mo-Ni interaction previously reported for medium-carbon steels(1).

## Conclusions

1. Molybdenum is about two times more effective than manganese and chromium, and about four times more effective than nickel in increasing the Jominy hardenability of 0.11%C constructional steels, especially when chromium in combination with large amounts of manganese (>1.0%) or manganese plus nickel, is also present.

2. Molybdenum is a little more than two times more effective than manganese and chromium in increasing the Jominy hardenability of 0.22%C Mn-Cr-Mo low-hardenability carburizing steels. Much of the molybdenum effectiveness in 0.22%C Mn-Cr-Mo low-hardenability carburizing steels results from a Cr-Mo-Mn three-way alloy interaction.

3. In high-hardenability 0.22%C carburizing steels molybdenum is more than two times more effective than nickel and slightly more effective than chromium in increasing the Jominy hardenability. The Mo-Ni two-way alloy interaction was confirmed.

4. Molybdenum is about eight times more effective than nickel in increasing the Jominy hardenability in 0.44%C chromium-free steels. The Mo-Ni two-way alloy interaction was confirmed.

5. No strong two-way interaction between manganese and chromium was detected in any of the steels studied.

## References

1. C.A. Siebert, D.V. Doane and D.H. Breen, The Hardenability of Steels – Concepts, Metallurgical Influence, and Industrial Applications, American Society for Metals, Metals Park, Ohio, 1977.

2. W.W. Cias and D.V. Doane, "Phase Transformation Kinetics and Hardenability of Alloys Medium-Carbon Steels," Metallurgical Transactions, October 1973, pp. 2257-2266.

3.  A.F. deRetana and D.V. Doane, "Predicting Hardenability of Carburizing Steels," Metal Progress, September 1971, pp. 65-66.

4.  G.T. Eldis and W.C. Hagel, "Effects of Microalloying on the Hardenability of Steel," Hardenability Concepts with Applications to Steel, D.V. Doane and J.S. Kirkaldy, eds., AIME, Warrendale, PA, 1978.

4.  T.M. Scoonover and H.L. Arnson, "Hardenability of Lowand Medium-Carbon Mn-Cr-Ni-Mo Steels," Report No. RP-46-78-01, January 20, 1982, Climax Molybdenum Company, 1600 Huron Parkway, Ann Arbor, Michigan 48105.

6.  J.M. Hodge and M.A. Orehoski, "Relationship Between Hardenability and Percentage Martensite in Some Low-Alloy Steels," Transactions of AIME 167 (1946), pp. 627-642.

7.  D.J. Carney, "Another Look at Quenchants, Cooling Rates and Hardenability," Transactions of ASM 46 (1954), pp. 882-927.

8.  M.A. Grossmann, Elements of Hardenability, American Society for Metals, Cleveland, Ohio, 1952.

# The Proeutectoid Ferrite Reaction

W.T. Reynolds, Jr.*, M. Enomoto** and H.I. Aaronson*

*Department of Metallurgical Engineering and Materials Science
Carnegie-Mellon University
Pittsburgh, PA  15213
**National Research Institute for Metals, Tsukuba Laboratories, 1-2-1,
Sengen, Sakura-Mura, Niihari-Gun, Ibaraki 305, Japan

A generation of progress made in understanding the morphology, kinetics and mechanisms of the proeutectoid ferrite reaction in steel is summarized. In view of increasing concern about interactive effects of small amounts of alloying and impurity elements upon the transformation, emphasis is placed upon studies conducted on high-purity Fe-C and Fe-C-X alloys. Measurements of nucleation kinetics of grain boundary allotriomorphs in Fe-C alloys have shown that these kinetics are markedly time-dependent. Interpretation of these data in terms of classical nucleation theory requires that the ferrite nuclei be largely or entirely coherent with one or both bounding austenite grains; only a small proportion of the austenite grain boundary area can support ferrite nucleation at the temperatures studied. Similar interpretation of measurements of allotriomorph nucleation kinetics in Fe-C-X alloys indicates that the influence of X upon the volume free energy change attending nucleation explains much of the X effect upon the nucleation rate when X = Si, Co and even Mo; reduction of both austenite grain boundary and austenite:ferrite boundary energies is required to understand the remaining effects observed. The broad faces of ferrite sideplates have been experimentally shown to have a partially coherent structure, incorporating closely spaced, coherent triatomic structural ledges and misfit dislocations in sessile orientation. This structure rules out participation of shear during plate thickening. Growth ledges provide the agency for plate thickening, and their erratic, time-dependent distribution is reflected in irregular thickening kinetics. Plate lengthening also appears to be controlled by growth ledges, but the generation kinetics of these ledges is remarkably uniform. Although grain boundary allotriomorphs have interphase boundaries incorporating both partially coherent and disordered areas, in proportions which are presently unknown, their growth kinetics are usually not much less than those predicted for purely carbon diffusion-control. During at least the early stages of growth in Fe-C-X alloys, the growth of ferrite allotriomorphs appears to take place under paraequilibrium conditions. However, significant deviations from the kinetics predicted on this basis are often found. A solute drag-type of effect appears to be the main kinetic factor causing these deviations. The local equilibrium model often fails to predict both the composition and the growth kinetics of ferrite allotriomorphs during the early stages of their growth.

## Introduction

This paper may be regarded as the successor to an extended review (1)
published in the proceedings of a symposium on the "Decomposition of Austenite by Diffusional Processes" more than a generation ago.  In that review,
primary emphasis was placed upon detailed morphological studies of the proeutectoid ferrite reaction.  A general theory of precipitate morphology was
also proposed and was shown to be capable of providing a qualitative accounting for the experimental observations.  This theory emphasized the role of
the boundary orientation-dependence of interphase boundary structure in
determining the growth mechanism, and thus the growth kinetics of these
boundaries.  At the time that review was written, however, transmission
electron microscopy (TEM) was still under development and the first observations of the structure of partially coherent interphase boundaries with this
technique had yet to be made.  A surprising amount of data on nucleation
kinetics of ferrite at austenite grain boundaries was available.  However,
the techniques for making these measurements were incompletely developed and
the theory for interpreting these data was not yet ready for even preliminary
application to this task.  The first data on the thickening kinetics of
precipitate plates in any diffusional transformation were reported and provided initial, though only tentative confirmation for the prediction of the
theory of morphology that this process would occur by a ledge mechanism.
Somewhat more data were available on the lengthening kinetics of ferrite
plates, and on the lengthening and thickening kinetics of ferrite allotriomorphs.  Development of the mathematical analyses of these processes was
being actively pursued.  Application of the analyses that were available had
just begun.  A substantial theoretical base for understanding the influence
of a single substitutional alloying element, X, upon the growth kinetics of
ferrite had already been developed, as reviewed by Kirkaldy (2) in the same
proceedings.  However, the limited experimental data on growth kinetics of
ferrite in alloy steels then available had been largely obtained with in-
sufficiently refined experimental techniques and serious efforts to apply
the theory had not yet begun.  Electronic techniques for microchemical
analyses were just then becoming available and had yet to be applied to the
proeutectoid ferrite reaction.

The 22 years which intervened between the publication of that proceedings

and the present one have seen the acquisition of rather extensive bodies of
data on the nucleation and the growth kinetics of grain boundary allotriomorphs in both Fe-C and Fe-C-X alloys and on the growth kinetics of ferrite
sideplates, especially in Fe-C alloys.  Quantitative studies have been made
of the interfacial structure of the broad faces of ferrite sideplates.
Nucleation theory has now developed sufficiently so that at least preliminary
analyses can be made of the data on allotriomorph nucleation kinetics.
Growth kinetics theory, though increasingly formidable mathematically, is
usually considerably simpler in a conceptual sense and as a result is now in
quite advanced stages of development.  The theory has been applied with some
care to the experimental data....and has helped to open up a new generation
of problems.  Considerable progress has thus been made.  However, understanding of the fundamental mechanisms of the proeutectoid ferrite reaction
still appears to be in a "divergent" rather than a "convergent" state of
development.

Because the "data base" of theory, particularly, but also of experiment
has expanded so much, it is no longer feasible to attempt a complete review
of all aspects of this subject within the compass of a single rather brief
article.  Emphasis will accordingly be placed upon the blocks of understanding
which have been acquired and major problems which have been uncovered through
selected comparisons between theory and experiment.  Growing concern over
alloying element effects upon the proeutectoid ferrite reaction has led to
division of the main body of this review into two parts.  The first deals
with studies conducted on Fe-C alloys of reasonably high purity (where other
elements are present at levels of no more than a few hundreths, and preferably not more than a few thousandths of a percent).  The second will be
concerned with Fe-C-X alloys of similar purity.

## The Dube Morphological Classification System (3)

As in the predecessor review (1), the various components of this system
are first defined with the assistance of Fig. 1, reproduced from that paper.
Since the present review will include very little description of the individual morphologies, some remarks of this type will be appended to the definitions as they are presented.

Grain Boundary allotriomorphs--crystals which nucleate at grain boundaries
in the matrix phase and grow preferentially, and more or less smoothly, along
these boundaries.

This morphology has been increasingly utilized, particularly in the proeutectoid ferrite reaction in steel but in a variety of other alloy systems
as well (4-7), as the prototype morphology whose interphase boundary structure is predominantly of the disordered type.  This assumption, though
originally reasonable, is becoming increasingly questionable.  An adequate
test of this assumption in steel, undertaken by means of TEM studies of
interphase boundary structure, would be exceedingly difficult because of
destructuion of nearly all of the untransformed austenite by martensite
formation during quenching.  However, the impressions easily secured from
studies of high-alloy Fe-C-X alloys that the broad faces of ferrite allotriomorphs are predominantly partially coherent (8-10), may be almost equally
misleading.  Sideplate formation in such alloys has often been inhibited (11),
probably by a strong solute drag-like effect (11-13), thereby permitting the
slowest growing boundary orientation (i.e., the one corresponding to the
partially coherent structure whose time-averaged density of ledges is
smallest) to occupy the largest proportion of the interphase boundary area
after sufficient thickening has occurred.

The view that ferrite allotriomorphs form at large-angle austenite grain boundaries (1) has since been confirmed by King and Bell (14) using a Kossel back-reflection X-ray microdiffraction technique. The association of grain boundary allotriomorphs with large-angle grain boundaries (which do not lie near a coincidence site lattice or some other orientation capable of yielding a low-angle boundary type of structure) now appears general, but only for fcc matrices (and precipitates having a different crystal structure); bcc matrices can produce allotriomorphs even at low-angle grain boundaries, for reasons not yet understood (15).

Widmanstatten Sideplates (or Sideneedles)--are plate- or needle-shaped crystals which develop into the interior of a matrix grain from the vicinity of the matrix grain boundaries.

Primary Sideplates (or Sideneedles)--(Fig. 1-(b) - (1))--grow directly from grain boundaries in the matrix phase.

The original finding that primary ferrite sideplates grow directly from small-angle austenite grain boundaries has been repeated in numerous alloy systems, but again only when the matrix is fcc and the precipitate has another structure (15,16); a bcc matrix (and also, curiously the fcc matrix in Al-Mg-Zn-type alloys (17)) can produce other morphologies at small-angle boundaries.

Secondary Sideplates (or Sideneedles)--Fig. 1-(b) - (2)--develop from crystals of another morphology (but of the same phase), usually part of the same single crystal as the grain boundary allotriomorph from which they evolved (18). Particularly because intragranular ferrite plates form in substantial numbers only at lower reaction temperatures in lower carbon alloys with a coarse austenite grain size (greater than ca. ASTM No. 0) (13, 19), most measurements of Widmanstatten plate thickening and lengthening kinetics have been performed on sideplates. A particularly striking aspect of ferrite sideplates is the repetitive degeneracy which they exhibit at intermediate reaction temperatures. The morphology of this feature has been documented in some detail (1) but has yet to be investigated at greater depth with modern techniques.

Widmanstatten Sawteeth--Fig. 1-(c)--have a triangular cross-section in the plane of polish and develop from the region of the matrix grain bound-aries.

Primary Sawteeth--Fig. 1-(c) - (1)--grow directly from the matrix grain boundaries.

The estimate that ferrite crystals of this morphology develop from inter-mediate-angle grain boundaries (16) has yet to be confirmed in steel but has been for the fcc→hcp transformation in an Al-Ag alloy (20). This mor-phology is as at least as scarce as primary sideplates.

Secondary Sawteeth--Fig. 1-(c) - (2)--develop from crystals of another morphology, usually from grain boundary allotriomorphs.

This morphology has attracted little further attention since it was first reported.

Idiomorphs--Fig. 1-(d)--are roughly equi-axed crystals.

Intragranular Idiomorphs--Fig. 1-(d) - (1)--form in the interiors of matrix grains.

This is a relatively rare ferrite morphology. The combination of this

circumstance with the inaccessibility of its interfacial structure to TEM
studies has evidently been responsible for the absence of quantitative
studies on this morphology, despite the possibilities which this precipitate
shape offers for simultaneous investigation of the growth kinetics of
differently oriented and thus structured partially coherent interphase
boundaries.

Grain Boundary Idiomorphs--Fig. 1-(d) - (2)--roughly equi-axed crystals
formed at matrix grain boundaries.

This morphology is rarely seen during the proeutectoid ferrite reaction,
but has been observed during diffusional transformations in other alloy
systems (15).

Intragranular Widmanstatten Plates (or Needles)--Fig. 1-(e)--Although
these plates likely nucleate on isolated dislocations or at small-angle
grain boundaries in austenite, the martensite transformation has so far
prevented a direct determination of the sites at which these plates are
formed.  Intragranular plates are often grouped in star-like and more complex configurations at lower reaction temperatures in the proeutectoid
ferrite region (1).  At still lower temperatures, where precipitation of non-
lamellar carbides converts the ferrite plates to bainite, intragranular
plates (as well as sideplates) frequently appear in sheaves.  These groupings
are considered to arise from sympathetic nucleation (21).  The details of
this process are only now being investigated in detail, and in Ti-base
alloys (22), where the "austenite" phase can be retained at room temperature.

For more detailed descriptions of these morphologies, the reader is
referred to ref. (1) and to confirmatory observations later reported by
Townsend and Kirkaldy (23).

## A General Theory of Precipitate Morphology (1,24)

The need to minimize the free energy of activation for critical nucleus
formation ensures that the orientation relationship of the nucleus with
respect to the matrix grain(s) with which it is in contact is such that interfacial energy per unit volume of the nucleating phase is minimized.
Strain energy minimization is simultaneously attempted but unless the crystal structures and lattice orientations of the two phases are the same (25),
interfacial energy minimization continues to dominate determination of the
orientation relationship.  From the standpoint of interphase boundary structure, a fully coherent boundary is expected at most if not at all orientations of the nucleus:matrix interface, particularly when two of the three
common metallic crystal structures, fcc, bcc and hcp, form the boundary (26).
The small size of the critical nuclei will usually make fully coherent interphase boundaries lower in interfacial energy than their partially coherent
counterparts (27,28).  Acquisition of misfit dislocations at these boundaries
will thus occur during the growth rather than the nucleation process.
However, whether such boundaries are partially or fully coherent, they will
be unable to migrate in the direction normal to themselves through diffusional, i.e., atom by atom processes because this would require that substitutional atoms temporarily occupy interstitial sites.  The probability
that such occupation would be long sustained is very small, and the atomic
jumps involved would be quickly retraced.  Only when the crystal  structure
and lattice orientation of both phases are the same can partially or fully
coherent interphase boundaries be displaced in this manner (25).  Displacement of partially or fully coherent boundaries between crystals differing
significantly in structure must accordingly be accomplished by growth ledges.
If the risers of these ledges are themselves partially or fully coherent,
as has been recently shown for some fcc:bcc interfaces (10), a ledge-on-

ledge mechanism must be invoked, if necessary unto n'th order, in order to accomplish boundary movement (24).

In its original form, this theory proposed that interphase boundaries between crystals with different structures, even when related by orientation relationships capable of yielding good matching of the two lattices at one boundary orientation, could produce matching close enough to correspond to partial coherency at few (or possibly no) other boundary orientations (1). At all other boundary orientations, the interphase boundary structure was taken to be of the disordered or incoherent type. Interfacial structures of this type are readily displaced by the essentially uncoordinated movement of individual atoms across the boundary. With a very short (ca. a lattice parameter) and constant diffusion distance and a high diffusivity, the growth kinetics of such an interfacial structure will normally be essentially unaffected by the details of this structure but will instead be controlled by such non-structural factors as the solute concentration gradient in the matrix phase, the diffusivity in the matrix and the curvature of the inter-phase boundary. TEM studies of the edges as well as the broad faces of pre-cipitate plates soon led, however, to the conlcusion that partial or full coherency obtains at more boundary orientations than originally anticipated (24). More recently, it was proposed that partial coherency may obtain during growth at a sufficient number of boundary orientations to enclose completely a precipitate crystal imbedded in a matrix of another crystal structure when structures involved are fcc, bcc or hcp (26). This proposal for growth is the counterpart of that previously noted for the presence of all round full coherency during nucleation. Hence the interfacial structure situation originally envisaged has now been effectively reversed. The central experimental (and theoretical) problem is now to locate the dis-ordered areas of interphase boundaries. At least some and perhaps most of these areas reside on the risers of ledges. However, particularly at high ($10^2$-$10^4$ nm.) growth ledges, it seems quite possible that only a very small fraction of the riser area may be disordered. At other boundary orienta-tions disordered areas may be even more difficult to find because (as was long ago postulated for the growth of crystals in liquid solutions) the dis-ordered areas, being highly mobile, will tend to "grow out", leaving the interphase boundary almost entirely fully or partially coherent. This tendency should be somewhat restrained in the case of grain boundary allot-riomorphs by the kinetic constraint that these crystals grow predominantly along their matrix grain boundaries.

Finally, an increasing ratio of the migration kinetics of disordered boundaries to that of partially or fully coherent boundaries (at a given, early reaction time) with increased undercooling was proposed as the factor responsible for concomitant increase in the acicularity of precipitate morphology under this circumstance. Adaptation of this view to the evidently more common situation in which the surviving boundary orientations all grow by means of the ledge mechanism probably follows from the tendency of boundaries containing closely spaced ledges to develop growth kinetics approaching those of disordered boundaries (29-31). However, considerably further experimental and especially theoretical studies will be required to test adequately this suggestion.

## Nucleation Kinetics of Ferrite Allotriomorphs at Austenite Grain Boundaries

Although a number of studies of this type have been attempted, as shown in a current review (32) most have been affected by problems in both measure-ment techniques and interpretation. The following is a brief summary of a recent investigation of the nucleation kinetics of ferrite allotriomorphs at austenite grain faces in which the experimental problems have been largely surmounted and some useful progress has been made in interpreting the data obtained (33).   Three Fe-C alloys, containing 0.13, 0.32 and 0.63 W/O C, respectively, were employed.   Specimens 0.025 cm. thick were austenitized at 1300°C in order to make nearly all of the austenite grian boundaries perpendicular to the specimen broad faces, on which (after grinding and polishing) the measurements were made.   Careful etching was utilized in order to distinguish allotiromorphs nucleated at austenite grain faces from those nucleated at grain edges.   Fig. 2 shows that, in agreement with predictions made by Cahn (34), nucleation occurs predominantly at grain edges when undercooling is small, but decreasingly so as the undercooling is increased.   Emphasis was placed in this study on grain face nucleation; hence considerably lower experimental values of nucleation were obtained through exclusion of edge nucleation.   Carbon concentration profiles in the plane of the grain boundaries about previously nucleated ferrite allotriomorphs were computed using the Atkinson analysis (35) for the growth of oblate ellipsoids of revolution.   On this basis, the total grain boundary length within which the nucleation of additional ferrite crystals would be significantly inhibited was determined.   The grain boundary length already occupied by allotriomorphs was also measured.   These two lengths were then deducted from the total grain boundary length per unit area on the plane of polish, $L_A$, determined from the experimentally measured number of grain boundary intercepts per unit length of random measurement line, $P_L$, and the relationship (36):

$$(1) \qquad L_A = (\pi/2) \, P_L$$

The Schwartz (37)-Saltykov (38) analysis for determining the number of spherical particles per unit volume from measurements of particle diameters on a random plane of polish was found to be applicable exactly to determination of the number of particles per unit area of unreacted and compositionally unaffected grain boundary area from measurements of particle length on the plane of polish when certain conditions are fulfilled.   The most important of these are that the grain boundaries be perpendicular to the plane of polish and that the cross-section of the allotriomorphs in the grain boundary plane be satisfactorily approximated as circles (a point independently demonstrated by means of the multiple sectioning technique).   Both SEM and TEM studies confirmed statistically that optical microscopy has sufficient resolution to reveal all of the allotriomorphs present.

Fig. 3 shows representative data on the allotriomorph number density per unit unreacted and unaffected austenite grain boundary area as a function of isothermal reaction time at the three reaction temperatures studied in the 0.13% C alloy.   The passage of these plots through a maximum is probably due more to the difficulty of revealing grain boundaries between impinged ferrite crystals (and grain growth within these aggregates) than to coarsening.   The variation of nucleation rate with time derived by taking slopes of the plots in Fig. 3 is shown in Fig. 4.   Note that only at the smallest undercooling was steady state nucleation achieved before the particle number density reached a maximum value (and measurement of slopes was terminated).

The nucleation rate data were examined by means of classical heterogeneous nucleation theory.   Since an appropriate $\gamma$-plot for the chemical interfacial energy of coherent fcc:bcc interphase boundary energies is not yet available, the critical nucleus shape cannot yet be rigorously deduced even for homogeneous nucleation, much less for nucleation at grain boundaries of unknown structure and energy.   Hence a more pragmatic approach to this problem, which is the central one in the application of nucleation theory at low and moderate undercoolings, was taken.   Models based upon the usual abutting spherical caps were first utilized.   Such shapes correspond to a disordered-type interfacial structure when the crystal structures of the two phases are different.   Truncation of first one and then both caps by a

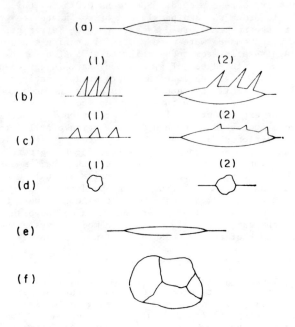

Fig. 1:   The components of the Dube (3) morphological classification system.

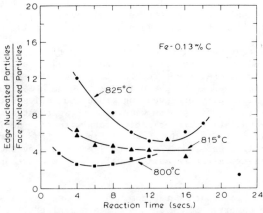

Fig. 2:   Variation of the ratio of edge-nucleated to face-nucleated ferrite
          crystals with isothermal reaction time at three temperatures in
          Fe-0.13% C (33).

planar facet was then introduced.   Sufficient data on activities of C and
Fe in austenite and on the energies of disordered-type austenite grain
boundaries and   austenite:ferrite boundaries are available to permit realistic evaluation of these models; the facet energy, though, had to be
treated parametrically.   The two unfaceted caps model, applied to the
Fe-0.13% C alloy reacted at 825°C, yielded a nucleation rate of about

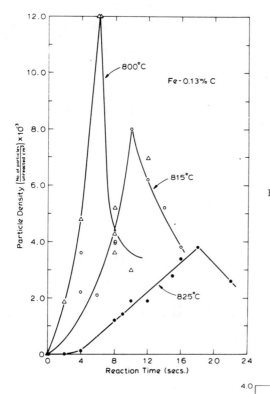

Fig. 3:  Number density of face-
nucleated ferrite
crystals vs. reaction
time at three tempera-
tures in Fe-0.13% C (33).

Fig. 4:  Nucleation rate of ferrite
at austenite grain faces
vs. reaction time at three
temperatures in
Fe-0.13% C (33).

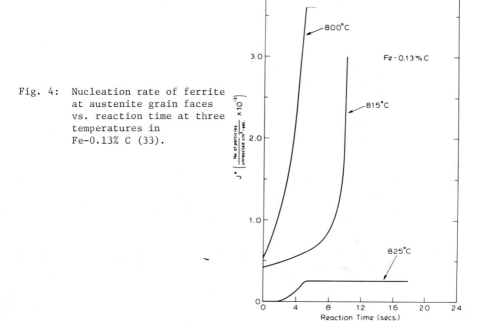

$10^{-1,000,000}$ cm.$^2$/sec. The faceted models yielded higher rates, but when the condition was applied that the facet energy could not be reduced below levels corresponding to physically reasonable sizes for the various models, the calculated nucleation rates were still more than ten orders of magnitude below the range of measured rates even at the lowest facet energy employed. Hence recourse was made to pillbox models, illustrated in Fig. 5. The model in Fig. 5a is based upon low energy broad faces and edges, whereas thos in Figs. 5b and 5c envisage replacement of one low energy (presumably coherent) broad face with a spherical cap in the austenite grain containing the nucleus and in the adjacent austenite grain, respectively.

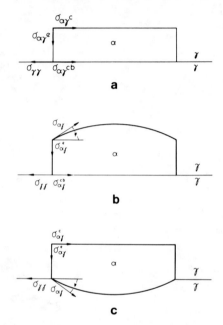

Fig. 5:   (a)  The coherent pillbox model of the ferrite critical nucleus, and (b) and (c) two variants thereof, with one incoherent interface each (33).

Note that the first two models, but especially the first, require an orientation relationship capable of yielding low energy facets between the ferrite nucleus and both austenite grains. The Fig. 5a design was found able to account for the steady state nucleation rate measured at the smallest undercooling in all three alloys when the edge energy, taken as the disposable parameter in this analysis, was ∿ 18 ergs/cm.$^2$ = √18mJ/m$^2$. This appears to be a reasonable value for the energy of a fully coherent α:γ boundary. The other two models produced similar results and it was not possible to distinguish among them. At larger undercoolings, though, the analysis predicted a much faster increase in nucleation kinetics than was found experimentally. Calculated and measured nucleation rates were brought into agreement over the temperature-composition region studied when the number of atomic nucleation sites (the number of substitutional atoms per unit area in contact with the grain boundary) was very greatly reduced below its normal value of $10^{15}$/cm.$^2$. This severe restriction is consistent with the crystallographic requirement that a {111}$_\gamma$ plane lie exactly in the "surface" of one austenite grain.

The finding of King and Bell (14) that many ferrite allotriomorphs exhibit a low energy orientation relationship between both austenite grains and the observation on their published micrograph that faceting occurs even in the apparent absence of such relationships provide generalized support for the models of Figs. 5a and 5b.  Equivalent observations have been made by means of selected area electron channeling on allotriomorphs of (hcp) $\xi_m$ formed at (bcc)$\beta$ grain boundaries during the massive transformation in an Al-26 A/O Ag alloy (39).

## Widmanstatten Ferrite

For the balance of this section on Fe-C alloys, attention will be focussed on crystallography, interfacial structure, and growth kinetics and mechanism.  Throughout these discussions, however, it should be recognized that growth occurs within the crystallographic framework provided by the orientation relationships  established during nucleation.

Widmanstatten sideplates and intragranular plates are considered first because their interphase boundary structure should be almost entirely of the partially coherent type.  Grain boundary allotriomorphs, being bounded by an as yet unevaluated mixture of partially coherent and disordered areas-- but having growth characteristics more nearly akin to those expected of dis-ordered boundaries--will be discussed next.

### Orientation Relationships

Mehl, Barrett and Smith (40) found that (presumably) intragranular ferrite plates in a plain carbon steel had the Kurdjumow-Sachs (K-S) (41) orientation relationship with respect to their parent austenite grain:

$$(111)_\gamma \text{ // } (110)_\alpha$$
$$(\bar{1}10)_\gamma \text{ // } [11\bar{1}]_\alpha$$

A reanalysis of their data by Young (42), however, yielded a rotation of $3.4°$ around the normal to the parallel planes; this rotation angle, $\theta$, may be defined as that between the directions $[1\bar{1}0]_\gamma$ and $[100]_\alpha$.  At the K-S relationship, $\theta = 5\text{-}1/4°$.  Watson and McDougall (43) used Kössel patterns to examine orientation relationships in an Fe-0.45% C alloy.  They found several degrees of scatter about K-S, whose average represents both a non-zero value of $\theta$ and also a finite value of $\emptyset$, defined as the angle between the $(111)_\alpha$ and the $(110)_\alpha$ planes.  The Kossel pattern data of King and Bell (44) on an Fe-0.47% C alloy, similarly heat treated, exhibited values of $\theta$ ranging from 0° to 7°.  At $\theta = 0°$, the Nishiyama (45)-Wassermann (46) orientation relationship obtains:

$$(111)_\gamma \text{ // } (110)_\alpha$$
$$[\bar{2}11]_\gamma \text{ // } [\bar{1}10]_\alpha$$

However, the mean $\theta$ was 3.3°, remarkably close to Young's reanalysis of the Mehl et al data.  Additionally, the mean value of $\emptyset$ was 1.1° and rose as high as 2.5°.  Briefly noting relevant results on fcc:bcc orientation rela-tionships in other alloys, Hall, Aaronson and Kinsman (47) found that $\theta$ varies between the values corresponding to N-W and K-S for bcc Cr-rich pre-cipitates formed in a Cu-rich Cu-Cr alloy; in a current study, Hall and Aaronson (48) used lattice fringe imaging to demonstrate that $\emptyset = 0$.  Rigsbee and Aaronson (49) observed variations in a $\theta$ from -3° to 3.5° in an Fe-0.62% C-2.0% Si alloy, and Southwick and Honeycombe (50) observed by means of Moire fringes that $\emptyset$ ranges from 0° to 5° in a duplex stainless steel.

Correlations have been observed between $\theta$ and the ratio of the fcc/bcc lattice parameter ratio (51, 52).  Attempts to rationalize such results in

terms of polar $\gamma$-plots based upon crude calculations of the misfit dislocation contribution to interphase boundary energy (10, 53, 54) have been extensively criticized (55, 56), particularly because orientation relationships
are fixed during nucleation, when the probable presence of full coherency
(56) makes the relevant $\gamma$-plot that of the orientation-dependence of the
chemical interfacial energy. An assumption that the two plots are intimately
related (10) requires proof which is not yet available. The circumstance
that the proeutectoid ferrite reaction and the formation of Cr-rich precipitates in Cu-Cr alloys yield similar $\theta$ but different $\emptyset$ values suggests that
other factors, such as the interaction of the nucleation site with the
critical nuclei and perhaps differences between interstitial and substitutional solid solutions play a role in determining the details of the
orientation relationships developed.

### Habit Planes

Belaiew (57) and Mehl, Barrett and Smith (40) evaluated the habit plane
of intragranular ferrite plates as $\{111\}_\gamma$. Using two Fe-C-Si alloys, Liu
et al (58) employed two-surface analysis to demonstrate that "the habit
plane of ferrite sideplates is neither unique nor $\{111\}_\gamma$. The habit planes
determined lay between 4° and 20° from $\{111\}$". Watson and McDougall (43)
demonstrated substantial scatter in the habit place of ferrite plates
formed "immediately adjacent to the free surface", with individual poles
lying well away from $\{111\}_\gamma$; less scatter and a mean value closer to $\{111\}_\gamma$
was found for "the tapered segment or 'tail'" associated with these plates.
"Some of the scatter is considered to be real..."; but in their discussion
of these results, Watson and McDougall stated that their measurements "show
that the habit planes do not scatter significantly from the average value
of $(0.5057, 0.4523, 0.7346)_F$". An association between a given variant of
the habit plane and a given variant of the lattice orientation relationship
was also noted. However, it seems clear that their data may have been
affected by transformation in association with a free surface, and thus the
generality of their findings seems doubtful. King and Bell (44), studying
plates formed in bulk alloys, found habit planes lying within 5° of $\{111\}_\gamma$
for well defined plates, but observed deviations of as much as 12° for
individual facets of less well formed plates. Inspection of their optical
micrographs suggests that a substantial fraction of the plates in their
specimens were sideplates. Also conducting studies on sideplates, but in an
Fe-C-Si alloy, Rigsbee and Aaronson (49) observed deviations of 9° to 18°
from $\{111\}_\gamma$. These data are consistent with those of Liu et al (58); the
differences with respect to the data of King and Bell (44) suggest that Si
may have had some effect upon the habit plane of ferrite sideplates. But it
seems relevant to note that deviations of 10° -20° from $\{111\}_{fcc}$ were observed in a Cu-Cr alloy (47).

### Three-Dimensional Shape of Widmanstatten Ferrite

Mehl, Barrett and Smith (40) concluded that the shape of Widmanstatten
ferrite in three dimensions is that of a plate. Published micrographs (59)
provided much of the basis for this statement. However, a section through
a slowly cooled 0.41% C, 0.68% Mn, 0.19% Si alloy deliberately cut parallel
to a $\{111\}_\gamma$ plane in a very large austenite grain revealed that ferrite
plates lying in the plane of polish have a roughly equi-axed, though highly
irregular cross-section. King and Bell (44) later confirmed this finding
on isothermally reacted specimens of a Fe-0.47% C alloy polished on two
orthogonal planes. They noted, consistent with the Mehl et al observation,
that the ferrite plates (probably sideplates) studied are by no means equi-
axed. Watson and McDougall (43) and Wayman (60) (the latter using an 0.4%
C steel of unstated alloy content), on the other hand, also used the two
plane-of-polish technique to demonstrate that ferrite plates have the three-

dimensional shape of laths.  As pointed out by King and Bell, however, in both of these investigations the Widmanstatten ferrite examined had formed at a free surface.  Watson and McDougall themselves noted that the shape of Widmanstatten ferrite so nucleated began to change as it developed into the interior of the specimen.

### Interphase Boundary Structure of Ferrite Plates

On the theory of morphology summarized in an earlier section, the broad faces of precipitate plates should be partially or fully coherent with the matrix crystal into which the plates develop.  This prediction has been confirmed with TEM in many transformations in a wide variety of alloy systems (24, 61).  Room temperature investigation of the broad face and edge structure of ferrite plates in Fe-C alloys is not possible--at least with the techniques presently available--because the undecomposed austenite is transformed to martensite during quenching.  However, this problem has been solved in an Fe-0.62% C-2.0% Si alloy (49).  Transformation at 425° - 475°C produced closely spaced ferrite sideplates.  The carbon concentration in the austenite regions trapped between these plates was sufficiently high, inhibition by Si of cementite formation at austenite:ferrite boundaries, (which would have seriously depleted the carbon stored in the austenite regions) was sufficiently complete, and the compressive stresses exerted by the ferrite plates on the entrapped austenite were great enough to permit full retention of the austenite.  This investigation was undertaken to test a model of partially coherent fcc:bcc interfaces developed by Hall et al (47) and Russell et al (62), and later further elaborated through computer modeling (63).  On this model, introduction of "structural ledges", from one to three atoms planes high, on $\{111\}_\gamma//\{110\}_\alpha$ boundaries, cause the limited regions of good matching between the two lattices to be repeated, laterally displaced, on the terraces of successive structural ledges. Introduction of a set of misfit dislocations, $a/2\langle110\rangle$ in the fcc phase and either $a/2\langle111\rangle$ or $a\langle100\rangle$ in bcc, usually of mixed edge-and-screw orientation, with the Burgers vector always parallel to the $\{111\}//\{110\}$ planes forming the terraces of the structural ledges, completes development of the partially coherent interface.  Fig. 6 is a schematic isometric portrait of such a boundary.

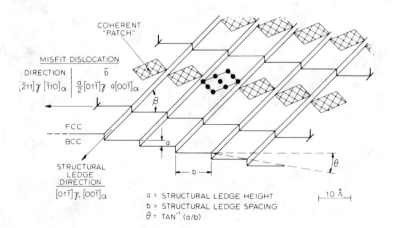

Fig. 6:    Isometric sketch of a partially coherent fcc:bcc interface with a Nishiyama-Wassermann (45, 46) orientation relationship operative (63).

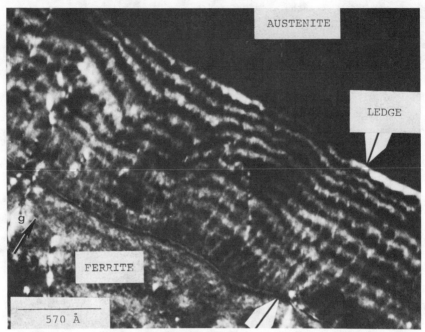

Fig. 7:  Weak-beam, dark-field TEM micrograph of the same area, showing:
(a) structural ledges; and (b) misfit dislocations (at D) as well
as structural ledges.  $\bar{g} = [110]; |\bar{s}| = 0.015$ Å$^{-1}$ (49).

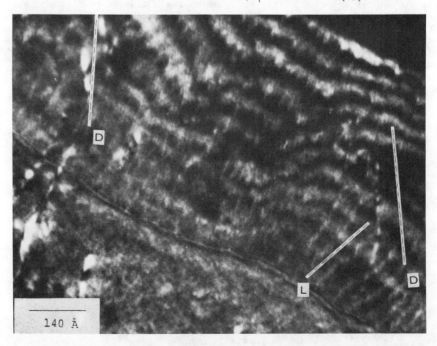

Figs. 7a and 7b were taken of the same area at a ferrite plate broad face at two different magnifications. In the lower magnification micrograph, Fig. 7a, diamonds at the ends of thickened white curves delineate one of a series of parallel structural ledges, about 3.7 nm. apart. These ledges were found to be three atoms planes high; computer modeling indicated that such ledges make more efficient use of the coherent regions of the boundary than do one- or two-layer high structural ledges (63). Fig. 7b shows, parallel to the dashed lines marked D, misfit dislocations with an average separation of about 1.5 nm. The interfacial structure of seven ferrite plates was analyzed in detail and shown to be in generally quite good agreement with the predictions of the computer model.

It is particularly important to take note of other experimental investigations of partially coherent fcc:bcc interfaces formed during solid-solid phase transformations. Howell et al (64, 65) have twice investigated the structure of austenite:ferrite boundaries in a duplex stainless steel (26% Cr, 5% Ni, 1.5% Mo). Partially coherent structures were observed; an O-lattice analysis yielded results in considerable disagreement with the experimental observations. No evidence for structural ledges was found.

Hall and Aaronson (48) have recently reinvestigated the interfacial structure of planar facets on Cr-rich precipitates in a Cu-Cr alloy. With one exception, they also found no evidence for structural ledges on the facets studied. The question thus arises as to the source of this difference among the three sets of results. An obvious possibility is that both phases involved in the Fe-C-Si alloy are interstitial solid solutions whereas those in the duplex stainless steel and Cu-Cr alloys are substitutional solutions. However, a mechanism through which this difference might affect the interphase boundary structure is not apparent. A less obvious source of difference is that both substitutional alloys were isothermally reacted for long times, sufficient to cause extensive coarsening, whereas the Fe-C-Si alloy was transformed for times usually less than a minute, and was quenched while growth was still actively in progress. It is possible that the interphase boundaries containing structural ledges represent more readily formed but higher interfacial energy structures than do those formed only with misfit dislocations. Indeed, Hall et al (66) have used the O-lattice analysis to show that this can be accomplished, but a comparison of the interfacial energies of the two types of structure has yet to be attempted.

Concern has been expressed about the effect upon the structural ledge model of $\emptyset \neq 0$, i.e., when $\{111\}_{fcc}$ is not exactly parallel to $\{110\}_{bcc}$ (44). However, such deviations can be readily accomodated by introduction of an additional set of misfit dislocations whose Burgers vector is perpendicular to the terraces of the structural ledges. In effect, these dislocations would superimpose a dislocation structure similar to that at a small-angle grain boundary upon the misfit dislocation and structural ledge components present when $\emptyset = 0°$.

Although growth ledges are usually an extrinsic component of partially coherent interphase boundaries, they are nonetheless vitally important. Because the martensite reaction makes difficult the identification of growth ledges on ferrite plates during studies conducted at room temperature, consideration of growth ledges is delayed until the next subsection, where it can be considered in conjunction with measurements of plate thickening kinetics.

<u>Thickening Kenetics of Ferrite Sideplates</u>

A detailed investigation of this topic has been reported on Fe-C alloys (67). Counterpart studies have not been made on intragranular ferrite plates, whose somewhat different morphology (broad faces more nearly parallel than those of sideplates) suggest the possibility of significant differences in kinetic behavior. Kinsman et al (67) used thermionic electron emission microscopy (THEEM) to measure the thickening kinetics of ferrite sideplates during transformation in Fe-C alloys containing 0.11, 0.22, 0.31 and 0.41% C at temperatures ranging (in different alloys) from 820° to 675°C. Measurements were made from motion picture film on which the growth of individual plates had been continuously recorded; a correction was made for the angle between the habit plane of the sideplates and the specimen surface.

Typical plots of plate half-thickness vs. isothermal reaction time are shown in Fig. 8. From data such as these the following conclusions were drawn. (i) Overall kinetic behavior is sufficiently erratic so that all effects of reaction temperature and carbon concentration in the alloy are effectively obscured. (ii) In a particular alloy reacted at a given temperature, linear, parabolic, stepped, undefinable and any combination of these behaviors was observed. (iii) Comparison with the dashed curves in Fig. 8, drawn to represent growth distance vs. isothermal reaction time for a planar, disordered interphase boundary, as computed from Atkinson's (68) analysis for a planar boundary with a composition-dependent diffusivity of carbon in austenite, makes immediately clear that plate thickening kinetics differ sharply from those for a disordered boundary. (iv) The stepped nature of some of these plots is consistent with growth by a ledge mechanism. A few examples were found of very high (ca. one micron) super-ledges, whose position as a function of reaction time could be photographed; their lengthening rates were constant and approximately these predicted for control by the volume diffusion of carbon in austenite. (v) Most plots of half-thickness vs. time displayed a tendency for the thickening rate to decrease from $5 \times 10^{-5 \pm 1}$cm./sec. to $1\text{-}30 \times 10^{-6}$cm./sec. after a few seconds of growth.

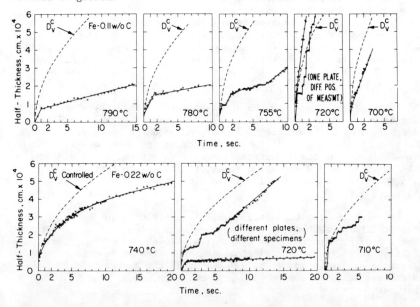

Fig. 8: Half-thickness vs. reaction time for individual ferrite sideplates in Fe-0.11% C (upper set) and Fe-0.22% C (lower set) (67).

Interpretation of these data requires a mathematical analysis of growth by the ledge mechanism. For the overall thickening rate, $G_t$, of an interphase boundary growing solely by this mechanism (i.e., only the risers of ledges are mobile; the terraces or broad faces of the ledges are taken to be wholly immobile), Cahn, Hillig and Sears (69) wrote:

(2)
$$G_t = hv/\lambda$$

where h = ledge height, v = lateral velocity of ledges and $\lambda$ = inter-ledge spacing (Fig. 8). From application of the flux balance equation to the riser of a ledge, Jones and Trivedi (70) obtained:

(3)
$$V = \frac{D(x_\gamma^{\gamma\alpha} - x_\gamma)}{\alpha\lambda(x_\gamma^{\gamma\alpha} - x_\alpha^{\alpha\gamma})} \equiv \frac{D\Omega_o}{\alpha\lambda}$$

where D = averaged (carbon) diffusivity in the (austenite) matrix, $x_\gamma^{\gamma\alpha}$, $x_\alpha^{\alpha\gamma}$, and $x_\gamma$ are the atom fractions of solute (carbon) in $\gamma$ at the $\gamma/\alpha(\alpha + \gamma)$ phase boundary, in $\alpha$ at the $\alpha/\alpha+\gamma)$ phase boundary and in the austenite prior to transformation, respectively, and $\alpha$ is the effective diffusion distance in front of the riser. To obtain $\alpha$, Jones and Trivedi rewrote Fick's Second Law for a moving, reduced coordinate system and solved it with the assistance of some approximations. In particular, the solute concentration along the riser was progressively diminished from its base upwards so as to ensure a constant concentration gradient in front of the riser (otherwise the upper portion of the riser would lengthen more rapidly than the lower) and the diffusion field associated with an individual ledge was terminated at finite, but angle-dependent distances from the riser (Fig. 9). From the relationship thereby obtained:

(4)
$$\Omega_o = 2p\alpha$$

where p, the Peclet number, is vh/2D; Jones and Trivedi provided graphs of the value of $\alpha$ as a function of p and $\Omega_o$.

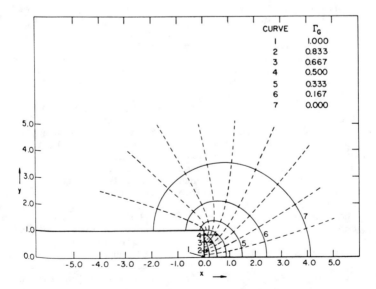

| CURVE | $\Gamma_G$ |
|-------|-----------|
| 1 | 1.000 |
| 2 | 0.833 |
| 3 | 0.667 |
| 4 | 0.500 |
| 5 | 0.333 |
| 6 | 0.167 |
| 7 | 0.000 |

Fig. 9:  Isoconcentrates (solid curves) and flux lines (dashed) in the diffusion field of an isolated ledge according to Jones and Trivedi (70).

Atkinson (71) subsequently mounted a different attack upon the problem.  He
too ensured the stability of the riser, but allowed the diffusion field
associated with it to extend indefinitely into the matrix.  Fig. 10 shows
that the anisotropy of the diffusion field about a riser which he obtained
is essentially the reverse of that of Jones and Trivedi.  However, his
p vs. α curve does not differ significantly from that of Jones and Trivedi
until p < 10$^{-3}$.  For the proeutectoid alpha reaction in Fe-C alloys, the
difference between the two treatments is unimportant as long as the ledges
are diffusionally isolated.  When they are not, the differences between
Figs. 9 and 10 leads to major differences in growth behavior.

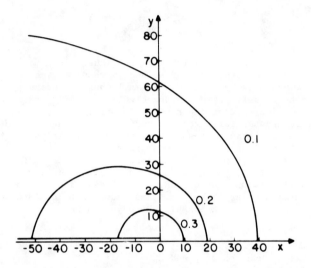

Fig. 10:  Isoconcentrates in the diffusion field of an isolated ledge
          according to Atkinson (71).

Kinsman et al applied the Jones-Trivedi analysis to their thickening
kinetics data and back-calculated the average inter-ledge spacing.  As
anticipated from Fig. 8, the scatter in the spacings thus obtained is large.
Fig. 11 shows the data obtained for their alloys.  During the initial
stages of growth, $\lambda$ is seen to vary in the range 1 - 10 x 10$^{-4}$ cm. and at
late stages from 1 - 5 x 10$^{-3}$ cm.  Inter-ledge spacings were also measured
with replication electron microscopy.  They, too, scattered appreciably, but
averaged about 2 x 10$^{-4}$ cm.  The agreement between the measured and back-
calculated $\lambda$'s is as good as one could hope to secure in view of the scatter
in both measurements during early reaction times.  At later times, however,
the back-calculated values are about an order of magnitude too large, suggesting that many of the growth ledges had migrated into boundary orientations
at which their risers contained a negligible proportion of disordered
structure.  Hence these results provide reasonable, though hardly complete,
support for the prediction of the theory of morphology that growth ledges
furnish the only growth mechanism for the thickening of precipitate plates.
Considerably more quantitative support for this conclusion has been obtained
from counterpart studies, employing hot-stage TEM, on Al-Ag alloys (72), and
from room temperature studies, correlated with measurements of inter-
ledge spacing, on θ plates in Al-4% Cu (73).  A more exacting test on plate
thickening during the proeutectoid ferrite reaction than that made by
Kinsman et al will be very difficult, since formation of the Widmanstatten
structure is largely absent in thin foils (74) due to the influence of
austenite grain size upon the evolution of sideplates (75).

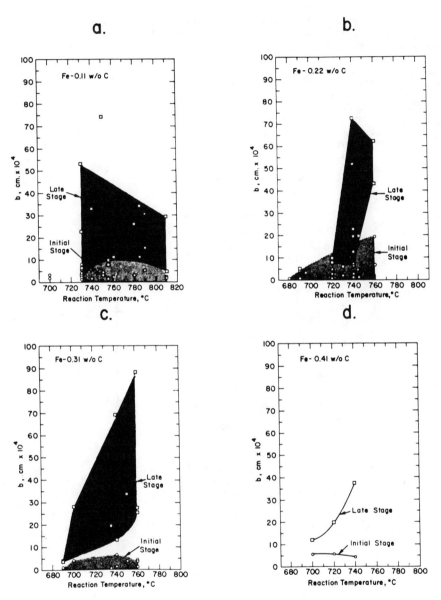

Fig. 11:  Ranges of average inter-ledge spacing vs. reaction temperature
         back-calculated through Jones-Trivedi (70) analysis from
         thickening kinetics data at initial and late stages of thickening
         in four Fe-C alloys (67).

Lengthening Kinetics of Ferrite Plates

Unlike thickening kinetics, measurement of lengthening kinetics is quite simple. Data of adequate accuracy are readily obtained by measuring the length of the longest plate in the plane of polish on specimens quenched to room temperature as a function of reaction time at a given isothermal reaction temperature. Three sets of lengthening rate data on ferrite (and bainite) plates are available in Fe-C alloys, due to Hillert (76, 77), Townsend and Kirkaldy (23) and to Simonen et al (78), respectively. These data, which are in quite reasonable agreement, always yield constant lengthening rates. Analysis of lengthening kinetics requires calculations of or preferably experimental data on the radius of plate edges. Simonen et al secured such information by means of replication electron microscopy; frequency histograms of the data were plotted, e.g., Fig. 12, and the radius appearing most frequently was taken as the true value of this quantity. No measurements of plate edge radii were included in the other two studies.

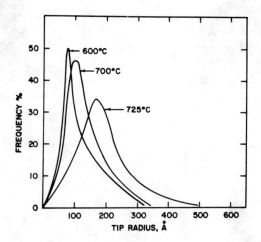

Fig. 12:   Smoothed curves drawn through frequency histograms for apparent radii of curvature of the edges of ferrite plates at three reaction temperatures in Fe-0.24% C. Simonen et al (78).

The most sophisticated analysis of plate lengthening kinetics presently available is due to Trivedi (79). He has also published an equivalent analysis for the lengthening kinetics of needles (80). For plate lengthening he obtained:

$$(5) \qquad \Omega_o = \sqrt{\pi p} \ e^p \ \mathrm{erfc} \sqrt{p} \ [1 + \frac{G_\ell}{G_c} \ \Omega_o \ S_1(p) + \frac{r_c}{r} \ \Omega_o \ S_2(p)]$$

$$(5A) \qquad G_c = \mu_o \ (x_\gamma^{\gamma\alpha} - x_\gamma)$$

$$(5B) \qquad r_c = \frac{x_\gamma^{\gamma\alpha}}{x_\gamma^{\gamma\alpha} - x_\gamma} \ [\frac{\gamma \bar{V}_{Fe}^\alpha}{(x_\gamma^{\gamma\alpha} - x_\alpha^{\alpha\gamma})RT}]$$

$$(5C) \qquad p = \frac{G_\ell r}{2D}$$

where $G_\ell$ = lengthening rate of a plate, $G_c$ = lengthening rate if growth were entirely controlled by interfacial reaction, $\mu_o$ = interface kinetics co-efficient, $r_c$ = radius at which growth is entirely prevented by capillarity, $\gamma$ = austenite:ferrite interfacial energy at the plate edge, $\bar{V}_{Fe}^{\alpha}$ = partial molar free volume of Fe in ferrite, R = gas constant and T = absolute temperature. $S_1(p)$ and $S_2(p)$ are complex functions of p, described and plotted against p by Trivedi (79); Bosze and Trivedi (81) later showed that $S_1(p) \approx 1/\pi p$ and $S_2(p) \approx 2/\pi p$. Equation(5) permits evaluation of $G_1$ as a function of r. Invoking Zener's (82) maximum growth rate principle yields:

$$(6) \quad \frac{\partial \Omega_o}{\partial r} = 0 = \frac{\Omega_o}{2p}(2p+1) - \frac{\Omega_o}{\sqrt{\pi p}\; e^p\;[1 - \mathrm{erf}\sqrt{p}]}$$

$$+ \sqrt{\pi p}\; e^p\; \mathrm{erfc}\sqrt{p}\; [\frac{G_\ell}{G_c} S_1'(p) + \frac{r_c}{r} S_2'(p) - \frac{r_c}{r} S_2(p)]$$

Where $S_1'(p)$ and $S_2'(p)$ are also shown as a function of p in Trivedi's paper. Solving equations (5) and (6) simultaneously provides unique values of both $G_1$ and r.

The lack of theoretical justification for assuming that $G_1$ is a maximum has long been troublesome. Experimental studies of dendrite formation during solidification of an organic compound have shown that the value of r adopted is much larger than that for maximum growth rate (83); instead, the tip radius is consistent with that which makes a growing sphere just stable against sinusoidal shape fluctuations. However, Trivedi (84) has pointed out that analyses of dendrite tip stability, such as that of Langer and Müller-Krumbhaar (85), take into account the formation of sidebranches. The presence of partial or full coherency on the broad faces of precipitate plates normally prevent development of sidebranches, and thus alter the basis of the analysis. Trivedi (84) has studied the stability of a circular cylinder (instead of the more appropriate parabolic cylinder, for which the analysis becomes exceedingly complex). He has shown that in the presence of the interfacial structure constraint against sidebranch formation the stability condition yields exactly the same result as the maximum growth rate hypothesis. Trivedi notes that analysis of the stability of eutectic structures, wherein the same type of constraint is operative, has repeatedly produced the same result (86-88).

In analyzing their experimental data on $G_1$ and r, Simonen et al first assumed that $\mu_o = \infty$, i.e., that no interface barrier to growth is present. Fig. 13 shows their plots of growth rate vs. driving force, calculated for two different values of $\gamma$: 200 ergs/cm.$^2$, the usual (89) estimate of the energy of a partially coherent austenite:ferrite boundary, and 800 ergs/cm.$^2$, an appropriate value for a disordered boundary of this type (90). The curves in this Figure were calculated; the individual points represent experimental data. Since the experimental data fell at higher growth rates than those calculated when $\gamma$ = 800 ergs/cm.$^2$, no mechanism is available through which calculated and measured kinetics can be brought into agreement. $\gamma$ = 200 ergs/cm.$^2$ was the highest interfacial energy at which this anomaly could be avoided. But even this value is seen in Fig. 13b to have been insufficient to produce agreement between calculated and measured kinetics at other temperatures. A finite, temperature-dependent $\mu_o$ accomplishes this in Fig. 13c. "Fortunately", the activation energy for the values of $\mu_o$ thus obtained, 11,800 cal./mole (49,400 J/mole) does not correspond to any readily identifiable mechanism which might participate in plate lengthening. Instead, the findings of $\gamma$ = 200 ergs/cm.$^2$ and a finite, temperature-dependent $\mu_o$ were taken to mean that lengthening as well as thickening of ferrite plates is controlled by a ledge mechanism. A still more complex analysis of ledge growth kinetics, now based upon three-dimensional diffusion because the length of a ledge at the edge of a plate is likely to be comparable to its height, is thus required to describe the lengthening kinetics of ferrite plates. This description must also include closely spaced ledges arrayed on the edges of a sharply curved plate. Data on the interledge spacing at the plate edges will thus be needed in order to effect a complete comparison between theory and experiment.

This type of conclusion for the mechanism of plate lengthening has now become nearly routine. The edges of $\theta'$ plates in Al-Cu (91,92), $\gamma$ plates in Al-Ag (93), $\eta$ plates in Al-Au (92) and $\alpha_1$ plates in Cu-Zn (94) have so far been shown to be partially coherent by TEM observations; hence these plates must lengthen by the ledge mechanism. Ledgewise lengthening of $\gamma$ Al-Ag plates has been observed with hot-stage TEM (72). Bosze and Trivedi (95) have made a preliminary attempt at solving the exceedingly difficult mathematical problem of describing plate lengthening by ledges through a study of the effects of anisotropy in $\mu_o$. However, a more direct attack upon this problem remains highly desirable.

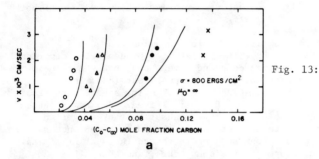

a

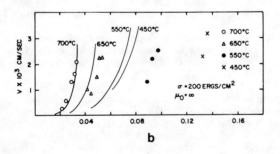

b

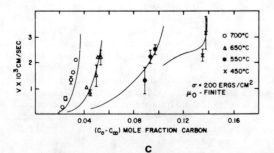

c

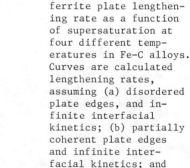

Fig. 13:  Data points represent ferrite plate lengthening rate as a function of supersaturation at four different temperatures in Fe-C alloys. Curves are calculated lengthening rates, assuming (a) disordered plate edges, and infinite interfacial kinetics; (b) partially coherent plate edges and infinite interfacial kinetics; and (c) partially coherent plate edges and finite interfacial kinetics. Simonen et al (78).

On the Role of Shear in the Growth of Ferrite Plates.

As a consequence of their morphology and the martensite-like relief effect which ferrite plates can produce when formed at a free surface, shear has long been considered to play an important role in the formation of ferrite plates.  A considerable range of growth processes has been postulated to result from a shear mode of transformation, ranging from slow, continuous shear (96-98), through short, fast shears, which are rapid enough to inherit essentially the full carbon concentration of the parent austenite, (99, 100), to long shears which swiftly form an entire plate (101). Several critical reviews have been made of such proposals for the formation of precipitate plates in general (1,24,102) and two reviews have dealt specifically with the growth of ferrite plates by shear (1, 67).  Hence the present discussion will focus attention upon the most important evidence bearing on the problem, emphasizing more recent items.

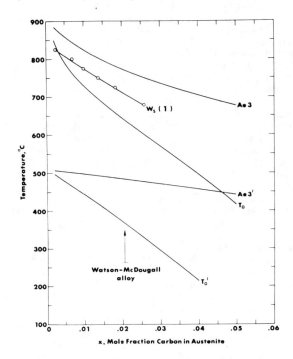

Fig. 14:   Ae3, Ae3', $W_s$ (experimental), $T_o$ and $T_o'$ curves for Fe-C alloys (103).

Consider first thermodynamic analyses of shear proposals.  Fig. 14 shows the experimentally determined highest temperature at which a significant number of ferrite sideplates can form, $W_s$, as a function of carbon concentration in Fe-C alloys (1).  This temperature is seen to lie significantly above the $T_o$ temperature, at which the free energies of austenite and ferrite of the same composition have the same free energy in the absence of strain energy (82).  Hence formation of ferrite plates with the same carbon concentration as the parent austenite is thermodynamically impossible over an appreciable temperature-composition region.  If shear strain energy is added to the free energy of ferrite, using a relationship due to Eshelby (104), the shear of 0.36 which Watson and McDougall (105)

measured on ferrite plates formed at a free surface, and an aspect ratio of
ferrite plates estimated from their micrographs a $T_o$ curve lying at much
lower temperature is obtained. Unsurprisingly on the view of Eshelby (104),
this curve is seen to approximate closely the $M_s$ temperature. If the
$\gamma/(\alpha + \alpha)$ or Ae3 curve is recalculated to incorporate the influence of (the
same amount of) shear strain energy, the resulting Ae3' curve lies well below the $W_s$ curve (103). A subsequent attempt (106) to recalculate the
Ae3' curve using a different value for the aspect ratio of ferrite plates
yielded a considerably higher Ae3' curve which, however, still lies well
below the $W_s$ curve; attempts to dismiss this gap (106) are readily
invalidated (107). These comparisons between the $W_s$ and Ae3' curves mean
that even though ferrite has only the carbon concentration corresponding to
the $\alpha/(\alpha + \gamma)$ curve or its metastable equilibrium extrapolation it cannot
form with the amount of strain energy corresponding to that attending
a shear transformation. On the other hand, if relaxation of some of the
shear strain energy through plastic deformation attends growth of the
ferrite plates (108), the Ae3' and $T_o$' curves will be raised, though the a-
mount of the increase, which is a sensitive function of the work hardening
which occurs during plastic deformation (109), has yet to be estimated for
this situation. However, Fig. 14 shows that even in the absence of strain
energy, inheritance of the full carbon concentration of the parent
austenite is out of the question and formation of equilibrium ferrite by
shear is questionable.

These thermodynamic considerations are consistent with electron probe
X-ray microanalysis experiments which show that the carbon concentration in
ferrite plates is not that of the parent austenite, but is less than 0.03
W/O, the limit of the experimental technique employed (110). The reaction
temperatures and alloy compositions of specimens on which such measurements
were made are indicated in Fig. 15.

The details of the structure of the broad faces of ferrite plates
shown in Figs. 7a and 7b, and the associated discussion previously presented
bear directly on the mechanistic feasibility of displacing these interfaces
by a shear mechanism. Since the Burgers vector of the misfit dislocations
was shown to lie in the terraces of the structural ledges and to range in
character from pure edge to mixed, but not pure screw, movement of these
interfaces in the normal direction by shear is mechanistically impossible.
Significant lateral displacement of the risers of structural ledges has
been predicted to be infeasible because of the disruption this would cause
in the atomic structure of the boundary (47, 63). Nonetheless, the proposal has been made that Olson-Cohon (111) partial dislocations with a very
small Burgers vector are present in the risers and provide a mechanism for
their lateral migration by shear (112). This proposal is readily tested by
means of data already summarized. The average spacing between structural
ledges is $10^{-7}$ cm. (63), whereas the average spacing between growth ledges
at early stages of growth is about $10^{-4}$ cm., later increasing to about
$10^{-3}$ cm. (67). Hence the structural ledges must be essentially immobile as
originally predicted.

Watson and McDougall (105) have made a careful study of the shape change,
orientation relationships and habit plane of ferrite plates formed at free
surfaces in an Fe-0.45% C alloy. On the evidence they provided, their
results are inconsistent with the phenomenological theory of martensite (113).
In particular, tent-shaped as well as invariant plane strain surface relief
effects were found, significant scatter appeared in the habit plane data and,
especially, the measured orientation relationship and shape strain were
found to be inconsistent. These findings contravene the dictum that accept-
ability as the product of a martensitic transformation requires not only
that all of the individual requirements of the theory be satisfied, but also

that all of the data sets be consistent with one another (114).

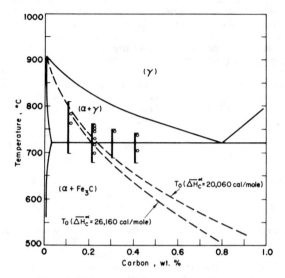

Fig. 15:  Temperatures at which electron probe measurements were made of
          carbon concentration in ferrite plates in four Fe-C alloys.  Two
          dashed curves shown represent extremum evaluations of $T_o$.  (103).

      The repeated finding that the relief effect at a free surface produced
by ferrite plates is tent-shaped or more complex rather than of the simple
invariant plane strain type (67) is a particularly prominent individual
inconsistency.  Claims that tent-shaped reliefs are actually formed by two,
back-to-back plates (115) are contradicted by a variety of experimental
studies (67, 116).  More recently, however, TEM evidence that ferrite plates
normally appear as pairs with different orientation relationships of the
same form has lent some support to the two-plate explanation (106).
However, the heat treatment technique applied to the Fe-0.41% C alloy used
in this study was inadequate to ensure isothermal reaction of even a richly
alloyed quaternary steel (117, 18); hence the possibility that the "second
plate" formed sympathetically during quenching from the transformation
temperature to room temperature cannot be discounted until the experiments
have been repeated with adequate technique.

Growth Kinetics of Grain Boundary Ferrite Allotriomorphs

      Measurement of the lengthening and thickening kinetics of grain bound-
ary allotriomorphs poses significant stereological problems.  Detailed consideration of these difficulties has led to the conclusion that using thin
specimens, austenitized for sufficient time at a high enough temperature
to allow the austenite grain boundaries to grow perpendicular to the broad
faces of the specimens, and measuring the length of the longest and the
thickness of the thickest allotriomorph on a plane of polish parallel to the
specimen broad face is the best solution presently available to these
problems (119).  These measurements should closely approximate the true
length and the true thickness of one of the first allotriomorphs nucleated
in the specimen.  Some interesting attempts have been made to circumvent
the stereological problem of relating apparent sizes on a plane of polish
to sizes in three dimensions.  A particularly incisive one was made by
Purdy and Kirkaldy (120), who used decarburization and slow cooling into

the austenite + ferrite region to establish essentially planar layers of
polycrystalline ferrite at both ends of a bar of an Fe-C alloy.  After a
homogenization anneal at a constant temperature in the austenite + ferrite
region to eliminate concentration gradients in both phases, the bar was
isothermally reacted at a higher temperature in order to grow austenite at
the expense of ferrite.  Periodic quenching and polishing on a plane parallel to the diffusion direction permitted accurate measurement of the true
austenite layer thickness as a function of isothermal reaction time.
Growth of the surface-nucleated ferrite through a number of austenite grains
during establishment of the ferrite layers was suggested to have eliminated
problems arising from orientation relationships between austenite and ferrite
capable of producing low energy, immobile boundaries at some orientations.
While the parabolic rate constant obtained from these measurements at a
single temperature did agree well with that calculated from the appropriate
constant-diffusivity solution to the diffusion equation, it is probably not
feasible to extend measurements of austenite (or ferrite) growth kinetics
to temperatures appreciably below that of the eutectoid with this technique because of the increased probability of nucleating transformation
ahead of the planar boundaries which it establishes.

Thermionic electron emission microscopy has also been used to measure allotriomorph growth kinetics in Fe-C alloys (12, 121).  Individual
data sets are very precise because this technique yields a metallographic-
like image of allotriomorphs which can be photographed at  short intervals
with a motion picture camera while growth is taking place.  However, this
technique yields much scatter among parabolic rate constants determined in
a single specimen during transformation at a given temperature (12, 121).
Stereological problems resulting from measuring the growth kinetics of an
allotriomorph on a plane lying above or below its "equator" and also from
grain boundaries lying at an angle to the specimen surface are likely responsible for some, though not all of this scatter (119).

Fig. 16 is a typical data plot obtained from measurements of growth
kinetics in an Fe-C alloy by the room temperature, optical metallographic
technique (122).  Replotting in log-log form yields a slope of one-half,
within experimental limits of error, provided that transformation did not
occur during either quenching to or quenching from the reaction temperature
(122).  Although a double spherical cap appears to be the best generalization
of the morphology of an idealized allotriomorph, an oblate ellipsoid is the
closest shape for which the diffusion problem has been solved.  As a consequence of the larger effect which the point effect of diffusion might
reasonably be expected to exert upon the growth kinetics of a double
spherical cap, the growth kinetics calculated for an oblate ellipsoid with
the same aspect (thickness/length) ratio should be somewhat slower than those
of the double cap.  Ham (123) and Horvay and Cahn (124) have solved the
ellipsoidal growth problem for a constant diffusivity in the matrix
phase.  Atkinson (125) has developed a computer-based solution for this
problem when the pertinent diffusivity varies with composition in a known
manner; this solution has been specialized to the particular case of ferrite
allotriomorphs (34).  Fig. 17 shows the variation of the ratio of the experimentally measured parabolic rate constants for thickening, $\alpha$, and
lengthening, $\beta$, to those calculated from this Atkinson analysis with temperature in the three Fe-C alloys in which measurements were made (122).
This ratio is seen to be less than unity under all conditions studied, and
to be particularly small when both the carbon concentration in the alloy
and the undercooling below the $\gamma/(\alpha+\gamma)$ temperature are lowest.

The results in Fig. 17 are not due, on the foregoing reasoning, to
the use of an oblate ellipsoid as a model for allotriomorphic growth.

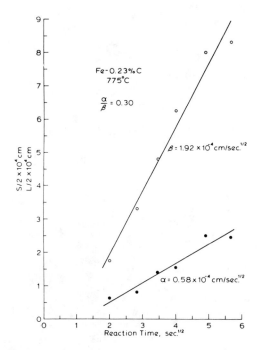

Fig. 16:  Typical plots of maximum half-thickness, S/2, and maximum half-
length, L/2, of grain boundary ferrite allotriomorphs vs.
(reaction time) 1/2 in Fe-0.23% C.  Bradley et al (122).

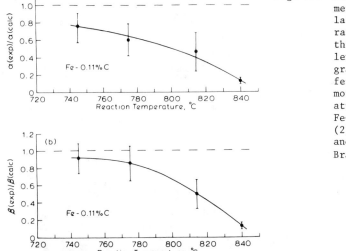

Fig. 17:  Ratio of experimental to calcu-
lated parabolic
rate constants for
thickening ($\alpha$) and
lengthening ($\beta$) of
grain boundary
ferrite allotrio-
morphs vs. temper-
ature in:   (1)
Fe-0.11% C;
(2) Fe-0.23% C;
and (3) Fe-0.42% C.
Bradley et al (122).

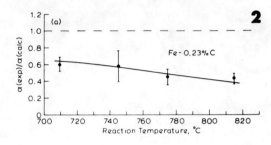

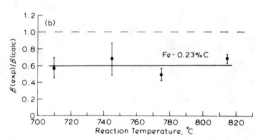

Fig. 17 (2)

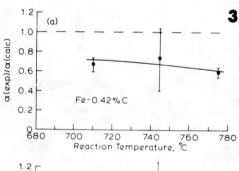

Fig. 17 (3)

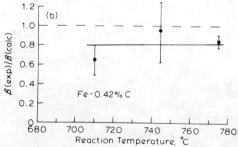

Instead, the slower than calculated growth kinetics are attributed to the presence of a distribution of partially coherent facets amongst the disordered-type interfacial structure of the allotriomorph faces. This is a consequence of low energy orientation relationships between the allotrio-morphs and one (14) (or more likely, all (14, 33)) of the bounding austenite grains. Such facets, growing by means of the erratically operating ledge mechanism, will normally migrate at lesser rates than those permitted by uniform atomic attachment and detachment at disordered austenite:ferrite boundaries, as discussed in the previous section. Fig. 18 is a schematic drawing of how the allotriomorph interface shape might change locally with growth time when a partially coherent facet has a disordered area on either side of it (1).

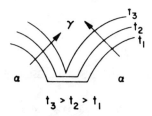

Fig. 18:   Mechanism for bypassing a partially coherent facet by disordered areas of an advancing interphase boundary (1).

Clearly, these growth kinetics will be somewhat different....and very much more complex...than those treated in the ellipsoidal analyses that was used. However, before a more exact treatment can be made of allotriomorph growth kinetics, it will be necessary to secure information on the propor-tions and distribution of partially coherent facets on the interphase boundaries of allotriomorphs, and also on the variation in the inter-ledge spacing of these facets with time. Destruction of the austenite matrix by the martensite transformation during quenching to room temperature will seriously interfere with securing reliable information of this type in Fe-C alloys; either a more highly alloyed steel will have to serve as a "stand in" or a high-resolution elevated-temperature technique which does not affect the transformation processes will have to be applied in order to acquire this information.

It should be noted, however, that Fig. 17 does show that allotriomorph growth kinetics are usually not very much slower than those predicted from oblate ellipsoidal kinetics assuming disordered interphase boundaries, and thus that allotriomorphs can serve, at least awhile longer, as a fairly good model for growth controlled by the migration of disordered interphase boundaries.

One further aspect of allotriomorph growth kinetics should be briefly noted. The aspect ratio of grain boundary ferrite allotriomorphs is roughly one-third, independently of reaction time, reaction temperature and carbon concentration in Fe-C alloys (112). This ratio, however, differs from the ratio of 0.55 which obtains when one or both phases are recrystal-lized by plastic deformation and then annealed at the original reaction temperature (126). Further, measurement of the dihedral angle at the edges of allotriomorphs in as-transformed specimens yields a value equiva-lent to an aspect ratio of 0.47 if the allotriomorphs were accurately representable by double spherical caps (122). Similar kinds of differences have been observed between as-transformed and "recrystallized" allotrio-morphs of the $\theta$ phase ($CuAl_2$) in an Al-4% Cu Alloy (127). Although a

number of possible explanations for these systematic deviations have been
considered (122), a satisfactory explanation for them remains to be devised.

## The Proeutectoid Ferrite Reaction in Fe-C-X Alloys

The amount of systematic research performed on the proeutectoid ferrite
reaction in high-purity Fe-C-X alloys remains surprisingly limited.
Quantitative investigations of ferrite formation in complex alloy steels,
such as are employed for industrial purposes, and in some instances even
qualitative studies, remain of uncertain value because interactions between
different substitutional alloying elements can be quite powerful and at the
present time are not readily predicted (128). These circumstances, and
space limitations, compel restriction of this section to two major topics;
the influence of X upon ferrite nucleation kinetics at grain boundaries, and
fundamental issues in the growth kinetics of grain boundary ferrite allotriomorphs. Attention is focussed on allotriomorphs rather than on
Widmanstatten plates in order to minimize entanglement of alloying element
effects with the idiocyncracies of the ledge mechanism.

### Nucleation Kinetics of Ferrite Allotriomorphs at Austenite Grain Boundaries in Fe-C-X Alloys

This problem has been recently investigated in high purity Fe-C-X
alloys (129). Four of the alloys studied contained ca. 0.5 A/O C and
3 A/O X, where X was successively Mn, Ni, Co and Si; a fifth alloy included a higher carbon concentration (0.84 A/O C) and a lower proportion of
Mo (2.5 A/O) because of the rapid constriction of the austenite region
engendered by Mo. The previously noted techniques employed to study
ferrite nucleation at austenite grain boundaries in Fe-C alloys (33) were
utilized in this study. Although nucleation at austenite grain edges as
well as at grain faces was observed (130), only the data on grain face
nucleation will be considered here. Unlike the data on allotriomorph
number density vs. isothermal reaction time collected in Fe-C alloys, those
obtained for Fe-C-X alloys did not exhibit a significant interval of non-
steady state nucleation kinetics--perhaps because the usually slower kinetics
involved permitted use of more widely spaced reaction times. Temperature
intervals over which these data could be obtained were again small, e.g.,
20-40°C, except in the Mo alloy, where a 300°C region was accessible.
Again  employing the pillbox-shaped critical nucleus of Fig. 5a, $J_s^*$ (the
steady state nucleation rate) was fitted to classical heterogeneous nucleation theory, and extrapolated over considerably wider temperature ranges
than could be covered experimentally so that nucleation kinetics in different
Fe-C-X alloys could be more conveniently compared. The results are
collected in Fig. 19, together with data on an Fe-0.6 A/O C alloy obtained
during the previous investigation. Since Si, Mo and Co raise the $\gamma/(\alpha+\gamma)$
temperature, $J_s^*$ is higher in these alloys than in the Fe-C alloy at the
higher reaction temperatures. The converse behavior occurs in the Mn and
Ni alloys relative to the Fe-C alloy because these alloying elements reduce
the $\gamma/(\alpha+\gamma)$ temperature. Note that below all of the maxima in $J_s^*$ vs.
temperature the order in which the alloying elements affect $J_s^*$ is largely
reversed. In order to minimize what is essentially a phase diagram effect
upon nucleation kinetics, $J_s^*$ is replotted as a function of the volume free
energy change driving nucleation in Fig. 20. A quite different situation
now emerges. The nucleation rate at driving forces less than those corresponding to the maxima in $J_s^*$ is usually most rapid in the Fe-C alloy. Both
Co and Si slightly depress $J_s^*$; surprisingly, the effects of Mo differ little
from those of Si, though it was not experimentally feasible to follow nucleation in the Mo alloy below the bay temperature. Only Ni and M· now emerge as
elements which markedly depress nucleation kinetics, with Mn still being the
most potent.

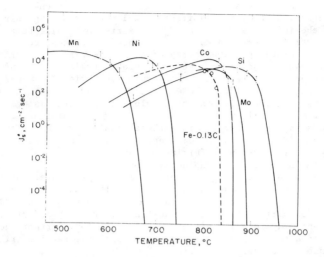

Fig. 19:   Steady state nucleation rate vs. reaction temperature for an Fe-C
           alloy and for the indicated Fe-C-X alloys, extrapolated from the
           experimental data shown via classical heterogeneous nucleation
           theory (129).

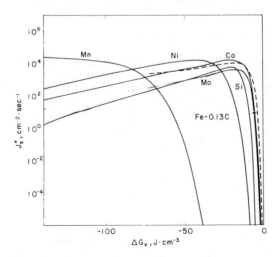

Fig. 20:   Fig. 19 replotted as a function of the volume free energy change
           driving nucleation (129).

Differential effects of X upon dilatational strain energy associated with nucleation can only be significant at low undercoolings; since Fig. 19 shows that the influence of X upon $J_s^*$ is exerted over wide temperature ranges, the strain energy factor need not be further considered. Influence of alloying elements upon the energy of austenite grain boundaries was e-valuated by means of scanning Auger microprobe studies (131) and an anal-ysis of interactive co-segregation to grain boundaries in Fe-C-X type alloys by Guttmann and McLean (132). Mo was found to reduce austenite grain boun-dry energy most effectively; the other elements are decreasingly effective in the order Si, Mn, Ni and Co, with the latter two being of minimal effectiveness in this regard. Since these results are inconsistent with those of Figs. 19 and 20, it becomes apparent that the effects of X upon the energies of the austenite:ferrite boundaries involved also play a sig-nificant role in determining the activation free energy for nucleation, the major factor affecting $J_s^*$. Unfortunately, experimental information on the energies of even disordered austenite:ferrite boundaries in alloys of interest is unavailable, and these data are really needed on fully coherent (or at least on partially coherent) austenite:ferrite boundaries, where their influence may be quite different. In addition, further theoretical study is needed to provide a more accurate model for ferrite critical nuclei at austenite grain boundaries.

## Composition and Growth Kinetics of Grain Boundary Allotriomorphs in Fe-C-X Alloys

### Basic Theory

The core problem associated with the migration of disordered austenite:ferrite boundaries in Fe-C-X alloys is that of balancing the fluxes of C and of X in the face of carbon diffusivities which typically exceed those of X by a factor of $10^4$-$10^5$ in the temperature ranges of interest. The two most important fully developed solutions to this problem are summarized in the following subsections.

#### Paraequilibrium (133, 134)

This hypothesis recognizes that below a certain limiting temper-ature, it is thermodynamically feasible for austenite to be transformed into ferrite with a lower carbon concentration but with the same atomic ratio, $(x_X / x_{Fe})$ as in the austenite phase. Gilmour, Purdy and Kirkaldy (135) have written the conditions for paraequilibrium as:

$$(7) \quad \bar{G}_C^{\alpha} = \bar{G}_C^{\gamma}$$

$$(8) \quad \bar{G}_X^{\gamma} - \bar{G}_X^{\alpha} = (\bar{G}_{Fe}^{\gamma} - \bar{G}_{Fe}^{\alpha}) \frac{x_{Fe}}{x_X}$$

where $\bar{G}_i$ is the partial molar free energy of the indicated species and $x_i$ is the atom fraction of i in the alloy. As sketched in Fig. 21 for an Fe-C-X alloy, the paraequilibrium boundaries of the $\alpha + \gamma$ region lie within those of the full equilibrium (now termed, following Hultgren (33), the "orthoequilibrium") phase boundaries. The growth kinetics of ferrite under paraequilibrium conditions are thus those of a binary Fe-C alloy, with only the phase boundary compositions modified to those appropriate to para-equilibrium. As indicated by Fig. 21, paraequilibrium growth can occur only when the supersaturation is sufficient so that the bulk composition of the alloy lies within the paraequilibrium $\alpha + \gamma$ region.

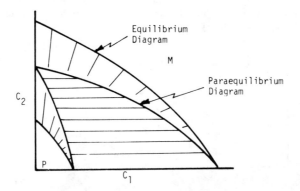

Fig. 21:   Orthoequilibrium ("Equilibrium Diagram") and paraequilibrium
           boundaries of the $\alpha + \gamma$ region of an Fe-C-X alloy; $C_1$ represents
           atom fraction of carbon; $C_2$ is atom fraction of X.   DeHoff (143).

Local Equilibrium (134, 136-139)

On this approach, the compositions of the ferrite and the austenite which are in contact at a disordered austenite:ferrite boundary lie at the ends of a tieline across the orthoequilibrium $\alpha + \gamma$ region of the phase diagram.  However, the bulk composition of the alloy does not lie on this tieline.  Instead, this composition is "relocated" so that one of the following two concepts may be utilized to solve the flux balance problem created because $D_C >> D_X$ (and diffusional interaction between C and X is ignored (139)):

(i)   at lower supersaturations, the bulk composition of the alloy lies on an isoactivity line for the activity of carbon in austenite which passes through the austenite terminus of the tieline;

This arrangement eliminates the carbon activity gradient, and also carbon concentration gradient in austenite; growth kinetics are thus controlled by the flux of X through austenite away from the advancing austenite:ferrite boundary.

(ii)   at higher supersaturations, the bulk composition lies on a line of constant X composition which passes through the ferrite terminus of the tieline;

Except for a thin "spike" in austenite near the austenite: ferrite boundary, this configuration eliminates the X concentration gradient needed to partition X between austenite and ferrite by making the X concentration in ferrite the same as that in the bulk alloy.  Hence growth kinetics are controlled by the flux of carbon through austenite away from the advancing austenite:ferrite boundary.  Hillert (140) has proposed an equivalent concept, couched in terms of carbon activities, for this situation.

Fig. 22 combines these two concepts at different levels of supersaturation, and shows that the composition regions in which they are applicable, termed respectively I and II, meet along a curve joining the $\alpha/(\alpha+\gamma)$ boundary in the Fe-X phase diagram at this temperature to the $\gamma/(\alpha+\gamma)$ boundary in the Fe-Fe$_3$C diagram. Coates (139,141,142) has considerably refined the Hillert (134) and the Kirkaldy school (136-138) treatments of the local equilibrium approach, and DeHoff (143) has more recently given an especially lucid review of it.

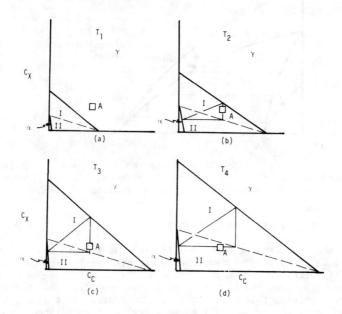

Fig. 22:  Path followed by the bulk composition of an Fe-C-X alloy along iso-carbon concentration ($T_2$ and $T_3$) and iso-X-concentration ($T_4$) lines appropriate to a given tieline across the $\alpha+\gamma$ region as a function of increasing supersaturation, i.e., decreasing temperature ($T_1 > T_2 > T_3 > T_4$). Dashed line represents boundary between regions I and II. DeHoff (143).

## Comparison of Basic Growth Theory with Experiment

### Ferrite Composition

The concentration of X in ferrite allotriomorphs after they had grown to a thickness not much greater than 2µ was determined as a function of reaction temperature, usually at two levels of carbon concentration, in a number of Fe-C-X systems by means of electron probe X-ray microanalysis (144 ). Partition of X between austenite and ferrite was not observed at any temperature studied, from as near to the Ae3 as it was feasible to approach down to the $M_s$, when X was Si, Mo, Co, Al, Cr and Cu. This finding was explained as a consequence of the paraequilibrium $\gamma/(\alpha+\gamma)$ boundary lying so close to the orthoequilibrium $\gamma/(\alpha+\gamma)$ boundary that the temperature range between these two boundaries was too small to permit sufficient undercooling to be achieved to permit formation of partitioned ferrite at detectable rates. In the Mn, Ni and Pt alloys, however, partition was observed above a characteristic critical temperature in each, as illustrated in Fig. 23 (144). Insufficient thermodynamic information is

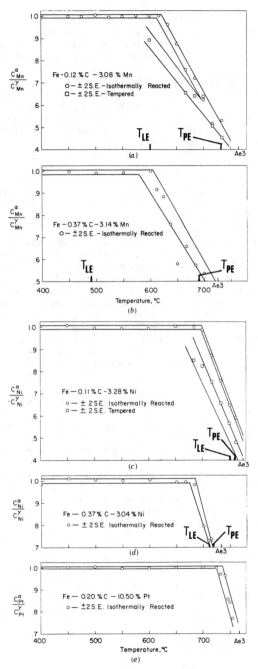

Fig. 23:  Ratio in five Fe-C-X alloys of atom fraction X in ferrite to that
in austenite in ferrite allotriomorphs, as determined by electron
probe microanalysis (144). $T_{PE}$ and $T_{LE}$ represent temperatures of
the paraequilibrium $\gamma/(\alpha+\gamma)$ boundary and of the I:II boundary,
respectively, calculated during the present study.

available on Fe-C-Pt to permit analysis of these data.  For the alloys
studied in the other two systems, however, the paraequilibrium $\gamma/(\alpha + \gamma)$
temperature ($T_{PE}$) and also the temperature of the I:II boundary in
Fig. 22 ($T_{LE}$) were computed.  These temperatures are marked in Fig. 23.

Considering first the paraequilibrium model, the circumstance that $T_{PE}$
lies above the critical temperature for the onset of partition in all four
alloys is consistent with the paraequilibrium growth model.  It is not
yet clear what the kinetic criterion of paraequilibrium is, though pene-
tration of X into austenite by a very small amount (as little as a lattice
parameter) has been suggested (139).  However, Fig. 24 introduces an important complication into this problem:  electron probe analysis shows that
partition of X from ferrite into austenite takes place by means of a
"rejector plate" mechanism ( 6 ), i.e., diffusion of X along austenite:
ferrite boundaries and thence along austenite grain boundaries before
passing via volume diffusion into the bounding austenite grains.  It is
readily shown that a concentration "spike" of X into austenite of physically viable thickness (i.e., ca. a lattice parameter) will occur at much
lower temperature-time products by diffusion from a stationary austenite
"rejector plate" than from a moving austenite:ferrite boundary.  Estimation
of the half-width of this spike at an austenite grain boundary at the four
critical temperatures yields values ranging from 3nm. in the Fe-0.11%
C-3.28% Ni alloy up to 8 nm. in the Fe-0.37% C-3.04% Ni alloy; spike widths
in austenite at the austenite:ferrite boundaries under these conditions are
all far less than one lattice parameter.  Hence the spike thickness
criterion applied to either type of boundary is insufficient to predict
the onset of paraequilibrium.  However, the data of Fig. 23 do make clear
that no bulk partition of X occurs over a substantial temperature-composition region in all five alloys in which partitioning was demonstrated at
higher temperatures.

On the local equilibrium approach, $T_{LE}$ should correspond to the highest temperature of no bulk partition of X between austenite and ferrite.
Fig. 23 shows that $T_{LE}$ falls significantly above the critical temperature
for partition in the two Fe-C-Ni alloys, in disagreement with this model.
The finding that $T_{LE}$ lies below the temperature for the beginning of partition in the two Fe-C-Mn alloys used could be dismissed as simply the consequence of fulfillment of the kinetic conditions for paraequilibrium; this
situation would automatically preclude the volume diffusion needed to
achieve local equilibrium conditions at the austenite:ferrite boundaries.
However, calculation of the concentration of X in ferrite for both alloys,
performed with the Local Equilibrium model using the orthoequilibrium phase
diagrams calculated from the Central Atoms Model (145), yields the results
shown in Fig. 25.  The partition of Mn is seen to be predicted to occur to
a substantial degree, in a nearly temperature-independent manner, well below
the critical temperature for partition in both alloys.  The Coates (142)
treatment of the effect of X upon the diffusivity of carbon in austenite
was incorporated in this analysis.  Inclusion of the strong ternary diffusional interaction in Fe-C-Mn effectively eliminated region II, in which no
X partition is required between austenite and ferrite.  Hence Figs. 24 and
25 provide substantial evidence that the local equilibrium model is in-
applicable to the Fe-C-Mn alloys studied.

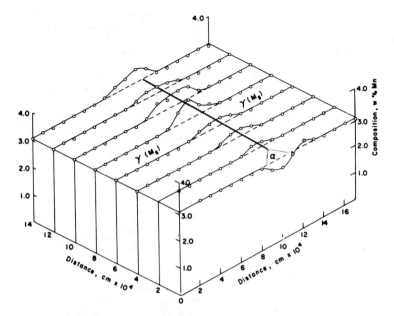

Fig. 24:  Electron probe concentration profiles taken normal to an austenite grain boundary (heavy solid line) leading to a grain boundary
ferrite allotriomorph (marked) formed in Fe-0.12% C-3.08% Mn
reacted 843,120 secs. at 717°C.  Transport of Mn along and away
from the austenite grain boundary by the "rejector plate"
mechanism is demonstrated.  (144).

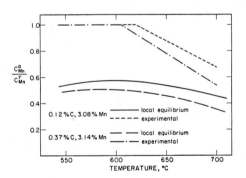

Fig. 25:  Ratio of atom fraction of Mn in α to that in γ vs. temperature in
two Fe-C-Mn alloys, determined experimentally (reproduced from
Fig. 23) (144) and calculated from the local equilibrium model.

## Growth Kinetics

Figs. 26 and 27 present the ratio of the parabolic rate constant
for the thickening kinetics of ferrite allotriomorphs determined experimentally to that calculated from the paraequilibrium and from the local

equilibrium models, respectively (146). (The experimental values were corrected for the influence of partially coherent facets on the basis of Fig. 17 at comparable levels of carbon concentration in the alloy and undercooling.) One of the Fe-C-Mn and one of the Fe-C-Ni alloys used in

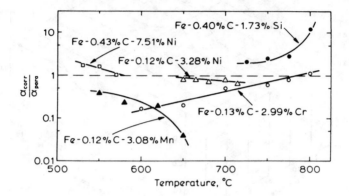

Fig. 26: Ratio of corrected experimental parabolic rate constant for allotriomorph thickening to that calculated from the paraequilibrium model vs. temperature in five Fe-C-X alloys (146).

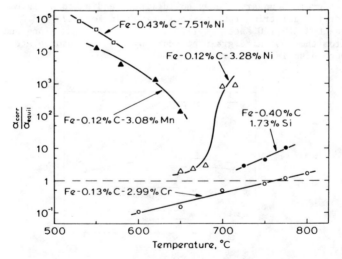

Fig. 27: Ratio of corrected experimental parabolic rate constant for allotriomorph thickening to that calculated from the local equilibrium model vs. temperature in five Fe-C-X alloys (146).

the partition experiments (Fig. 23) was employed in these studies. The local equilibrium model is seen to have resulted in values of $\alpha$ from $10^{-2}$ to $10^{-4}$ of those obtained experimentally in the Fe-C-Mn alloy. At two of the five temperatures at which growth kinetics measurements were made in the Fe-0.11% C-3.28% Ni alloy this ratio was $10^{-3}$; in the 0.43% C, 7.51% Ni alloy the local equilibrium model is seen to have resulted in values as low as $10^{-5}$ of that secured experimentally. Agreement to within an order of magnitude was obtained for the other three temperatures used in the lower Ni alloy and in an Fe-C-Si and an Fe-C-Cr alloy. It is possible, however, that in these alloys Region II conditions obtained, thereby yielding results close to those of the paraequilibrium model.

Figure 26 shows that the paraequilibrium model did predict values of $\alpha$ usually within an order of magnitude of those obtained from the corrected experimental data on thickening kinetics. Thus kinetic evidence is added to partition data supporting the basic soundness of this model at sufficiently low reaction temperatures. Romig and Salzbrenner (147) have recently published STEM data on alloying element partition in an Fe-C-V-Si alloy purporting to disprove the paraequilibrium model by demonstrating the presence of local equilibrium. However, the use of very high reaction temperatures, long reaction times and very small austenite grain sizes makes this test invalid; a more detailed discussion of their experiments is in preparation (148).

## Effects Upon Paraequilibrium Growth Kinetics

Fig. 26 shows that the corrected experimental parabolic rate constant for the thickening kinetics of grain boundary ferrite allotriomorphs agrees well with that calculated from the paraequilibrium model only for Fe-C-Ni alloys. The slower than predicted growth kinetics in Fe-C-Mn and Fe-C-Cr alloys and the faster than expected kinetics in the Fe-C-Si alloy studied require further explanation, albiet in a paraequilibrium context.

## Interfacial Structure

The rather empirical corrections made to the measured values of $\alpha$ in Fe-C-X alloys on the basis of comparable data in Fe-C alloys emphasize our present ignorance as to the details of the interfacial structure of ferrite allotriomorphs. The present presumption is that this structure is a mixture of partially coherent and disordered areas. As Trivedi (84) has pointed out, the growth kinetics of such an interface, with two different growth mechanisms and hence two different sets of kinetics simultaneously operative, have yet to be treated. Before such an analysis is attempted, however, it would appear very desirable to make TEM observations on the interphase boundaries of allotriomorphs in order to secure estimates not only of the proportions of the two types of structure* but also of the morphologies in which they are combined. This information would permit more realistic
*Presumably the areas of these boundaries whose interfacial structure cannot be resolved and which do not appear to be accurately planar may be described as disordered--though with the obvious danger of overestimating the proportion of the total interphase boundary area which does have this type of structure even if lattice fringe imaging or atomic resolution TEM are used.
modeling of the growth processes. Data on the time-dependence of the interledge spacing on the partially coherent facets would also be important. A detailed study of this type would significantly improve the accuracy with which deviations from paraequilibrium can be assessed.

## Influence of Carbide Precipitation

Interference with grain boundary motion by precipitates and inclusions is a long familiar effect. Much interest has developed during the past decade in carbide precipitation at austenite:ferrite boundaries. Two morphologies of such precipitation are particularly prominent. One is periodic precipitation of isolated carbides at many locations along austenite:ferrite boundaries; these row-like arrays are known as "interphase boundary carbides"

(149). The other is a fibrous or needle-like version of pearlite (149).
"Fibrous carbides" as this morphology is named, have been deduced always to
precipitate on disordered austenite:ferrite boundaries and interphase boundary carbides to do so in some situations (150-152); hence these carbides
are said to precipitate on moving interphase boundaries (152). Accordingly,
such modes of carbide precipitation ought to affect boundary migration
kinetics, especially when they are closely spaced. However, the finding that
neither the interphase boundary carbide nor the fibrous carbide morphologies
develop in an Fe-C-Mo alloy in which proeutectoid ferrite has been recrystallized in the austenite + ferrite region prior to (further) transformation in
the ferrite + carbide region, even though such morphologies are common when
this alloy is conventionally heat treated at the same reaction temperatures
(153), casts considerable doubt upon the validity of these deductions. A
theoretical analysis of precipitation at interphase boundaries has similarly
indicated that critical nucleus formation is usually infeasible at moving
boundaries (154). Hence it does not presently appear likely that carbides
will interfere with the migration of austenite:ferrite boundaries in the
manner with which they would affect grain growth. On the other hand, accepting the earlier conclusion that precipitation of interphase boundary carbides
is confined to the terraces of growth ledges (155), an effect upon the migration kinetics of adjacent areas of disordered boundary can be expected because the carbon concentration in austenite in contact with the carbide will
be changed to that of the extrapolated $\gamma/(\gamma + carbide)$ phase boundary (igno-
ring, for simplicity, the good possibility that areas of the carbide:austenite
boundaries are partially coherent). Hence the driving force for the growth
of these disordered austenite:ferrite boundary areas will be altered.
If the bulk composition of the alloy and the reaction temperature intersect
within the region between the extrapolated $\gamma/(\alpha + \gamma)$ and $\gamma/(\gamma + carbide)$
boundaries, the driving force for ferrite growth into austenite (assuming
"semi-infinite bar" diffusion conditions) will be increased. Outside this
region, however, the driving force for the growth of ferrite will be diminished
(156). The dashed curves in Fig. 28 show that the ratio of the ledgewise
growth rate of ferrite in the presence to that in the absence of carbides
(assuming the same, constant inter-ledge spacing in both situations) can vary
by more than an order of magnitude. It should thus be noted that in the
time-temperature envelopes used to acquire the growth kinetics data for
Figs. 26 and 27, carbide precipitation at interphase boundaries was observed
only in the Fe-C-Si alloy (157). On Fig. 28, however, it appears that such
precipitation should diminish $\alpha$, whereas $\alpha_{corr}$ actually exceeded by up to an
order of magnitude the value calculated from the paraequilibrium model (Fig.
26).

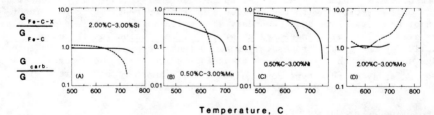

Fig. 28:    Dashed curves:  ratio of ledgewise growth rate in the presence to
            that in the absence of carbides at $\alpha:\gamma$ boundaries. Solid curves:
            ratio of the ledgewise growth rate in Fe-C-X alloys to that in
            Fe-C alloys with the same A/O C. Both curves assume constant inter-
            ledge spacing which is the same in each pair of alloys being com-
            pared. Bradley et al (156).

## Solute Drag-Like Effect

Although not yet described by quantitative theory and evaluated experimentally only by difference between observed kinetics and those predicted by the paraequilibrium model, the solute drag-like effect (SDLE) is now the leading candidate for explaining most of the deviations found experimentally from paraequilibrium behavior. This effect was conceived some years ago by Kinsman and Aaronson (12), and subsequently by Hillert (140). The ideas of the former authors have since undergone some changes as well as elaboration and are presented here in their present form (158). Transformation is assumed to take place in the temperature region wherein microchemical (electron probe or STEM) analysis establishes that the concentration of X in ferrite is indistinguishable from that in austenite, and thus that paraequilibrium growth conditions are operative. If there is significant size misfit between X atoms and Fe atoms, particularly in austenite but also in ferrite, X atoms will accumulate in disordered areas of austenite:ferrite boundaries during growth. (This "sweeping up"-by-a-moving-boundary mechanism of X segregation is a basic difference with respect to the familiar solute drag effect operative during grain growth and recrystallization (159), wherein X atoms diffuse to a stationary boundary with volume diffusion-controlled kinetics.) If X atoms are strongly attracted to carbon atoms, the high carbon concentration in austenite at disordered austenite:ferrite boundaries provides an additional driving force for retention of X atoms at the boundaries during growth. However, calculations based upon the Guttmann-McLean (160) theory of consegregation of interstitial and substitutional alloying elements to grain boundaries indicate that the latter, chemical type of driving force can be an order of magnitude less than that of strain energy (158).

Since there is a discontinuity in the partial molar free energies of X and of Fe across disordered autenite:ferrite boundaries, the segregation of X on the ferrite side of the boundary region will not be that which permits the activity of carbon in austenite in contact with the boundary to retain its paraequilibrium value computed without reference to austenite:ferrite boundary segregation. If X reduces the activity of carbon in austenite, segregation of X under this circumstance should depress the carbon activity in contact with the boundary and thereby diminish the driving force for growth. Conversely, if X increases carbon activity, the driving force for growth will be larger, leading to an inverse SDLE. The absence of X segregation in bulk austenite adjacent to the interphase boundary and the change in growth kinetics through alteration in the activity of the carbon in bulk austenite in contact with the boundary by segregation of X to the boundary through a "sweeping up" mechanism rather than by the dragging of X along with the boundary by volume diffusion through austenite are additional differences between the solute drag-like effect discussed here and the solute drag effect applicable to grain boundary migration.

Development of a minimum in the parabolic rate constant for thickening at the temperature of the bay in the TTT-curve for initation of ferrite and bainite formation in an Fe-C-Mo alloy (Fig. 29) is explicable in terms of the SDLE (11). Mo segregation to austenite:ferrite boundaries should increase with decreasing reaction temperature. The additional postulate is now made that the boundaries become saturated with Mo at the bay temperature. The continued increase in the driving force for growth at lower temperatures then reduces the SDLE at these temperatures. The deeper bay in the TTT-diagram produced by Mo than by Cr is consistent with the larger size misfit of Mo; the Wagner interaction parameter for C-X is somewhat more negative (attractive) than for Mo at temperatures of interest here (160). It has also been suggested that the incomplete transformation phenomenon occurs when the SDLE is particularly strong (161).

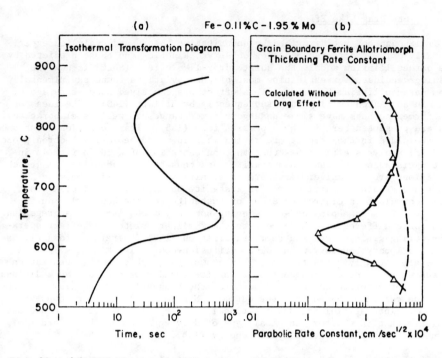

Fig. 29: (a) TTT-diagram for beginning of transformation in an Fe-0.11% C-1.95% Mo alloy. (b) Calculated (dashed) and measured (solid) parabolic rate constant for thickening of allotriomorphs vs. temperature in the same alloys. Boswell et al (11).

## Concluding Remarks

Compared to the level of understanding attained at the time a predecessor review of the proeutectoid ferrite reaction was published (1), considerable progress has been made in understanding the growth of ferrite in Fe-C and Fe-C-X alloys, and a beginning has been achieved in understanding ferrite nucleation at austenite grain boundaries. Extensive further studies of interphase boundary structure and chemistry with very high-resolution electron microscopy—a peculiarly difficult task when investigating the proeutectoid ferrite reaction because of transformation of the remaining austenite to martensite during quenching to room temperature—will be required, however, in order to make further advances in studies of growth kinetics. Advances in understanding of the nucleation kinetics of ferrite are crucially dependent upon more rigorous modeling of critical nucleus shape, particularly at grain boundaries with well defined structures, and upon experimental measurements of interfacial segregation and energies at types of austenite grain boundaries and austenite:ferrite boundaries likely to participate in critical nucleus formation.

## Acknowledgments

W.T.R. Jr. expresses appreciation for support of his contribution to preparation of this review to the Army Research Office, extended through Contract DAAG-29-80-0071. H.I.A. also thanks the ARO, as well as the National Science Foundation for support through individual Grant DMR-81-16905 from the Division of Materials Research and also through participation in an NSF-funded Materials Research Laboratory funded through Grant DMR-81-19507.

# References

1.  H.I. Aaronson, Decomposition of Austenite by Diffusional Processes, p. 387, Interscience, NY (1962).
2.  J.S. Kirkaldy, ibid, p. 39.
3.  C.A. Dube, Ph.D. Thesis, Carnegie Institute of Technology, Pittsburgh, PA (1948); C.A. Dube, H.I. Aaronson and R.F. Mehl, Rev. de Met., 55, 201 (1958).
4.  H.I. Aaronson, Trans. AIME, 224, 693 (1962).
5.  E.B. Hawbolt and L.C. Brown, Trans. TMS-AIME, 239, 1916 (1967).
6.  H.B. Aaron and H.I. Aaronson, Acta Met., 16, 789 (1968).
7.  A. Pasparakis, D.E. Coates and L.C. Brown, Acta Met., 21, 991 (1973) 21, 1259 (1973).
8.  R.W.K. Honeycombe, Met. Trans. A, 7, 915 (1976).
9.  R.W.K. Honeycombe, Met. Sci., 14, 201 (1980).
10. P.R. Howell and R.W.K. Honeycombe, Proceedings of an International Conference on Solid-Solid Phase Transformations, p. 399, TMS-AIME, Warrendale, PA (1983).
11. P.G. Boswell, K.R. Kinsman, G.J. Shiflet and H.I. Aaronson, unpublished research (1968).
12. K.R. Kinsman and H.I. Aaronson, Transformation and Hardenability in Steels, p. 39, Climax Molybdenum Co., Ann Arbor, MI (1967).
13. H.I. Aaronson, The Mechanism of Phase Transformations in Crystalline Solids, p. 270, Institute of Metals, London (1969).
14. A.D. King and T. Bell, Met. Trans. A, 6, 1428 (1975).
15. H.B. Aaron and H.I. Aaronson, Met. Trans., 2, 23 (1971).
16. H.I. Aaronson, The Mechanism of Phase Transformations in Metals, p. 47, Institute of Metals, London (1956).
17. P.N.T. Unwin and R.B. Nicholson, Acta Met., 17, 1379 (1969).
18. K.R. Kinsman and H.I. Aaronson, unpublished research, Ford Motor Co., Dearborn, MI (1968).
19. H.I. Aaronson, Trans. TMS-AIME, 212, 212 (1958).
20. J.B. Clark, High-Temperature, High-Resolution Metallography, p. 347, Gordon and Breach, NY (1967).
21. H.I. Aaronson and C. Wells, Trans. AIME, 206, 1216 (1956).
22. E.S.K. Menon, unpublished research, Carnegie-Mellon University, (1983).
23. R.D. Townsend and J.S. Kirkaldy, Trans. ASM, 61, 605 (1968).
24. H.I. Aaronson, C. Laird and K.R. Kinsman, Phase Transformations, p. 313, ASM, Metals Park, OH (1970).
25. H.I. Aaronson, Phase Transformations, Vol. I, P. II-1, Institution of Metallurgists, Chameleon Press, London (1979).
26. H.I. Aaronson, Trans. Indian Inst. of Metals, 32, 1 (1979).
27. J.H. van der Merwe, Jnl. App. Phys., 34, 117 (1963).
28. J.H. van der Merwe, ibid, 34, 123 (1963).
29. C. Atkinson, K.R. Kinsman and H.I. Aaronson, Scripta Met., 7, 1105 (1973).
30. M. Enomoto, H.I. Aaronson, J. Avila and C. Atkinson, Proceedings of an International Conference on Solid-Solid Phase Transformations, p. 567 TMS-AIME, Warrendale, PA (1983).
31. R.D. Doherty and B. Cantor, ibid, p. 547.
32. W.F. Lange III and H.I. Aaronson, Int. Met. Revs., in press.
33. W.F. Lange III and H.I. Aaronson, Met. Trans., in press; Met. Trans. A, 10, 1951 (1979).
34. J.W. Cahn, Acta Met., 4, 449 (1956).
35. C. Atkinson, K.R. Kinsman, H.B. Aaron and H.I. Aaronson, Met. Trans., 4, 783 (1973).
36. E.E. Underwood, Quantitative Microscopy, p. 149, McGraw-Hill, NY (1968).
37. H.A. Schwartz, Metals and Alloys, 5, 139 (1934).
38. S.S. Saltykov, Stereometric Metallography, 2nd ed., Metallurgizdat, Moscow (1958).

39.   M.R. Plichta and H.I. Aaronson, Acta Met., 28, 1041 (1980).
40.   R.F. Mehl, C.S. Barrett and D.W. Smith, Trans. AIME, 105, 215 (1935).
41.   G. Kurdjumow and G. Sachs, Zeit. Phys., 64, 325 (1930).
42.   J. Young, Phil. Trans. Roy. Soc., 238A, 393 (1939).
43.   J.D. Watson and P.G. McDougall, Acta Met., 21, 961 (1973).
44.   A.D. King and T. Bell, Met. Sci. J., 8, 253 (1974).
45.   Z. Nishiyama, Sci. Rep. Tohoku Univ., 23, 638 (1934).
46.   G. Wassermann, Arch. Eisenhuttenwesen, 16, 647 (1933).
47.   M.G. Hall, H.I. Aaronson and K.R. Kinsman, Surface Sci., 31, 257 (1972).
48.   M.G. Hall and H.I. Aaronson, unpublished research (1983).
49.   J.M. Rigsbee and H.I. Aaronson, Acta Met., 27, 365 (1979).
50.   P.D. Southwick and R.W.K. Honeycombe, Met. Sci., 14, 253 (1980).
51.   L.A. Bruce and H. Jaeger, Phil. Mag., 38, 223 (1978).
52.   D.A. Smith, K.M. Knowles, H.I. Aaronson and W.A.T. Clark, Proceedings
      of an International Conference on Solid-Solid Phase Transformations,
      p. 587, TMS-AIME, Warrendale, PA (1983).
53.   R.C. Ecob and B. Ralph, Proc. Natl. Acad. Sci., USA, 77, 1749 (1980).
54.   R.C. Ecob and B. Ralph, Acta Met., 29, 1037 (1981).
55.   G.D.W. Smith, discussion to ref. 10, p. 424.
56.   H.I. Aaronson and K.C. Russell, Proceedings of an International Confer-
      ence on Solid-Solid Phase Transformations, p. 371, TMS-AIME, Warrendale,
      PA (1983).
57.   N.T. Belaiew, J. Inst. Metals, 29, 379 (1923).
58.   Y.C. Liu, H.I. Aaronson, K.R. Kinsman and M.G. Hall, Met. Trans., 3,
      1318 (1972).
59.   H. Hanemann and A. Schrader, Atlas Metallographicus, Vol. 5 (1927).
60.   C.M. Wayman, Phase Transformations, p. 59, ASM, Metals Park, OH (1970).
61.   H.I. Aaronson, Jnl. of Microscopy, 102, 275 (1974).
62.   K.C. Russell, M.G. Hall, K.R. Kinsman and H.I. Aaronson, Met. Trans.,
      5, 1503 (1974).
63.   J.M. Rigsbee and H.I. Aaronson, Acta Met., 27, 351 (1979).
64.   P.R. Howell, P.D. Southwick and R.W.K. Honeycombe, Jnl. of Microscopy,
      116, 151 (1979).
65.   P.R. Howell, P.D. Southwick, R.C. Ecob and R.A. Ricks, Proceedings of
      an International Conference on Solid-Solid Phase Transformations,
      p. 591, TMS-AIME, Warrendale, PA (1983).
66.   M.G. Hall, J.M. Rigsbee and H.I. Aaronson, submitted to Acta Met.
67.   K.R. Kinsman, E. Eichen and H.I. Aaronson, Met. Trans. A, 6, 303 (1975).
68.   C. Atkinson, Acta Met., 16, 1019 (1968).
69.   J.W. Cahn, W.B. Hillig and G.W. Sears, Acta Met., 12, 1421 (1964).
70.   G.J. Jones and R. Trivedi, Jnl. App. Phys., 42, 4299 (1971).
71.   C. Atkinson, Proc. Roy. Soc., London, A378, 351 (1981).
72.   C. Laird and H.I. Aaronson, Acta Met., 17, 505 (1969).
73.   R. Sankaran and C. Laird, Acta Met., 22, 957 (1974).
74.   G.R. Purdy, Acta Met., 26, 477 (1978).
75.   P.R. Krahe, K.R. Kinsman and H.I. Aaronson, Acta Met., 20, 1109 (1972).
76.   M. Hillert, unpublished research (see ref. 77).
77.   L. Kaufman, S.V. Radcliffe and M. Cohen, Decomposition of Austenite by
      Diffusional Processes, p. 313, Interscience, NY (1962).
78.   E.P. Simonen, H.I. Aaronson and R. Trivedi, Met. Trans., 4, 1239 (1973).
79.   R. Trivedi, Met. Trans., 1, 921 (1970).
80.   R. Trivedi, Acta Met., 18, 287 (1970).
81.   W.P. Bosze and R. Trivedi, Met. Trans., 5, 511 (1974).
82.   C. Zener, Trans. AIME, 167, 550 (1946).
83.   S.C. Huang and M.E. Glicksman, Acta Met., 29, 701 (1981).
84.   R. Trivedi, Proceedings of an International Conference on Solid-Solid
      Phase Transformations, p. 477, TMS-AIME, Warrendale, PA (1983).
85.   J.S. Langer and H. Müller-Krumbhaar, Acta Met., 26, 1681 (1978).
86.   S. Strassler and W.R. Schneider, Phys. Condensed Matter, 17, 153 (1974).

87. V. Datye and J.S. Langer, Phys. Rev., $\underline{24B}$, 4155 (1981).
88. G.E. Nash, J. Crystal Growth, $\underline{38}$, 155 (1977).
89. H. Knapp and U. Dehlinger, Acta Met., $\underline{4}$, 289 (1956).
90. L.H. Van Vlack, Trans. AIME, $\underline{191}$, 251 (1951).
91. C. Laird and H.I. Aaronson, Trans. TMS-AIME, $\underline{242}$, 1391 (1968).
92. R. Sankaran and C. Laird, Phil. Mag., $\underline{29}$, 179 (1974).
93. C. Laird and H.I. Aaronson, Acta Met., $\underline{15}$, 73 (1967).
94. K. Chattopadhyay and H.I. Aaronson, unpublished research, Carnegie-Mellon University (1982).
95. W.P. Bosze and R. Trivedi, Acta Met., $\underline{23}$, 713 (1975).
96. T. Ko and S.A. Cottrell, J.I.S.I., $\underline{172}$, 307 (1972).
97. T. Ko, ibid, $\underline{175}$, 16 (1973).
98. J.W. Christian, Decomposition of Austenite· by Diffusional Processes, p. 371, Interscience, NY (1962).
99. J.M. Oblak and P.F. Hehemann, Transformation and Hardenability in Steels, p. 15, Climax Molybdenum Co., Ann Arbor, MI (1967).
100. E. Eichen and J.W. Spretnak, Trans. ASM, $\underline{51}$, 454 (1959).
101. E.S. Deavenport and E.C. Bain, Trans. AIME, $\underline{90}$, 117 (1930).
102. H.I. Aaronson, J.K. Lee and K.C. Russell, Precipitation Processes in Solids, p. 31, TMS-AIME, Warrendale, PA (1978).
103. H.I. Aaronson, M.G. Hall, D.M. Barnett and K.R. Kinsman, Scripta Met., $\underline{9}$, 705 (1975).
104. J.D. Eshelby, Proc. Roy. Soc., London, $\underline{241A}$, 376 (1957).
105. J.D. Watson and P.G. McDougall, Acta Met., $\underline{21}$, 961 (1973).
106. H.K.D.H. Bhadeshia, Acta Met., $\underline{29}$, 1117 (1981).
107. H.I. Aaronson, G.J. Shiflet, S.K. Liu, W.T. Reynolds, Jr. and D.A. Smith, unpublished research.
108. J.K. Lee, Y.Y. Earmme, H.I. Aaronson and K.C. Russell, Met. Trans. A, $\underline{11}$, 1837 (1980).
109. Y.Y. Earmme, W.C. Johnson and J.K. Lee, Met. Trans. A, $\underline{12}$, 1521 (1981).
110. K.R. Kinsman, E. Eichen and H.I. Aaronson, Met. Trans., $\underline{2}$, 346 (1971).
111. G.B. Olson and M. Cohen, Acta Met., $\underline{27}$, 1907 (1979).
112. H.K.D.H. Bhadeshia, Scripta Met., $\underline{17}$, 1475 (1983).
113. C.M. Wayman, Metallography, $\underline{8}$, 105 (1975).
114. C.M. Wayman, Phase Transformations, p. 59, ASM, Metals Park, OH (1970).
115. C.M. Wayman and G.R. Srinivasan, The Mechanism of Phase Transformations in Crystalline Solids, p. 310, Institute of Metals, London (1969).
116. K.R. Kinsman, R.H. Richman and J. Verhoeven, Materials Science Sympos-Abstracts, p. 41, ASM, Metals Park, OH (1974).
117. H.K.D.H. Bhadeshia and D.V. Edmonds, Met. Trans. A., $\underline{10}$, 895 (1979).
118. S.K. Liu, W.T. Reynolds, Jr., H. Hu, G.J. Shiflet and H.I. Aaronson, Met. Trans., submitted.
119. J.R. Bradley and H.I. Aaronson, Met. Trans. A, $\underline{8}$, 317 (1977).
120. G.R. Purdy and J.S. Kirkaldy, Trans. TMS-AIME, $\underline{227}$, 1255 (1963).
121. K.R. Kinsman and H.I. Aaronson, Met. Trans., $\underline{4}$, 959 (1973).
122. J.R. Bradley, J.M. Rigsbee and H.I. Aaronson, Met. Trans. A, $\underline{8}$, 323 (1977).
123. F.S. Ham, Quart. Appl. Math., $\underline{17}$, 137 (1959).
124. G. Horvay and J.W. Cahn, Acta Met., $\underline{9}$, 695 (1961).
125. C. Atkinson, Trans. TMS-AIME, $\underline{245}$, 801 (1969).
126. N.A. Gjostein, H.A. Domian, H.I. Aaronson and E. Eichen, Acta Met., $\underline{14}$, 1637 (1966).
127. H.B. Aaron and H.I. Aaronson, Acta Met., $\underline{18}$, 699 (1970).
128. J.S. Kirkaldy, Met. Trans., $\underline{4}$, 2327 (1973).
129. M. Enomoto and H.I. Aaronson, to be submitted to Met. Trans.
130. M. Enomoto and H.I. Aaronson, to be submitted to Met. Trans.
131. M. Enomoto and H.I. Aaronson, to be submitted to Met. Trans.
132. M. Guttmann and D. McLean, Interfacial Segregation, p. 261, ASM, Metals Park, OH (1979).

133. A. Hultgren, Trans. ASM, 39, 915 (1947).
134. M. Hillert, Internal Rep., Swedish Institute for Metals Research (1953).
135. J.B. Gilmour, G.R. Purdy and J.S. Kirkaldy, Met. Trans., 3, 3213 (1972).
136. J.S. Kirkaldy, Can. Jnl. Phys., 36, 907 (1958).
137. G.R. Purdy, D.H. Weichert and J.S. Kirkaldy, Trans. TMS-AIME, 230, 1025 (1964).
138. M.P. Puls and J.S. Kirkaldy, Met. Trans., 3, 2777 (1972).
139. D.E. Coates, Met. Trans., 4, 2313 (1973).
140. M. Hillert, The Mechanism of Phase Transformations in Crystalline Solids, p. 231, Institute of Metals, London (1969).
141. D.E. Coates, Met. Trans., 3, 1203 (1972).
142. D.E. Coates, Met. Trans., 4, 1077 (1973).
143. R.T. DeHoff, Proceedings of an International Conference on Solid-Solid Phase Transformations, p. 503, TMS-AIME, Warrendale, PA (1983).
144. H.I. Aaronson and H.A. Domian, Trans., TMS-AIME, 236, 768 (1966).
145. C.H.P. Lupis and J.F Elliott, Acta Met., 15, 265 (1967).
146. J.R. Bradley and H.I. Aaronson, Met. Trans. A, 12, 1729 (1981).
147. A.D. Romig, Jr. and R. Salzbrenner, Scripta Met., 16, 33 (1982).
148. J.R. Bradley, G.J. Shiflet and H.I. Aaronson, unpublished research.
149. M. Mannerkoski, Acta Polytech. Scand., Ch. 26, p. 7 (1964); K. Relander, ibid, Ch. 34, p. 7.
150. P.R. Howell, J.V. Bee and R.W.K. Honeycombe, Met. Trans. A, 10, 1213 (1979).
151. P.R. Howell, R.A. Ricks, J.V. Bee and R.W.K. Honeycombe, Phil. Mag., 41, 165 (1980).
152. R.A. Ricks and P.R. Howell, Acta Met., 31, 853 (1983).
153. T. Obara, G.J. Shiflet and H.I. Aaronson, Met. Trans. A, 14, 1159 (1983).
154. H.I. Aaronson, M.R. Plichta, G.W. Franti and K.C. Russell, Met. Trans. A, 9, 363 (1978).
155. A.T. Davenport and R.W.K. Honeycombe, Proc. Roy. Soc. London, 322, 191 (1971).
156. J.R. Bradley, G.J. Shiflet and H.I. Aaronson, Proceedings of an International Conference on Solid-Solid Phase Transformations, p. 819, TMS-AIME, Warrendale, PA (1983).
157. G.J. Shiflet, H.I. Aaronson and J.R. Bradley, Met. Trans. A, 12, 1743 (1981).
158. W.T. Reynolds, Jr., unpublished research, Carnegie-Mellon University, Pittsburgh, PA (1984).
159. J.W. Cahn, Acta Met., 10, 789 (1962).
160. J.S. Kirkaldy, B.A. Thomson and E.A. Baganis, Hardenability Concepts with Applications to Steel, p. 82, TMS-AIME, Warrendale, PA (1978).
161. H.I. Aaronson, The Mechanism of Phase Transformations in Crystalline Solids, p. 270, Institute of Metals, London (1969).

# THE PEARLITE TRANSFORMATION

N. Ridley

University of Manchester/UMIST
Department of Metallurgy and Materials Science,
Grosvenor Street, Manchester M1 7HS, England.

After summarising well established features of pearlite formation in plain carbon steels, theories of pearlite growth are outlined. Methods used to meaure interlamellar spacings, growth rates and partitioning data which are necessary to test growth theories, are discussed. Crystallographic information relating to pearlite nucleation and growth is also reviewed. Unified measurements of temperature, growth velocity and spacing for Fe-C alloys show that the relationship betwen the parameters is identical for pearlite growth under isothermal and forced velocity conditions. For isothermal growth a linear relationship between reciprocal spacing and undercooling is generally observed, but for steels containing alloying additions there is little evidence of the predicted inflexion corresponding to a temperature at which alloy partitioning ceases. For Fe-C alloys isothermal, forced velocity and continuous cooling data leads to the conclusion that volume diffusion of carbon in austenite is the dominant rate controlling process for pearlite growth, although a discrepancy between measured and calculated growth rates remains to be explained. Experimental data on spacing, growth rate and partitioning is reviewed for Fe-C-X alloys, where X is Ni, Mn, Co, Si, Cr or Mo. While it is likely that local equilibrium models of growth can successfully predict pearlite growth rates for additions of Ni, Mn, Co and Si, this is not so for the strong carbide forming elements where growth is probably retarded by solute drag effects, particularly for lower reaction temperatures.

## Introduction

Pearlite is a lamellar product of eutectoid decomposition and is probably the most familiar of all metallographic structures. It can form in steels and also in a range of non-ferrous alloys during transformation under isothermal, continuous cooling and forced-velocity growth conditions. Figure 1, taken from the early work of Villela, Guellich and Bain (1), shows pearlite formed isothermally in a plain carbon steel. Many important characteristics of the pearlite reaction were established prior to and at the time of the 1960 Symposium on the Decomposition of Austenite by Diffusional Processes (2,3). Since 1960 various features of the reaction have been reviewed from time to time and, in particular, theoretical and experimental aspects of the pearlite reaction were the subject of a comprehensive review by Puls and Kirkaldy in 1972 (4-6). Recently Hillert (7) has analysed the effects of alloy additions on pearlite growth, while the author has reviewed the data on the interlamellar spacing of pearlite (8).

Figure 1 - Lamellar pearlite formed isothermally at 705°C in a plain carbon eutectoid steel.

In the present review the well established features of pearlite formation will be outlined and attention will be concentrated on some of the more recent studies on eutectoid steels, particularly those in which the pearlitic carbide is cementite rather than an alloy carbide. The areas to be covered will include a summary of theories of pearlite growth, methods for obtaining experimental data, crystallographic aspects of pearlite formation, pearlite growth in Fe-C and Fe-C-X alloys transformed under isothermal and forced velocity growth conditions, and partitioning of alloying additions.

## Main Features of Pearlite Formation in Fe-C Alloys

To set the scene some of the more important features of the pearlite reaction in isothermally transformed Fe-C alloys will be outlined. Figure 2a shows the appropriate section of the Fe-Fe$_3$C phase diagram. The area bounded by the extrapolated Ae$_3$ and A$_{cm}$ phase boundaries represents the region, or range of temperatures and compositions, for which the parent austenite phase is simultaneously saturated with respect to both ferrite and cementite. At higher reaction temperatures below Ae$_1$ the cooperative growth of these phases leads to the characteristic lamellar structure of pearlite. An isothermal transformation diagram for an alloy of eutectoid composition is shown schematically in Figure 2b. At temperatures below the nose of the curve the reaction product is bainite, while for a range of temperatures in the region of the nose both pearlite and bainite are the products of competitive reactions. Since pearlite and bainite represent

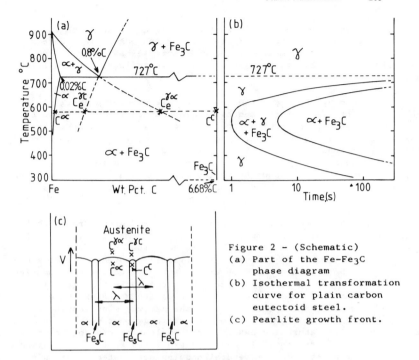

Figure 2 - (Schematic)
(a) Part of the Fe-Fe₃C phase diagram
(b) Isothermal transformation curve for plain carbon eutectoid steel.
(c) Pearlite growth front.

two different modes of eutectoid decomposition, it is believed that each reaction product will have its own C-curve such that the isothermal transformation curve shown in Figure 2b is the result of overlapping C-curves, one for pearlite and one or more for bainite. Evidence in support of this view has been summarised and discussed recently by Kennon (9,10).

Pearlite nucleation frequently occurs on austenite grain boundaries and prior to impingement of growing nodules, growth often takes place on a spherical front. Each pearlite nodule is composed of colonies within which the lamellae are essentially parallel, but differ in orientation from lamellae in adjacent colonies. The colony structure is clearly evident in Figure 1. Repeated sectioning studies have shown that pearlite is a branched structure such that each colony consists of two interwoven crystals, one of ferrite and the other of cementite (11).

The austenite-pearlite reaction front is shown schematically in Figure 2c. The important growth parameters are V, the radial growth velocity and $\lambda$, the interlamellar spacing. From Figure 2a it can be seen that ferrite, or $\alpha$-phase, has a low solubility for carbon, <0.02 wt pct, whereas Fe₃C contains a large amount of carbon, ~6.68 wt pct. Hence, during pearlite growth there will be a considerable redistribution of carbon at or near the transformation front. In the case of alloy steels the substitutional elements may also redistribute between the product phases at the growth front.

Theories of pearlite growth assume that growth rate and interlamellar spacing are constant with time, and although this is a reasonable assumption for plain carbon eutectoid alloys it is not always true for alloyed steels. The wide range of spacings seen in Figure 1 is due to the varying angles at which the colonies are intersected by the plane of section. Pearlite growth rate increases with decreasing reaction temperature before passing through a maximum, while most alloying additions with the exception of cobalt retard growth rate. Interlamellar spacing reflects the diffusion geometry at the transformation front and decreases as the reaction temperature is decreased. The spacing is also influenced by alloying additions (2).

Although pearlite nucleation rate has always proved difficult to measure, the studies of Mehl and his collaborators (2) have shown the nucleation rate increases with time, decreases with temperature before passing through a maximum, and, like growth rate, is slowed down by most alloying additions. However, it is a structure sensitive parameter and is affected by austenite grain size, inhomogeneities and impurities. The effect of the latter is shown by the observation that in steels of very high purity nucleation at high transformation temperatures occurs mainly on the free surface.

It might be expected that both nucleation and growth rates would contribute to overall reaction rates and this is implicit in several relationships which have been proposed to describe the progress of transformation with time (2). However, Cahn (12,13) pointed out that since nucleation is heterogeneous and usually restricted to grain boundaries there will be a tendency, particularly when the nucleation rate is relatively rapid, for all the nucleation sites to be saturated early in the reaction. Apart from temperatures near the $Ae_1$ where nucleation rates are low, site saturation is believed to cover a large part of the pearlite range. Once site saturation has occurred nucleation can no longer play any role in the transformation. It has been predicted that if nucleation is sufficiently rapid to produce several nodules per austenite grain early in the reaction i.e. in a time of <0.1 d/V, where d is the austenite grain diameter, then the reaction is finished in the time, $t_f$, required for the growth front to reach the grain centres, where $t_f \sim 0.5$ d/V.

Overall reaction rates have been derived by Cahn for site saturation transformation on grain surfaces, edges and corners (12). If site saturation is operative then to change the overall transformation rate it is necessary to change d or V. From a practical point of view the most satisfactory and controllable way of changing the overall rate, and hence the hardenability, is to use alloy additions to change V.

## Theories of Pearlite Growth

Theories of pearlite growth assume that the reaction product grows into the parent austenite phase at a constant velocity, V, and maintains a constant interlamellar spacing, $\lambda$. During growth the redistribution of solute may occur ahead of the transformation front by volume diffusion through the austenite, and/or by diffusion along the austenite-pearlite interface. Attention will be concentrated on theories required for the interpretation or testing of experimental data to be presented in later sections.

The Zener-Hillert model (14,15) which assumes that the supersaturated parent phase is in local equilibrium with the product phases at the transformation front and that volume diffusion of carbon in austenite is rate controlling leads to the relationship

$$V = \frac{D}{a} \cdot \frac{\lambda^2}{\lambda^\alpha \lambda^c} \cdot \frac{C_e^{\gamma\alpha} - C_e^{\gamma c}}{C^c - C^\alpha} \cdot \frac{1}{\lambda} \left[ 1 - \frac{\lambda_c}{\lambda} \right] \qquad (1)$$

where D is the volume diffusion coefficient for carbon in austenite, a is a geometric parameter equal to 0.72 for an Fe-C eutectoid alloy; $C_e^{\gamma\alpha}$, $C_e^{\gamma c}$, $C^c$ and $C^\alpha$ are carbon concentrations in austenite in equilibrium with ferrite and cementite lamellae, and in cementite and ferrite, respectively (Figure 2a); $\lambda$ is the interlamellar spacing, $\lambda^\alpha$ and $\lambda^c$ are the respective thicknesses of ferrite and cementite lamellae, and $\lambda_c$ is the critical spacing for which the growth rate is zero.

The pearlite interface cannot be perfectly flat otherwise no concentration gradient would exist to drive the diffusional processs, except at the $\alpha/Fe_3C/\gamma$ junction where there would be a discontinuous change of composition. The $\alpha/\gamma$ and $Fe_3C/\gamma$ interfaces must be curved with the composition of the parent phase varying with curvature along the interface and with no discontinuity at the 3-phase junction. The curvature will, however, reduce the concentration gradient because of the Gibbs-Thomson effect. In equation (1) the maximum value of the concentration term $(C_e^{\gamma\alpha} - C_e^{\gamma c})$ is multipled by $(1 - \lambda_c/\lambda)$ to account for the effect of curvature, and this reduces the effective concentration gradient driving the growth process. Hillert incorporated the Gibbs-Thomson condition into his analysis of pearlite growth and this allowed detailed calculations of the interface shape to be made (15).

However, there is no unique solution of equation (1) since for a given reaction temperature, or fixed undercooling, $\Delta T$, there will be a range of spacings and growth rates which satisfies the relationship. To obtain a unique solution it is necessary to impose further conditions by means of an optimisation procedure. Zener predicted that the system stabilises at a spacing for which the growth rate is a maximum and this criterion leads to $\lambda = 2\lambda_c$. Support for the criterion is found in studies on eutectics where growth under forced constant velocity conditions occurs close to minimum undercooling (16), which Tiller has shown is equivalent to growth at maximum velocity (17). Trevedi (18) has recently considered experimental observations on Widmanstatten plates in a number of alloy systems and has concluded that the lengthening of the precipitates is consistent with the maximum growth rate criterion. He has argued that lamellar structures such as eutectics and eutectoids should also grow under conditions equivalent to maximum growth rate.

The critical spacing is related to undercooling by the relationship

$$\lambda_c = \frac{2\sigma^{\alpha c} T_E}{\Delta H_v . \Delta T} \qquad (2)$$

where $\sigma^{\alpha c}$ is the surface energy of the $\alpha/Fe_3C$ interface, $\Delta H_v$ is the enthalpy change per unit volume between the parent and product phases, and $T_E$ is the eutectoid temperature. Combining equation (2) with the Zener criterion gives

$$\lambda = 2\lambda_c = \frac{4\sigma^{\alpha c} T_E}{\Delta H_v . \Delta T} \qquad (3)$$

The maximum rate of entropy production criterion proposed by Kirkaldy (4,19), which has its theoretical basis in irreversible thermodynamlcs, and is supported by perturbation tests leads to

$$\lambda = 3\lambda_c = \frac{6\sigma^{\alpha c}T_E}{\Delta H_v . \Delta T} \qquad (4)$$

Since the terms $\Delta H_v$ and $\sigma^{\alpha c}$ are relatively independent of temperature equations (3) and (4) make the important prediction that interlamellar spacing and undercooling are inversely related.

Pearlite growth involving boundary diffusional control has been examined by various workers including Shapiro and Kirkaldy (20), Sundquist (21,22) and Hillert (23). Their analyses lead to similar results. The models assume that local equilibrium exists across the tranformation front. To achieve this some volume diffusion is required ahead of the moving interface, although the dominant mechanism of solute redistribution is by boundary diffusion. The relationship proposed by Hillert (23) for an Fe-C eutectoid alloy is

$$V = 12KD_B\delta . \frac{\lambda^2}{\lambda^\alpha \lambda^c} . \frac{C_e^{\gamma\alpha} - C_e^{\gamma c}}{C^c - C^\alpha} . \frac{1}{\lambda^2} \left[ 1 - \frac{\lambda_c}{\lambda} \right] \quad (5)$$

where K is the boundary segregation coefficient, $D_B$ is the boundary diffusion coefficient and $\delta$ is the boundary thickness. This equation may also be used for boundary diffusional growth involving substitutional elements by replacing the carbon concentration and diffusion terms with values appropriate to the substitutional element. For growth control by boundary diffusion the maximum growth rate and maximum rate of entropy production criteria lead to $\lambda = 3/2 \lambda_c$ and $\lambda = 2\lambda_c$, respectively. These relationships may be combined with equation (2) as illustrated earlier.

In the Sundquist model (21) of pearlite growth controlled by inter-facial diffusion an alternative optimisation criterion is adopted. The pearlite is assumed to grow with a maximum spacing such that any spacing larger than this results in interface instability and to the nucleation of a new lamella of cementite in the centre of each ferrite lamella at the transformation front. The configurations associated with what Sundquist, following the earlier observations of Hunt and Jackson (24) on eutectics, has termed the "upper catastrophic limit" are shown schematically in Figure 3. The Sundquist model was based on the view that experimental growth rates in plain carbon steels were much greater than those which could result from volume diffusion and that measured interlamellar spacings were considerably larger than those predicted by the maximum growth rate criterion (13).

AUSTENITE

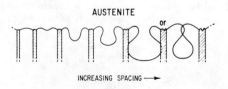

INCREASING SPACING ⟶

T = const.

Figure 3 - Predicted shapes of pearlite growth front (Ref. 20).

Equations (1) and (5) give the relationship between growth velocity, spacing, concentration gradient and diffusivity. Since the concentration term, $C_e^{\gamma\alpha} - C_e^{\gamma c}$, is approximately proportional to undercooling (Figure 2a) and undercooling is proportional to reciprocal spacing, $1/\lambda$, (Equations 3 and 4) then for volume diffusion control of growth, equation (1) may be rewritten as

$$V\lambda^2 = K_1 D \qquad (6)$$

while for boundary diffusion control equation (5) may be rearranged to give

$$V\lambda^3 = K_2 D_B \qquad (7)$$

where $K_1$ and $K_2$ are constants. Hence, logarithmic plots of $V/D$ or $V/D_B$ versus $1/\lambda$ should give straight lines with slopes of 2 or 3, depending on whether growth is controlled by volume or boundary diffusion. Equations (6) and (7) are sometimes written as

$$V\lambda^2 = \text{constant} \qquad (8)$$

or
$$V\lambda^3 = \text{constant} \qquad (9)$$

and although these relationships may be applicable for narrow ranges of temperature, diffusion is usually strongly temperature dependent and this should be taken into account.

As will be seen, equations (6)-(9) are frequently used in attempts to interpret the significance of kinetic data obtained from forced-velocity growth studies. In their studies on directionally aligned lamellar structures Carpay and Van den Boomgaard (25) who assumed that growth was controlled by boundary diffusion, proposed a further relationship

$$V\lambda^4 = \text{constant} \qquad (10)$$

although their model has been criticised on the grounds that it had neglected temperature dependent parameters (26).

### Experimental Methods

To examine the theories of pearlite growth it is necessary to have reliable measurements of pearlite growth velocities and interlamellar spacings, while for steels containing alloying additions information on the way in which these elements redistribute, or partition, at the growth front is also required.

### Forced-Velocity Growth

The bulk of growth data has been obtained using the long established isothermal technique in which undercooling is fixed and growth rate is the free variable. More recently forced-velocity growth has been used as an alternative method of studying pearlite formation. In this technique a specimen, usually in rod form, is translated at a constant velocity relative to a temperature gradient which establishes a single transformation interface and which is sufficiently steep to prevent nucleation ahead of this front. The procedure is similar to that used to produce aligned lamellar structures in eutectic alloys. The growth rate is fixed by the imposed velocity and transformation temperature is the free variable. The technique was first applied to Fe-C eutectoid alloys by Bramfitt and Marder (27) and Bolling and Richman (28), to several non-ferrous eutectoid alloys by Carpay (29), and to Cu-Al eutectoid by Livingston (30).

While measurements of the temperature of the transformation front have only occasionally been reported, the relationship between the interlamellar spacing of the aligned product and translation velocity, assumed to be equal to the growth velocity, can be analysed to determine the rate controlling process for pearlite growth using equations (6)-(9). It is important to measure the spacing of fully aligned pearlite because lamellae lying at an angle will have grown more rapidly than the imposed velocity in order to maintain a planar growth front. Experiment has shown that if pearlite is forced to grow at velocities greater than the maximum isothermal growth rate then the structure tends to become increasingly degenerate (28).

## Measurement of Growth Rate

Most growth rate measurements are made on isothermally transformed specimens using the maximum nodule radius method. The technique is difficult to apply to rapidly transforming specimens and cannot be used once impingement of growing nodules has become significant. It also gives a maximum, rather than an average, growth rate. Despite these limitations the method appears to be capable of giving results which are similar to the average growth rate method proposed by Cahn and Hagel (31,32).

A hot stage technique incorporating cine photomicrography and thermal analysis has been used by Marder and Bramfitt (33-35) to measure pearlite growth rates under continuous cooling conditions. The procedure enables in-situ measurements to be readily made simultaneously on several rapidly growing nodules in a single specimen. However, it is not easy to relate the growth rate to reaction temperature because transformation takes place over a range of temperatures, and although the temperature range decreases with increased cooling rate, recalescence can make the determination of temperature difficult at higher cooling rates or lower reaction temperatures. Nevertheless the results obtained show reasonable agreement with those measured using the maximum nodule radius method.

## Measurement of Interlamellar Spacing

Methods for measuring interlamellar spacing have been recently reviewed (8). The lack of resolution associated with the optical techniques developed by Mehl et al. (36) has led to them being largely superceded by methods involving scanning, replication and transmission microscopy. The most versatile and most widely used technique is probably that involving the examination of shadowed carbon replicas taken from fully or partially transformed specimens, in which measurements are made on electron microgaphs or on the fluorescent screen of the microscope (32). The distance measured is usually the average minimum observed spacing, $\lambda_0$ or $\lambda_{min}$, taken from several pearlite colonies, or the mean intercept spacing, $\bar{\ell}$. The former parameter is a perpendicular distance across two consecutive lamellae i.e. ferrite and cementite, while the mean intercept spacing is an average distance across two successive lamellae taken from test lines randomly applied to the plane of polish.

There is experimental evidence for both ferrous and non-ferrous pearlites produced under isothermal and forced-velocity growth conditions that the true spacing is not the minimum spacing observed in a metallographically prepared surface (8). The true spacing does not have a constant value but is a distribution of spacings about a mean true value, $\bar{\lambda}_0$. Provided the pearlitic structure is predominantly lamellar, values of $\bar{\lambda}_0$ may be obtained from the mean true intercept, $\bar{\ell}$, using the relation-

ship: $\bar{\ell} = 2\bar{\lambda}_o$ (37). It is clear that the factor which relates the minimum
and mean true spacing is not constant and may vary from system to system,
and with temperature in a given system. For isothermally transformed plain
carbon steels the factor is frequently less than that of 1.65 which was
proposed by Mehl and his collaborators (36). It is because of the
uncertainty in the relationship between the minimum and mean true spacing
that the former parameter, which can be both readily and reproducibly
measured, is frequently used to characterise the structure. However, it
would in fact seem more satisfactory to characterise the microstructure in
terms of the mean true spacing, $\bar{\lambda}_o$.

In some medium and high alloy eutectoid steels where transformation
times are long, the growth velocity can decrease with time and the interlamellar spacing at the reaction front may decrease. The resultant
microstructure has been termed divergent pearlite and has been studied in a
5.2 wt pct Mn eutectoid steel (31) and in a hypereutectoid silicon steel
(38). It has also been reported for a 13 wt pct Cr eutectoid steel (39).
In the former the apparent spacings at the reaction front in partially
transformed specimens were converted to an average reciprocal square
spacing in order to weight the smaller spacings in the same way as it was
anticipated that the growth rate should average them, while for the silicon
steel the interlamellar spacing was characterised by the minimum value at
the transformation front.

## Partition studies

For steels containing alloying additions, partitioning of the
substitutional elements between the pearlitic phases may occur at and
behind the transformation front. Carbide forming elements such as Mn, Cr,
Mo would be expected to partition to pearlitic cementite whereas Si, Ni and
Co would concentrate in the ferritic component. Earlier studies of alloy
partitioning involved the chemical analysis of extracted carbides (40-42).
However, the techniques gave an average composition so it was not readily
possible to distinguish between partitioning which had occurred at the
transformation front from that which may have occurred behind the front.

Most of the recent studies of alloy partitioning have been carried out
using analytical electron microscopy of thin foils or carbide extraction
replicas. During analysis the intensities of characteristic X-ray lines
from the specimen are recorded and, after applying appropriate absorption
and/or fluorescence corrections, are converted to weight concentrations.
The procedures involved and the practical problems to be overcome to obtain
quantitative data have been discussed by Ridley and Lorimer (43). The
analysis results are often expressed as a partition coefficient, $K^{cem}_{\alpha}(X)$,
which is defined as the ratio of the weight percent of X in cementite to
that in ferrite. The partition coefficient can be readily computed when
measurements are made on thin foils, but for measurements on replicas it is
necessary to calculate the concentration of alloy elements in ferrite by
using a mass balance procedure (44).

There are obvious advantages with thin foils particularly if the
reaction front can be located, when analyses can be made at the pearlite/
austenite interface and at various positions behind it. However, in
practice when the foil is prepared by electropolishing the probability of
obtaining an interface in a thin region is low. Another limitation of the
foil technique is that of resolution. Analysis of cementite in a foil can
be a problem because in a specimen of mean interlamellar spacing, 100 nm,
the cementite width will be ~12.5 nm. Even with a probe diameter of 10 nm,

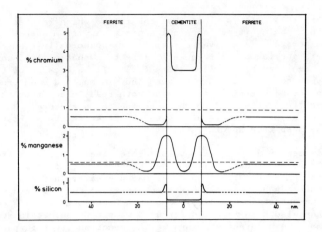

Figure 4 - Schematic distribution of alloy elements
in pearlite determined by atom-probe microanalysis. (Ref. 47)

beam spreading into the adjacent ferrite could be appreciable unless the
foil was exceptionally thin.  The carbide extraction replica does not
suffer the limitations due to resolution since only one phase is present,
and the transformation front is easy to locate.

Atom-probe microanalysis has recently been applied to the study of
partitioning in pearlitic steels.  The design and performance of atom probe
field ion microscopes has been described by Smith et al. (45).  The
analytical procedure carried out under UHV conditions involves applying
high voltage pulses to a thin wire specimen.  Layers of atoms are removed
from the tip of the specimen and the resulting ions pass through an
aperture and are analysed in a mass spectrometer.  Spatial resolution is
2-5 nm and is defined by the probe aperture, whereas depth resolution is of
the order of the interplanar spacing (~0.2 nm).  Hence, the technique is
capable of very fine scale analysis as is indicated by the data shown
schematically for a 0.6 wt pct C, 0.85 wt pct Cr, 0.66 wt pct Mn, 0.26 wt
pct Si steel transformed isothermally to pearlite at 597°C in a time of 120
seconds, followed by cold drawing to fine wire (Figure 4).

The atom-probe technique has so far only been used to a limited extent
and has been applied to fully transformed specimens so that the effects of
partitioning behind the interface have to be separated from those at the
interface (46-48).  Experimentally it could prove very difficult to locate
a transformation front in a partially transformed specimen, but the
information which could be gained on solute distribution in this region
would be of great interest.  However, it seems likely that future work on
pearlite partitioning will rely mainly on analytical electron microscopy
since with the best STEM instruments currently available it is possible for
a skilled operator using a very thin foil (~20 nm) produced by ion-beam
thinning to obtain quantitative data from regions of ~2 nm diameter (49).

## Crystallographic Aspects of Pearlite Formation

Hillert (11) has pointed out that the nucleation of pearlite is not a
well defined event but is a gradual process involving the development of a
favourable geometry between ferrite and cementite, which leads to rapid
co-operative growth of the two phases at a high mobility incoherent

interface.  The experimental observation that nucleation rate increases
with time at constant temperature is consistent with the gradual
development of co-operative growth (2).  A low degree of co-operation
results in degenerate pearlitic structures.  Depending on composition
either phase can be the initial or active nucleus, but since pearlite is a
duplex structure two phases must nucleate.  Following the nucleation of the
second phase and the development of co-operative growth, a pearlite colony
expands sideways and undergoes branching so as to maintain an interlamellar
spacing characteristic of the reaction temperature at the growth front.
There is some evidence that in the early stages of formation of a pearlite
colony sidewise growth by repeated nucleation, a process envisaged by Mehl
and his collaborators (2) may also occur to a limited extent (50).

According to the Smith-Hillert hypothesis (11,51) the essential
pre-requisite for co-operative growth is that the lattice orientations of
ferrite and cementite are random with respect to the austenite grain into
which they are growing provided that orientations which lead to partial
coherency at the transformation front, and hence reduced interface
mobility, are avoided.  Measurements of ferrite/cementite orientation
relationships have been reported by a number of workers (50,52-57) and the
two most important appear to be the Pitsch-Petch and the Bagaryatski
relationships.

| Pitsch-Petch | Bagaryatski |
|---|---|
| $[100]_c$ 2-3° from $[13\bar{1}]_\alpha$ | $[100]_c \parallel [0\bar{1}1]_\alpha$ |
| $[010]_c$ 2-3° from $[113]_\alpha$ | $[010]_c \parallel [1\bar{1}\bar{1}]_\alpha$ |
| $(001)_c \parallel (5\bar{2}\bar{1})_\alpha$ | $(001)_c \parallel (211)_\alpha$ |

Both relationships may be observed in the same specimen.

Dippenaar and Honeycombe (50) used T.E.M. to study crystallographic
and morphological aspects of pearlite formation in a hypereutectoid alloy
of composition 13 wt pct Mn, 0.8 wt pct C which was partially transformed
to pearlite at temperatures in the range 450-650°C.  Unlike plain carbon
and low alloy steels the parent austenite phase in this alloy is retained
on cooling to room temperature so enabling orientation relationships
between ferrite, cementite and austenite to be examined (Figure 5).

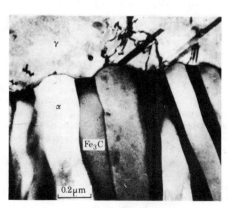

Figure 5 - Austenite-pearlite growth front in high Mn steel (Ref. 50).

Pearlite which was nucleated on a clean austenite boundary showed the Pitsch-Petch relationship. Both pearlitic phases were rationally oriented with respect to the austenite grain into which they were not growing, i.e. the true parent grain; the ferrite had the Kurdjmov-Sachs relationship and the pearlitic cementite the Pitsch relationship for Widmanstatten cementite (58). These relationships between ferrite/austenite and cementite/ austenite lead to the Pitsch-Petch relationship observed experimentally. Both of the pearlitic phases were unrelated to the austenite grain in which they were growing.

The Bagaryatski relationship was found for pearlite nucleated on grain boundary proeutectoid cementite. The pearlitic cementite was continuous, and hence crystallographically identical with the grain boundary cementite, but was unrelated to the austenite grain into which it was growing. The cementite was however, related to the adjacent austenite grain by the Pitsch relationship for Widmanstatten cementite (58). The crystallography of the pearlitic ferrite was determined entirely by the proeutectoid cementite on which it nucleated and with which it had the Bagaryatski relationship. The pearlitic ferrite was not related to either the grain in which it was growing or to the adjacent grain. It was proposed that the Pitsch-Petch relationships would predominate at compositions close to the eutectoid while the Bagaryatski would become increasingly important in hyper- and hypo-eutectoid steels as the composition deviated from the eutectoid composition.

In Figure 5 it is interesting to note the presence of facetting on the pearlite/austenite transformation front. This indicates that despite the absence of crystallographic relationships between the pearlitic phases and the austenite, some degree of partial coherency at the moving interface can exist, although this may be associated with the very slow rate at which pearlite grows in this highly alloyed steel.

The fact that pearlite, which in a plain carbon eutectoid steel only contains about 12% volume of cementite, has a lamellar rather than a rod morphology has been attributed to a marked anisotropy of interfacial energy (59). There is also evidence of partial coherency across the $\alpha/Fe_3C$ interface from field-ion microscopy and lattice resolution electron microscopy (50,57), but the interfacial planes are not well established. Various workers using electron microscopy have proposed that the cementite habit plane is $(001)_c$ (52,55), while Hillert has reported that Modin, using polarized light microscopy, observed that the plane deviated from $(001)_c$ and the deviation involved rotations about $[010]_c$ (11). The latter proposals are supported by field-ion microscopy studies (57). There is clearly scope for further work to ascertain the habit plane or planes of the cementite plates and the structure of the $\alpha/Fe_3C$ interfaces.

## Pearlite Growth in Fe-C Eutectoid Steels

Unified measurements of transformation temperature, growth velocity and interlamellar spacing have been reported for high purity alloys transformed under isothermal (60,61) and continuous cooling conditions (34), while recently Verhoeven and Pearson (62) have measured the three parameters for forced-velocity growth. The significance of this and other data will be examined with respect to the relationship between interlamellar spacing and undercooling, identification of the appropriate optimisation criterion and the rate controlling process for pearlite growth.

## Interlamellar Spacing Measurements

Measurements of minimum interlamellar spacings made on isothermally

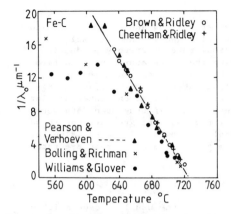

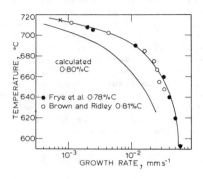

Figure 6 – Reciprocal minimum spacing spacing versus temperature for Fe–C alloys.

Figure 7 – Measured and calculated growth rates for Fe–C alloys.

transformed high purity Fe–C eutectoid steels by several workers (28,60,61,63) are shown in Figure 6 as a plot of reciprocal spacing versus transformation temperature. The recent results of Verhoeven and Pearson (62) for alloys transformed under forced-velocity growth conditions are also included. The results show good agreement for temperatures above 650°C and lie on a straight line in accordance with the predictions of Zener (equation 4). Below 650°C the measurements of Bolling and Richman (28) and Williams and Glover (63) show an increasing deviation from linearity and have clearly been influenced by recalescence. A regression analysis on the combined data of Ridley et al (60,61) and Verhoeven and Pearson (62) predicts a eutectoid temperature of 727.5°C which is close to the accepted value of 727°C, and to the relationship $\lambda_0\Delta T$ = 6.19 $\mu$m°C. A logarithmic plot of the spacing versus undercooling gives a straight line as predicted by equation (2), of slope very close to -1 (60).

Marder and Bramfitt (34) measured minimum spacings on alloys transformed to pearlite during continuous cooling. The spacings showed some scatter and tended to be slightly lower than those lying close to the straight line in Figure 6, but were in broad general agreement. From all of the data on minimum spacing available at the time Marder and Bramfitt obtained $\lambda_0\Delta T$ = 8.02 $\mu$m°C. It is important to note that the above values of $\lambda_0\Delta T$ are very different from that calculated from the line chosen by Cahn and Hagel (13) to represent the earlier mean interlamellar spacing data of Pellisier et al (36) made on commercially based plain-carbon eutectoid steels for which $\overline{\lambda}_0\Delta T$ = 18.8 $\mu$m°C.

In principle it should be possible to use the experimental value of $\lambda_0\Delta T$ with enthalpy and surface energy data to identify the optimisation criterion. Making the assumption of volume diffusion control of growth and substituting $\lambda_0\Delta$ = 6.19 $\mu$m°C into equations (3) and (4) with the enthalpy data of Kramer et al (64), $\Delta H_v$ = 607 MJ m$^{-3}$, the maximum growth rate criterion gives $\sigma^{\alpha c}$ = 0.94 Jm$^{-2}$, while the maximum rate of entropy production criterion predicts $\sigma^{\alpha c}$ = 0.63 Jm$^{-2}$. Of these two values the latter would seem the more realistic since the boundary involved must be at least semi-coherent. However, both values fall within the range 0.70±0.30 Jm$^{-2}$ obtained by Kramer et al (64), although recently Kirchner et al (65) have proposed a lower value of $\sigma^{\alpha c}$ = 0.50±0.34 Jm$^{-2}$.

Alternatively, if the above mean values of the surface energy are substituted in a generalised form of equations (3) and (4) they lead to values of $\lambda/\lambda_c$ of 2.7 and 3.7, respectively. While the available data appear to give some support to the maximum rate of entropy production criterion it will not be possible, using the above approach, to determine with confidence which, if either, of the criteria examined is appropriate until a much more precise value of $\alpha/Fe_3C$ interlamellar surface energy becomes available. It is also important to note that the interlamellar spacings in Figure 6 are minimum values so that if true mean data were available then the values of $\lambda/\lambda_c$ predicted would be greater than those above. For boundary diffusional control of growth calculated values of surface energy of 1.24 $Jm^{-2}$ and 0.83 $Jm^{-2}$ are obtained. Both are rather high, the former unrealistically so.

## Rate Controlling Process for Pearlite Growth

To identify the rate controlling process the usual procedure is to substitute measured values of interlamellar spacings into the Zener-Hillert relationship (equation 1) and to calculate growth rates consistent with these spacings, and then to compare calculated and measured growth rates. The concentration terms can be obtained from the extrapolated phase boundaries of the Fe-Fe$_3$C phase diagram. However, because of its curvature the $\gamma/\alpha + \gamma$ phase boundary cannot be extrapolated by geometrical means, and an acceptable extrapolation must be based on thermodynamic considerations. An equation proposed by Aaronson et al (66) based on a thermodynamic treatment by Kaufman et al (67) can be used to obtain $C^{\gamma\alpha}$ below the eutectoid temperature. The $\gamma/\gamma + Fe_3C$ boundary may be extrapolated geometrically. The volume diffusion coefficient can be obtained from the data of Wells et al (68) for the average carbon content of austenite at the interface.

The approach has been adopted by several workers to compare calculated and measured growth rates and has led to the conclusion that the data are not inconsistent with volume diffusion control. Puls and Kirkaldy (4) and Cheetham and Ridley (61) have shown that by evaluating the diffusion coefficient at an average carbon content of $(7/8 \ C^{\gamma\alpha} + 1/8 \ C^{\gamma c})$ reduces the difference between the measured and calculated growth rates to between 2 and 3 from the very much higher discrepancies, ~50 or more, reported previously (13,21). Figure 7 compares measured growth rates with calculated values for an arbitrarily selected value of $\lambda/\lambda_c = 3$. It has been suggested that because measured growth rates are greater than those calculated, the partial involvement of interfacial diffusion cannot be ruled out. Calculations cannot be carried out for interfacial diffusional control because of the lack of diffusion data. Verhoeven and Pearson (62) recently carried out similar calculations for their forced velocity growth data and observed discrepancies of 3 to 4 times between calculated and measured diffusivities, which is consistent with the above observations.

The forced-velocity growth technique provides an alternative way of examining the rate controlling process. A detailed study of forced velocity growth was made by Bolling and Richman (28) who examined the relationship between minimum spacing and velocity over five orders of magnitude. The data fitted the relationship $V\lambda^n$ = constant, where n = $2.3\pm0.1$. It was noted that "good spacings" were rarely obtained for speeds >50 m s$^{-1}$, which is not far from the maximum growth rate observed isothermally. Chadwick and Edmonds (69) and Cheetham and Ridley (61) obtained n = $2.70\pm0.11$ and 2.0, in their respective studies.

In their recent work Verhoeven and Pearson (62) found that forced

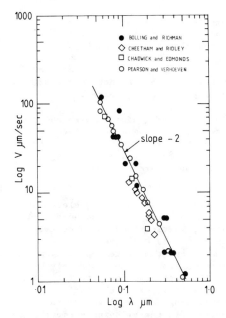

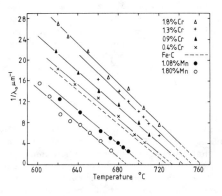

Figure 8 - Spacing-velocity
relationship for Fe-C eutectoid
alloys.

Figure 9 - Minimum spacing data for
Mn and Cr eutectoid steels.

velocity pearlite was limited to a maximum growth rate ~100 $\mu$m s$^{-1}$. At low
velocities, <1 $\mu$m s$^{-1}$ temperature oscillations led to oscillatory pearlite
interface motion. This was believed to have caused reduced minimum spacing
values and would result in an increase in the exponent of $\lambda$. Since a
similar experimental arrangement was used in the previous work of Bolling
and Richman (28) and Chadwick and Edmonds (69) it was suggested that the
results of these workers had been similarly affected at low velocities.
For forced velocity growth between 100 and 1 $\mu$m s$^{-1}$ Verhoeven and Pearson
obtained an exponent of $\lambda$ equal to 2.07. It was proposed that over this
range of forced velocities all of the available spacing and velocity data
showed satisfactory agreement (Figure 8) and gave the relationship:

$$V\lambda^2 = \text{constant}$$

Puls and Kirkaldy (4) have pointed out that the diffusion coefficient
in austenite at the transformation front, which is strongly dependent on
composition, is essentially constant at ~1.3 x 10$^{-8}$ cm s$^{-1}$ for temperatures
down to 600°C so that the above relationship is fully consistent with
volume diffusion control. Hence, these observations when combined with
those from the isothermal studies now provide strong support for volume
diffusion being the dominant rate controlling mechanism.

### Effect of Alloying Additions on Pearlite Growth

The redistribution of alloy additions between the pearlitic phases at
the austenite-pearlite reaction front in Fe-C-X alloys is a thermodynamic
requirement for growth at low undercoolings. Below a composition dependent
temperature, the no-partition temperature, it has been proposed that growth
can occur without the necessity of partitioning. Clearly, the rate
controlling process for pearlite growth will be different at temperatures

where the diffusion of substitutional elements is necessary from that at lower temperatures where only the partitioning of carbon is required.  As was indicated previously, the carbide forming elements such as Cr, Mo and Mn would be expected to partition to cementite, while Si, Ni and Co, would concentrate in the pearlitic ferrite.  Cahn and Hagel (13) have predicted that the pearlite interlamellar spacing should show a distinct change of slope in the region of the temperature where partitioning ceases.

Puls and Kirkaldy (4) have proposed that alloying additions which raise the eutectoid temperature, particularly the strong carbide forming elements which have a marked effect on hardenability, should be forced by thermodynamics to show partitioning over a wide range of temperatures, whereas additions which lower the eutectoid temperature might show a no-partitioning temperature.

In the temperature range where partitioning of the alloying element has to occur there is also simultaneous diffusion of carbon.  Mathematical treatments of growth under these complicated conditions have been attempted by Hillert (70) and by Bolze, Puls and Kirkaldy (71).  However, it may be assumed that diffusion of the substitutional element is rate controlling, and that the activity of carbon near the growth front is approximately uniform.  The diffusion path must be along the austenite-pearlite interface since volume diffusion of substitutional elements is extremely small. Growth rates can be then predicted by substituting values of the diffusion coefficient and concentration terms appropriate to the alloying element into Equation 5.

At low temperatures where the growing phases inherit the alloy content of the parent austenite, growth occurs under carbon diffusion control.  Two ways in which the no-partition reaction can occur are termed para-equilibrium and local equilibrium no-partitioning, respectively.  For the former, growth occurs without redistribution of the alloying element which is uniformly distributed across the growth front.  In local equilibrium no-partition the growing phases have the same concentration of alloying additions as the parent austenite, but remain in complete chemical equilibrium (orthoequilibrium) at the interface (23,72).

The latter model involves the existence of a pile-up, or spike, of alloying elements ahead of the transformation front.  Depending on the solute, carbon activity in the interface spike can be markedly influenced and this can lead, in effect, to a marked change in the carbon concentration gradient in the austenite driving the growth process. Paraequilirium and local equilibrium no-partion phase boundaries can be calculated for Fe-C-X alloys providing the necessary thermodynamic data is available.  Carbon concentration terms taken from these diagrams can be used with Equation 1 to predict growth rates.  The procedures involved have been described in detail forthe Fe-Cr-C system by Sharma, Purdy and Kirkaldy (73,74).

Most of the earlier measurements of Mehl and his collaborators (2) were on steels of commercial purity, and the data have been reviewed by Cahn and Hagel (13).  More recently measurements of transformation temperature, growth velocity, interlamellar spacing and partitioning at the transformation front, combined in some cases with microstructural observations, have been made for several Fe-C-X alloys.  The results of these studies will be examined bearing in mind that Hillert (7) has recently made a detailed analysis of the effect of nickel, manganese and silicon on pearlite growth.

## Effect of Manganese

Razik et al (75) measured pearlite growth rates and minimum
interlamellar spacings, and used analytical electron microscopy of thin
foils to examine partitioning at the interface for steels containing
nominally 1 and 2 wt pct Mn. Manganese was observed to partition to
pearlitic cementite at higher reaction temperatures but not at lower
temperatures, a result consistent with the earlier studies of Picklesimer
et al (41) who chemically analysed extracted carbides. Composition
dependent no-partition temperatures measured were in reasonable agreement
with values calculated for the local equilibrium no-partition model.

The spacing data is included in Figure 9 and the experimental points
lie on reasonably straight lines which extrapolate to give $Ae_1$ temperatures
close to those predicted by the equation of Andrews (76). There is no
change in the slope of the reciprocal spacing data in the region of the
no-partition temperatures, and the curves are similar in slope to that for
Fe-C alloys. As can be seen from Figure 9 that manganese increases the
spacing for a given reaction temperature. This observation is consistent
with earlier measurements on commercially based manganese steels reviewed
by Cahn and Hagel (13). Flugge et al (77) reported for commercially based
steels that manganese additions caused a decrease in spacing, but these
steels also had small but different chromium concentrations.

Growth rates calculated using the local equilibrium no-partition model
were consistent with diffusion of carbon as the rate controlling process,
although it was not possible to distinguish between volume and interfacial
diffusion. The calculations support the earlier predictions of Hillert
(78) and Sundquist (22).

Recent data obtained for a 1.0 wt pct Mn steel is summarised in Figure
10. The growth rate and spacing data are in excellent agreement with the
previous work. However, the growth rate measurements have been extended to
lower temperatures and show a maximum in the same temperature range as that
in the corresponding isothermal transformation curve. Partitioning data
was obtained using replicas and the higher resolution of the technique led
to generally higher distribution coefficients and to a lower value of the
no-partition temperature, $T_o^P$, i.e. ~670°C compared with 683°C previously
reported. Growth rates calculated assuming volume diffusion control are in
excellent agreement with experiment, and are shown in Figure 10 for $\lambda/\lambda_c$ =
3. At temperatures above 670°C it is clear that pearlite growth has been
markedly retarded by the necessity for manganese to partition at the
reaction front. The partitioning data shows that the distribution
coefficient decreases steadily with falling temperature until the
no-partition region is reached. This behaviour is indicative of a
relatively smooth transition between the two rate controlling mechanisms.

Cahn and Hagel (31) observed during isothermal transformation of a 5.2
wt pct Mn steel that the growth rate decreased with time and the spacing
increased, producing what was termed divergent pearlite. The observations
were attributed to a continuously changing carbon concentration ahead of
the reaction front and to the formation of orthopearlite with a higher
average carbon content than the alloy. Hillert has explained these
observations using a schematic phase diagram (7). He has also predicted
that partitioned pearlite with a constant spacing, which he termed constant
orthopearlite could occur, and has constructed ternary isothermal sections
identifying regions in which parapearlite, constant orthopearlite and
divergent orthopearlite can form. For the alloy examined by Cahn and Hagel
(31) it was believed that the growth rate was controlled by carbon

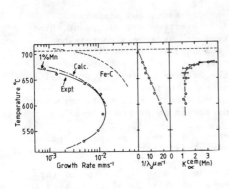

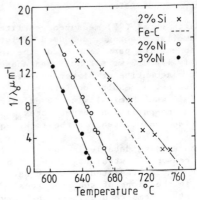

Figure 10 - Growth rate, minimum spacing and partitioning data for 1% Mn eutectoid steel.

Figure 11 - Minimum spacing data for Ni and Si eutectoid steels.

diffusion through the depleted zone, while boundary diffusion of manganese controlled the spacing.

Effect of Nickel

Measurements of minimum spacings and pearlite growth rates for high purity eutectoid alloys containing 0, 2 and 3 wt pct Ni were reported by Brown and Ridley (60). The reciprocal spacings shown in Figure 11 lie on reasonably straight lines which for the nickel steels extrapolate to temperatures of 678°C and 655°C, compared with values of 689°C and 672°C, respectively, calculated for the eutectoid temperatures. To explain the apparent temperature discrepancies it was recently suggested that the extrapolated values might relate to parapearlite, and that investigation of the region above these temperatures might provide evidence for partitioned pearlite (8). However, Hillert has predicted that pearlite in the nickel steels examined always forms as parapearlite (7), and using the local-equilibrium no-partition model has calculated limiting temperatures for its formation in agreement with the extrapolated values from Figure 11.

Partitioning studies currently in progress on extracted carbides from a 2 wt pct Ni steel confirm that pearlite formed at lower temperatures is unpartitioned, but some partitioning occurs at higher temperatures (Figure 12). Since nickel diffuses away from the cementite and concentrates in the ferrite the diffusion path is relatively short, and it is conceivable that some nickel redistribution can accompany the rate controlling process involving carbon diffusion. Even at the highest reaction temperature examined nickel does not partition very strongly.

It can be seen from Figure 11 that nickel additions, like manganese, increase the pearlite spacing for a given reaction temperature, although the reciprocal spacing plot is steeper than for the Fe-C alloy. For both nickel alloys $\lambda_0 \Delta T = 4.0$ $\mu m$ °C, which is approximately two-thirds of the product for Fe-C. It is not known whether this reflects a reduction in the $\alpha/Fe_3C$ interfacial energy.

Effect of Chromium

Measurements of minimum interlamellar spacing and growth rates of pearlite formed in high purity Cr eutectoid steels have been made by Ridley et al (1.3 and 1.41 wt pct Cr) and Sharma et al (0.4 and 1.8 wt pct Cr) and show good agreement (74,79,80).  Growth rates are progressively decreased and this is consistent with the well established effect of chromium on isothermal transformation and hardenability.  The reciprocal spacing data in Figure 9 lie on reasonably straight lines, which have slopes similar to that for Fe-C, and extrapolate to temperatures close to calculated Ae$_1$ values.  It can be seen that chromium additions progressively reduce the pearlite spacing for a given reaction temperature.

Partitioning data for the 1.41 wt pct Cr steel was obtained from extraction replicas and showed that chromium segregated to pearlitic cementite at the transformation front for the full range of temperatures examined, 720-550°C (Figure 13).  Atom probe studies on a commercially based 0.9 wt pct Cr steel (Figure 4) also showed partitioning at 597°C to an extent similar to that seen in Figure (13).  Previous work on thin foils gave a generally lower extent and narrower range of partitioning.  Bainite formed at the lower reaction temperatures is unpartitioned, and this supports the early measurements of Hultgren (40).

Although a no-partitioning temperature was not observed experiment-ally, the thermodynamic data of Sharma et al (73) enabled a local equilibrium no-partition temperature of 684°C to be calculated.  If this temperature is valid then partitioning below 684°C should merely accompany the rate controlling step for growth which will be carbon diffusion. Assuming that interfacial diffusion of chromium is rate controlling at higher temperatures then pearlite growth rates can be calculated from equation 5, modified for substitutional diffusion, using measured spacings, partition data, the thermodynamic data of Sharma et al (73) and the interfacial diffusion coefficient of Fridberg et al (81).  Calculated growth rates were found to show excellent or good agreement with measured values for a wide temperature range and are shown in Figure 14 for $\lambda/\lambda_c = 2$.

Growth rates calculated for temperatures below 684°C using the local equilibrium no-partition model and assuming volume diffusion of carbon to be rate controlling (Equation 1) were close to those predicted for interfacial diffusion, and in reasonable agreement with experiment except

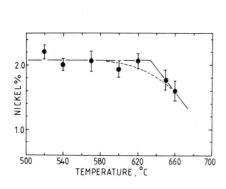

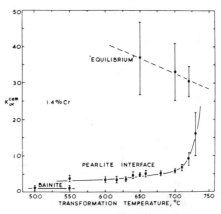

Figure 12 - Nickel concentration in cementite versus reaction temperature for a 2wt pct Ni eutectoid steel.

Figure 13 - Partition coefficient for chromium versus reaction temperature for a 1.4 wt pct Cr eutectoid steel.

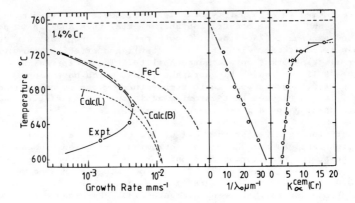

Figure 14 - Growth rate, minimum spacing and partitioning measurements for
a 1.4 wt pct Cr eutectoid steel.   Calc. B - boundary diffusion of Cr;
Calc. L - local equilibrium no-partition.

for the lowest temperatures.   At the lowest temperatures which correspond
to the upper part of the bay region in the isothermal transformation curve
(Figure 15), measured growth rates were becoming much slower than those
calculated and were probably controlled by solute drag effects.   A spiky
pearlite growth front develops in the bay region and may be associated with
localised inhibition of growth leading to a non-smooth front with
projecting lamellae.   Over the full temperature range examined, where up to
three different rate controlling processes could be operating, the
reciprocal spacing curve was essentially linear (Figure 14).

For the higher and intermediate temperatures the growth studies were
in good agreement with the finding of Sharma et al (74) for Cr eutectoid
steels.   These workers used their thermodynamic data to predict the
variation of spacing with temperature from the relationship

$$\lambda_c = \frac{2\sigma^{\alpha c} V_m}{\Delta G_o} \qquad (11)$$

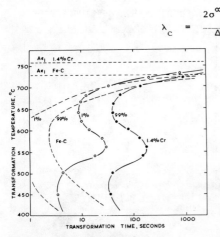

Figure 15 - Isothermal
transformation diagram
for 1.4 wt pct Cr eutectoid
steel.

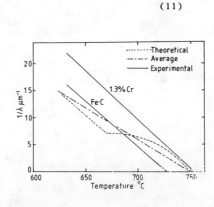

Figure 16 - Spacing-temperature
relationship predicted for 1.3 wt
pct Cr eutectoid steel.

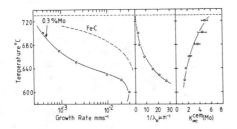

Figure 17 - Growth rate, minimum spacing and partitioning data for 0.3 wt pct Mo eutectoid steel

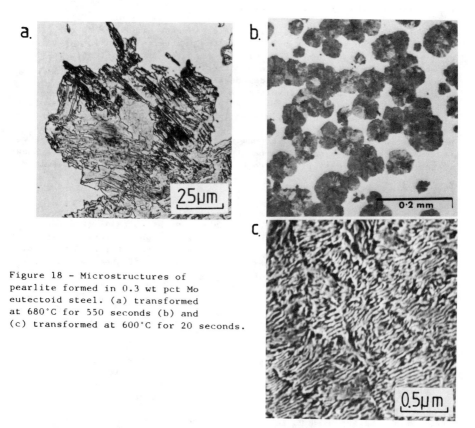

Figure 18 - Microstructures of pearlite formed in 0.3 wt pct Mo eutectoid steel. (a) transformed at 680°C for 550 seconds (b) and (c) transformed at 600°C for 20 seconds.

where $V_m$ is molar volume and $\Delta G_0$ the free energy of transformation. However, the agreement between theory and experiment was not good (Figure 16). It was suggested that if nature chooses a smooth transition between mechanisms, as is apparent for the growth rate data, the reciprocal spacing could be represented by a straight line. The remaining discrepancy would be removed if chromium reduced the surface energy possibly by changes in lattice parameter which reduce interfacial misfit.

Effect of Molybdenum

Minimum spacing, growth rate and partitioning measurements made for a

0.3 wt pct Mo eutectoid steel are shown in Figure 17. The reciprocal
spacing lies on a curve that is distinctly concave downwards which reflects
the variation of pearlite morphology with transformation temperature. At
high temperatures the pearlite is substantially lamellar and growth occurs
on an approximately spherical front. For temperatures in the range
680-620°C the pearlite is markedly degenerate, the microstructure shows few
areas of lamellar pearlite, and the growth front is irregular. At 600°C
the pearlite is again uniformly lamellar and for temperatures down to
500°C, and below, when pearlite has become largely replaced by bainite,
growth occurs on a spherical or smooth transformation front. Typical
microstructures are illustrated in Figure 18. The pearlite formed at
intermediate temperatures appears to be the 'wrinkled pearlite' reported by
Hultgren (40), while Parcel and Mehl (82) have also commented on the
degenerate morphology of pearlite in molybdenum steels.

The temperature range in which the degenerate structure forms
corresponds to an incipient bay in both the isothermal transformation and
growth rate curves (Figure 17). Similar features can be detected in the
growth and/or transformation data of other workers who have examined
molybdenum steels (82-84). The degenerate pearlite and associated
phenomena are probably the result of a solute drag effect on the growth
front which has hindered the co-operative growth of pearlite. This effect
is superimposed on the already marked influence that molybdenum has on the
transformation as can be seen from its effect at temperatures above and
below the incipient bay. Molybdenum partitioned to cementite at the growth
front over the full range of temperatures examined although it is not known
whether this was necessary for growth at the lowest temperatures examined.

### Effect of Silicon

Minimum spacing, growth rate and partitioning measurements have been
made for an isothermally transformed 2 wt pct Si alloy (85). (Figure 19).
The reciprocal spacing data versus temperature lie on a relatively straight
line with a slope which is about three-quarters of that for Fe-C with $\lambda\Delta T =$
8.3 $\mu m°C$. It is uncertain whether recalescence has affected the spacings
at lower reaction temperatures. Pearlite growth rate is retarded at
temperatures below 700°C and, if a comparison is made as a function of
undercooling, at all undercoolings. Recalescence does not appear to have
influenced the partitioning of silicon which segregates to pearlitic
cementite at the transformation front to a decreasing extent as the
reaction temperature is reduced.

Hillert (7) has recently predicted the temperature ranges over which
various types of pearlite can form in silicon steels, and has also
constructed ternary isothermal sections identifying regions where
parapearlite, constant orthopearlite, and divergent orthopearlite similar
to that studied by Fridberg and Hillert (38) in a hypereutectoid silicon
steel, might be expected to form. It was proposed that for most of the
temperature range examined the kinetic data in Figure 19 was consistent
with constant orthopearlite, the rate of formation of which is controlled
by interfacial diffusion of silicon. Silicon has a low equilibrium solid
solubility in cementite, and even though partitioning occurs the growth
velocity is not markedly retarded because the silicon content of the
growing cementite can be reduced by diffusion over a relatively short
distance. Using the spacing and growth rate data in Figure 19, Hillert (7)
has calculated concentration profiles for silicon across cementite and
ferrite lamellae and has shown that silicon can be removed from the
interiors of cementite lamellae while the interiors of ferrite lamellae are
still unaffected. The calculated profiles have the same general form as

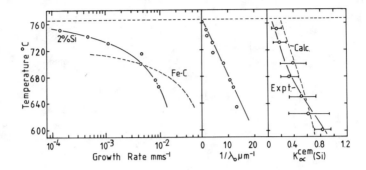

Figure 19 - Growth rate, minimum spacing and partitioning
data for 2 wt pct Si eutectoid steel.

those observed in Figure 4 for silicon partitioning in the atom probe study
of Smith et al (47). The calculated partition coefficients for the 2 wt
pct Si steel are shown in Figure 19 to be in good agreement with those
measured.

At reaction temperatures where partitioning was not complete at the
reaction front a further rapid segregation of silicon occurred behind the
front so that within a short time, <1 minute, near to equilibrium
partitioning was achieved over the full range of temperatures examined.
The variation of the partition coefficient as a function of distance from
the reaction front is shown in Figure 20 for a specimen transformed at
650°C. Partitioning of silicon behind the transformation interface to the
adjacent cementite must occur by volume diffusion through cementite, and
the fact that it occurs rapidly reflects the short diffusion path involved
and also the relatively high rate of volume diffusion of silicon in
cementite.

## Effect of Cobalt

Mellor and Edmonds (86,87) carried out detailed kinetic studies of
pearlite growth in high purity eutectoid steels containing 1 and 2 wt pct

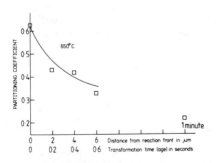

Figure 20 - Partition coefficient
as a function of distance from
growth front for 2 wt pct Si
eutectoid steel.

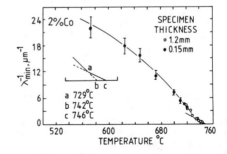

Figure 21 - Minimum spacing versus
reaction temperature for 2 wt pct Co
eutectoid steel.

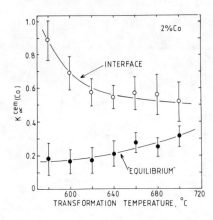

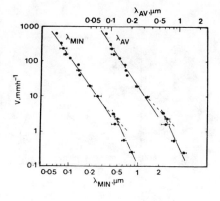

Figure 22 - Partition coefficient versus reaction temperature for 2 wt pct Co eutectoid steel.

Figure 23 - Velocity-spacing relationship for 2 wt pct Co eutectoid steel.

Co transformed under isothermal and forced velocity growth conditions. Minimum spacing data from isothermally transformed specimens is shown in Figure 21 for the 2 wt pct Co alloy. For both alloys the reciprocal spacing data appear to show a small change of slope near the respective $Ae_1$ temperatures and it was suggested that this might reflect a transition from partitioning to no-partitioning.

Partitioning measurements on the 2 wt pct Co alloy examined by Mellor and Edmonds (86) have recently been made by Ridley and Burgess (88) and the results are shown in Figure 22. Cobalt partitioned to pearlitic ferrite to a decreasing extent as the transformation temperature was decreased but a no-partition temperature could not be experimentally identified. However, cobalt only partitions relatively weakly and calculations based on the equilibrium partitioning data (Figure 22) suggest that the no-partition temperature will be very close, ~2°C, to the $Ae_1$ temperature. Hence, the partitioning observed experimentally for a wide range of temperatures could, because of the short diffusion path involved, have accompanied the rate controlling step which is likely to be carbon diffusion. Estimates of the diffusion distances for cobalt atoms at the transformation front support this view (88).

In an attempt to overcome problems associated with rapid isothermal growth rates Mellor and Edmonds also carried out forced-velocity growth studies (86). The relationship obtained between imposed velocity and spacing is shown in Figure 23 for the 2wt pct Co alloy. The data for both steels fitted the relationship

$$V\lambda^n = constant$$

where n = 2.7±0.1 and n = 2.8±0.1 for the 1 and 2 wt pct Co alloys, respectively. Alternatively it was proposed that the data could be represented by lines of different slopes with n ~ 4 at very slow growth rates and n ~ 3 at higher growth rates with a narrow transition region of intermediate slope, n ~ 2. It was proposed that n ~ 4 might reflect a temperature range where partitioning of cobalt was rate controlling, while the exponents of 3 and 2 would correspond to interfacial and volume diffusion of carbon, respectively, as the rate controlling processes for growth.

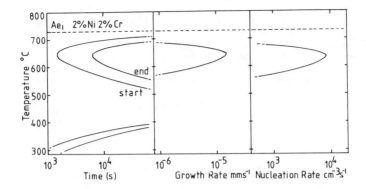

Figure 24 - Isothermal transformation rate, growth and nucleation
rates, versus temperature for 2 wt pct Ni 2 wt pct Cr eutectoid steel.

Following the recent observations of Verhoeven and Pearson (62) it is
possible that the spacing measurements made by Mellor and Edmonds (86) at
high growth velocities may have been influenced by degeneracy, and those at
low growth velocity by temperature oscillations leading to reduced spacing
spacings and an increased value of the exponent, n.  If attention is
concentrated on the velocity range 3.6-360 mm h$^{-1}$ (1-100 $\mu$m s$^{-1}$) the data
in Figure 23 may be represented by a single line giving a value of n =
2.7.  If the diffusion coefficient for carbon in austenite is constant over
the temperature range involved, as for Fe-C alloys, the exponent lies
between the values for volume and interfacial diffusion.  In view of the
observations reported earlier for Fe-C alloys an exponent closer to 2 might
have been expected.  For most, if not for all, of the pearlite region
carbon diffusion is likely to be rate controlling and the increase in
growth rate brought about by cobalt, as reported by Hawkes and Mehl (89)
will primarily be associated with an increase in the eutectoid temperature
leading to an increse in the carbon concentration gradient driving the
growth process.

Fe-C-X-Y alloys

Kinetic studies have been made on a commercially based 2 wt pct Ni, 2
wt pct Cr eutectoid steel for which the T.T.T. diagram showed two C-curves
separated by a deep bay of austenite stability (32).  The upper curve
represents pearlite formation and measurements of nucleation and growth
rate in the early part of the transformation for this region are shown in
Figure 24.  Both rates increase with decreasing temperature and then pass
through a maximum at a temperature corresponding to that of the nose of the
C-curve.  The presence of the austenite bay is likely to be due to a solute
drag effect associated with the presence of the strong carbide forming
element and results in a pearlite nose at a much higher temperature,
~640°C, than that encountered in an Fe-C alloy, ~550°C.

Al-Salman et al (44) have measured minimum spacings, growth rates and
partitioning at the reaction front for a 1 wt pct Mn, 1 wt pct Cr steel.
The isothermal transformation diagram again showed two C-curves separated
by an austenite bay, and attention was concentrated on the upper curve for
which the transformation product was primarily pearlite.  The measured
growth rate curve mirrored the pearlite C-curve, with both showing a
maximum rate at ~640°C.  Both Cr and Mn partitioned preferentially to
cementite at the reaction front for temperatures down to 600°C, the

chromium to a generally greater extent. The partitioning of manganese was probably due to the relatively slow rates of pearlite growth at all temperatures and particularly those attributed to solute drag in the lower part of the pearlite C-curve. The reciprocal spacing versus temperature plot showed some degree of scatter but was essentially a straight line despite the involvement of at least two different rate controlling processes for pearlite growth over the temperature range involved.

## General Discussion

### Transformation Mode

The unified measurements of transformation temperature, growth velocity and interlamellar spacing, now available for high purity Fe-C eutectoid alloys transformed under isothermal and forced velocity growth conditons (60-62), show that, apart from minor differences, the relationship between the three parameters is identical for the two modes of transformation. The three parameters have also been measured for continuous cooling transformation and, although the velocity-spacing relationship is the same as that in Figure 8 for the other two modes of transformation, the growth rate-temperature curve is displaced to higher undercoolings (34). This behaviour is consistent with the effect of continuous cooling in displacing the start of isothermal transformation to lower temperatures, and is probably due to delayed nucleation.

### Crystallography and Morphology

The studies of Dippenaar and Honeycombe (50) on pearlite nucleation show strong support for the Smith-Hillert hypothesis, although crystallographic effects are apparent at the $Fe_3C$-$\gamma$ interface in Figure 5, despite the absence of orientation relationships.

A morphological feature of forced velocity growth which is less evident in the two alternative transformation modes is the presence of pearlite with a "zig-zag" or "herring-bone" structure (27,28,61) (Figure 25). The proportion of this structure appears to increase as the imposed velocity is increased and as the temperature gradient is decreased (61,90). The transformation front is cusped and it has been suggested that the morphology may be due to a change to a different variant of the $\alpha/Fe_3C$ orientation relationship which enables the lamellae to grow as normal as possible to the local curvature of the interface (61,90).

In general, the alignment of pearlite colonies with the growth direction in the forced velocity method is not as good as that obtained for eutectic growth. However, observations made on a single surface can be misleading since Mellor and Edmonds (86) have shown, using two-surface analysis, that for Co eutectoid steels 90% of the pearlite colonies had lamellae which lay within 12° of the growth direction.

### Interlamellar Spacing; Optimisation Principle

For Fe-C eutectoid alloys the reciprocal minimum spacing showed a linear variation with undercooling (Figure 6). However, because of the uncertainty in the existing values of $\alpha/Fe_3C$ interfacial energy, it was not possible to use the slope of this line with Equations 3 and 4 to identify the appropriate optimisation principle. An alternative approach is to study the shape of the pearlite/ austenite interface and to compare this with theoretically predicted shapes for various values of $\lambda/\lambda_c$ (15,20,21). For volume diffusion control of growth $\lambda/\lambda_c > 3$ would be associated with

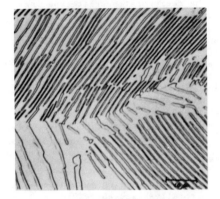

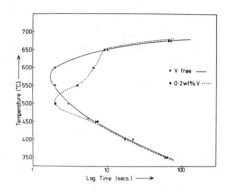

Figure 25 - Herring-bone or
zig-zag morphology in forced-
velocity pearlite (Ref. 27).

Figure 26 - Start times for isothermal
transformation in eutectoid steels
containing 0 and 0.2 wt pct V.

bending back of the interface and Hillert (5) has suggested that this is
not normally observed.  It was pointed out by Puls and Kirkaldy (4) that no
systematic examination of the growth front appears to have been made, and
this is clearly an area which should be pursued.

There appears to be little support for the optimisation principle
proposed by Sundquist (21).  His theory was based on an earlier analysis
which showed a large discrepancy between measured growth rates and those
calculated for volume diffusion control (13), and which also involved the
use of the spacing data of Pellisier et al (36).  Recent measurements of
pearlite spacing are much smaller than those previously reported, while
analysis of pearlite growth has reduced the discrepancy between prediction
and experiment to a factor of ~3.  The Sundquist theory would also require
appreciable bending back of the interface (Figure 3).

A recent literature review on the interlamellar spacings of ferrous
and non-ferrous pearlites formed under isothermal and forced velocity
growth conditions (8) confirms the observations of Pellisier et al (36)
that pearlite can grow with an essentially constant velocity with a range
of true spacings.  The spacings are grouped about a mean true value, $\lambda_0$,
and it would seem more appropriate to use this parameter to characterise
the structure rather than $\lambda_0$.  The main advantage of $\lambda_0$ is that it can be
rapidly and reproducibly measured, and for some applications, such as the
testing of growth theories, the values of $\lambda_0$ and $\bar{\lambda}_0$ are not significantly
different.  Apart from the earlier measurements, there is little
information about the range of true spacings which exist during isothermal
growth.  This information could be obtained using an analysis developed by
Cahn and Fullman (91) in which the volume fraction distribution function of
true spacings in a lamellar structure can be calculated from the lineal
fraction distribution function of random intercept spacings ($\ell$) measured
experimentally on the plane of polish.  Limited studies by Roosz et al
(92,93) on a low alloy steel indicate that the spread of spacings decreases
as the transformation temperature is decreased.  Mellor and Edmonds (86)
made detailed studies for forced-velocity pearlite in Co eutectoid steels
and proposed that the spacings were grouped about a mean true value equal

to that given by the maximum rate of entropy production criterion.

For steels containing alloy additions the reciprocal pearlite spacings show an essentially linear variation with undercooling, except for the Mo steel where degeneracy led to anomalous behaviour. For steels containing Mn or Cr more than one process is likely to be rate controlling over the range of temperatures examined. However, for Mn steels there were no apparent inflexions or gradual changes of slope in the reciprocal spacing data during the change from growth controlled by interfacial diffusion of manganese to that involving volume diffusion of carbon. The experimental curves lay essentially parallel to that for Fe-C. Similar observations were made for the Cr steels where two or even three processes could be successively rate controlling as the temperature was decreased. These unexpected observations require further investigation.

A further area which should be examined concerns the role of austenite grain size on pearlite spacing. Hyzack and Bernstein (94) have reported a statistically significant decrease of spacing with increasing austenite grain size for isothermally transformed commercially-based eutectoid steels. However, this trend is not apparent in the corresponding results of Flugge et al (77) on commercial steels, and Marder and Bramfitt (95) have reported that interlamellar spacing is independent of prior austenite grain size.

Pearlite Growth in Fe-C Alloys

Analyses of the isothermal and forced-velocity growth data for Fe-C alloys leads to the conclusion that volume diffusion of carbon is the dominant rate controlling process for pearlite growth (4,6,62). Similar conclusions were reached by Marder and Bramfitt (34) for their results obtained from continuous cooling transformation studies. However, the discrepancy between measured and calculated growth rates remains to be explained. Earlier data had been interpreted as showing an increasing involvement of interfacial diffusion of carbon with decreasing temperature (4,28).

Verhoeven and Pearson (62) suggest that diffusivity is accelerated due to strain in the austenite adjacent to the reaction front which arises because of differences in the specific volume of growing pearlite nodules and the parent austenite phase, and between the two pearlitic phases. Elastic strains could enhance diffusivity as would dislocations introduced by local plastic deformation. Evidence for the existence of elastic strains ahead of the growth front is to be found in the electron emission microscopy studies of Rathenau and Baas (96). Hillert (5,15) has also discussed the significance of plastic deformation but has pointed out that this would reduce the free energy available to create interfaces and to drive the growth process, and hence would retard rather than accelerate pearlite growth.

The maximum rate at which forced velocity pearlite grows is in reasonable agreement with the maximum isothermal rate. However, the relationship $V\lambda^2$ = constant, or $V\alpha(\Delta T)^2$, which describes the data does not predict that the growth rate will pass through a maximum. The continuous cooling growth rate data of Marder and Bramfitt (34) shows a maximum in the temperature range 600-550°C, and also the isothermal growth rates of Frye et al (97) seen in Figure 7, and the earlier measurements of Hull et al (98), seem to be approaching a maximum at ~600°C. A well defined maximum growth rate at temperatures where carbon diffusion is rate controlling is seen in Figure 10 for a 1% Mn steel.

A departure from the parabolic relationship requires that either the concentration gradient ($C^{\gamma\alpha}-C^{\gamma C}$) falls, which is unlikely according to theoretical models of the $\alpha+\gamma/\gamma$ phase boundary considered by Aaronson et al (66), or the diffusion coefficient falls.  It has been suggested by Verhoeven and Pearson (62) that at lower temperatures the proposed strain enhanced diffusion of carbon in austenite might decrease because the austenite flow stress increases.  On the other hand the prediction of a constant diffusion coefficient at higher temperatures could be in error. The diffusion data is obtained by extrapolation from measurements made at temperatures of 750-1150°C (68), and the theories assume a diffusivity in austenite related to an average interface composition (14,15).

## Effect of Alloying Additions

Hillert (7) in his recent review considered primarily the effects of additions of Ni, Mn and Si and concluded that agreement between predictd and experimental growth rates is good if local equilibrium is assumed. Coates has reached similar conclusions (99).  A problem associated with local equilibrium is that it requires a spike of solute of width D/V at the reaction front.  However, the significance of the spike at lower temperatures and faster growth rates is uncertain since it would have a calculated width which is considerably less than atomic dimensions.

Steels containing additions of 2-3% Ni form unpartitioned pearlite at all temperatures.  The retardation of pearlite growth rate is not marked and is probably due primarily to constitutional effects, since nickel depresses the $Ae_1$.  Manganese additions have a more marked effect on growth rate, and on hardenability, and this is partly attributable to constitutional effects.  Manganese is the only addition which shows an experimental composition dependent no-partition temperature in agreement with theoretical prediction.  Above the no-partition temperature manganese diffusion will be rate controlling while below it carbon diffusion will be rate controlling.  However, Mn reduces carbon activity and so reduces the growth rate.  Calculated growth rates for local equilibrium no-partition conditions are in excellent agreement with measured rates (Figure 10).

Silicon partitions by interfacial diffusion over a wide range of temperatures.  For additions of 2 wt pct the effect on growth rate is not appreciable because the requirement for silicon to reduce its concentration in cementite can be achieved by diffusion over short distances.  Hence, growth proceeds relatively rapidly, and the effect of silicon on hardenability is not as substantial as partitioning of substitutional elements would normally lead to.  It is likely that steels containing additions of cobalt form unpartitioned pearlite except at temperatures close to the $Ae_1$ where the position is uncertain.  The small increase in growth rate due to cobalt additions, and the decrease in hardenability, are primarily due to constitutional effects since cobalt raises the $Ae_1$.

Both Cr and Mo partition at the pearlite reaction front over a wide range of temperatures.  For chromium steels at higher reaction temperatures calculations which assume interfacial diffusion of chromium to be rate controlling give good agreement with measured growth rates (80).  At lower temperatures the situation is more complicated since growth rates calculated for interfacial diffusion of chromium, and for local equilibrium no-partition give similar values, which become increasingly different from the measured rates as the temperature falls (Figure 14).  The divergence is attributed to "solute drag" which retards interface movement.  The solute drag effect is associated with the development of an austenite bay in the isothermal transformation curve (Figure 15) which pushes the pearlite nose to higher temperatures and shorter times. The bay does not necessarily separate the pearlite and bainite regions since pearlite forms in the upper part of the lower C-curve in Figure 15. The combined effects of growth controlled by interfacial diffusion of chromium at higher reaction temperatures and solute drag at lower temperatures results in a considerable increase in hardenability.

Solute drag effects appear to be operating at intermediate temperatures in the 0.3 wt pct Mo steel (Figure 17), where an incipient bay is formed, and also may be seen in the isothermal transformation start curve for a 0.2 wt pct V steel shown in Figure 26. Both steels in Figure 26 were austenitised at 1050°C when ~0.1 wt pct V would be in solution in austenite. The 0.3 wt pct Mo addition inhibits cooperative growth in the bay region giving a degenerate pearlite (Figure 18), while in Cr steels pearlite nodules develop increasingly spiky interfaces with decreasing temperature in the bay region. The spikes are not bainite and do not have a leading phase. They consist of ferrite and cementite lamellae which have undergone partitioning at the reaction front to a similar but slightly lesser extent that occurs at the smoother interfaces from which they project.

It is clear that for the chromium steels, and is probably true for other steels containing strong carbide formers, that the local equilibrium no-partition model cannot predict pearlite growth rates in the bay region, a view supported by Coates (99). Solute drag effects have been reported by Aaronson and his collaborators for pro-eutectoid ferrite (100,101).

The mechanism of solute drag is uncertain but it may be due to segregation of strong carbide forming elements to the C-rich boundary regions ahead of the growing ferrite lamellae where they drastically reduce C activity in austenite and slow down growth (23). An alternative suggestion by Sharma and Purdy (102) is that clusters of strong carbide formers and carbon exist in austenite and exert an inhibiting effect on growth because of the necessity of carbon to dissociate from the clusters before ferrite can grow. In the low alloy Ni-Cr eutectoid steel examined by Brown and Ridley (32) it was observed that solute drag effects retarded both nucleation and growth rates to similar extents. In view of the important effects that solute drag appears to have on pearlite and ferrite growth, and consequently on hardenability, it is important that urgent attention is directed towards understanding and modelling the effect.

## Summary

Unified measurements of transformation temperature, growth velocity and interlamellar spacing, available for high purity Fe-C eutectoid steels transformed under isothermal and forced-velocity growth conditions, show that the relationship between the three parameters is identical for the two modes of transformation. It is clear that pearlite can grow with an essentially constant velocity with a range of true interlamellar spacings. Interlamellar spacing measurements for high purity Fe-C alloys are appreciably lower than the earlier measurements reported for steels of commercial purity. Measurements of minimum reciprocal spacing show a linear variation with undercooling, but it is not possible to use the data to identify an optimisation principle because of the uncertainty in the available values for energy of the $\alpha/Fe_3C$ lamellar interfaces. However, it is unlikely that criteria which require $\lambda/\lambda_c > 3$ will be appropriate.

There is experimental evidence which shows that although the pearlitic phases are crystallographically related they do not have a lattice

orientation with the austenite grain into which they are growing.  Further
work is required to determine the structure of the $\alpha/Fe_3C$ interface and the
habit plane of the $Fe_3C$ plates.

Analyses of pearlite growth under isothermal, forced-velocity and
continuous cooling lead to the conclusion that volume diffusion of carbon
in austenite is the dominant process controlling pearlite growth in pure
Fe-C alloys.  However, the reason for the persistent observation that the
measured growth rate is greater than that calculated remains to be
identified.

Most alloying additions retard pearlite growth rate.  The strong
carbide formers Cr and Mo partition to pearlitic cementite at the reaction
front over a wide range of temperatures.  For chromium, the interfacial
diffusion of the element controls growth at higher reaction temperatures.
The marked retardation of growth at lower temperatures and the associated
development of an austenite bay in T.T.T. diagrams is probably due to a
solute drag effect.  Silicon partitions to pearlitic ferrite at the
reaction front for a wide range of temperatures but diffusion is rapid and
the diffusion path is short so that retardation of pearlite growth is not
appreciable.  Manganese partitions to pearlite cementite at high reaction
temperature but not at lower temperatures, and is the only element which
shows a composition-dependent no-partition temperature which agrees with
theoretical prediction.  Above the no-partition temperature manganese
diffusion is rate controlling and below this temperature carbon diffusion
controls the rate of pearlite growth.  The effect of Ni or Co in
respectively decreasing or increasing the rate of pearlite growth is due
primarily to constitutional effects i.e. to their effect on the $Ae_1$
temperature.  In both cases carbon diffusion is rate controlling and the
partitioning of these elements observed experimentally has accompanied the
rate controlling process.

For steels containing small additions of alloying elements the
reciprocal spacing usually shows an approximatly linear variation with
undercooling, even when more than one process is rate controlling over the
range of temperatures involved.  In more highly alloyed steels the
isothermal growth rate may decrease with time and the interlamellar spacing
increase.  This behaviour has been attributed to a continuously changing
carbon concentration ahead of the growth front, and to the formation of
pearlite with a higher average carbon content than that of the alloy.

## Acknowledgements

The author is grateful to Dr. S.A. Al-Salman, Denise Burgess and Dr.
John Chance for permission to use the results of unpublished work.

## References

1.    J.R. Villela, G.E. Guellich and E.C.Bain, Trans.ASM, 24 (1936)
      pp 225-52.

2.    R.F. Mehl and W.C. Hagel, Prog.Metal Phys., 6 (1956) pp. 74-134.

3.    Decomposition of Austenite by Diffusion Processes, V.F. Zackay
      and H.I. Aaronson, eds.; Interscience Publishers, N.Y., 1962.

4.    M.P. Puls and J.S. Kirkaldy, Metall.Trans., 3 (1972) pp. 2777-96.

5.    M. Hillert, in Chemical Metallurgy of Iron and Steel, pp. 241-47,
      Iron and Steel Institute, London, 1973.

6.    N. Ridley, in Heat Treatment '76, pp. 201-208, The Metals Society,
      London, 1976.

7.    M. Hillert, in Solid-Solid Phase Transformations, pp. 789-806,
      H.I. Aaronson, D.E. Laughlin, R.F. Sekerka and C.M. Wayman, eds.;
      TSM-AIME, Warrendale, PA., 1982.

8.    N. Ridley, in Symposium on "Establishment of Microstructural
      Spacing during Dendritic and Cooperative Growth", paper presented
      at TMS-AIME Annual Meeting, Atlanta, Ga, March, 1983.  To be
      published in Metall.Trans., 1984.

9.    N.F. Kennon, Metall.Trans.A, 9A (1978) pp. 57-66.

10.   N.F. Kennon and N.A. Kaye, Metall.Trans.A, 13A (1982) pp. 975-80.

11.   M. Hillert, in Decomposition of Austenite by Diffusional Processes,
      pp. 197-237, V.F. Zackay and H.I. Aaronson, eds., Interscience
      Publishers, N.Y., 1962.

12.   J.W. Cahn, Acta Met., 4 (1956) pp. 449-59.

13.   J.W. Cahn and W.C. Hagel, in Decomposition of Austenite by
      Diffusional Processes, pp. 131-92, V.F. Zackay and H.I. Aaronson,
      eds.; Interscience Publishers, N.Y., 1962.

14.   C. Zener, Trans.AIME, 167 (1946) pp. 550-83.

15.   M. Hillert, Jerkont.Ann., 141 (1957) pp. 757-89.

16.   K.A. Jackson and J.D.Hunt, Trans.TMS-AIME, 236 (1966) pp. 1129-42.

17.   W.A. Tiller in Liquid Metals and Solidification, pp. 276-318, ASM
      Cleveland, 1958.

18.   R. Trevedi, in Solid-Solid Phase Transformation, pp. 477-502,
      H.I. Aaronson, D.E. Laughlin, R.F. Sekerka and C.M. Wayman, eds.;
      TMS-AIME, Warrendale, PA., 1982.

19.   J.S. Kirkaldy, in Decomposition of Austenite by Diffusional
      Processes, pp. 39-123, V.F. Zackay and H.I. Aaronson, eds.;
      Interscience Publishers, N.Y., 1962.

20.   J.M. Shapiro and J.S. Kirkaldy, Acta Met., 16 (1968) pp. 579-85.

21.   B.E. Sundquist, Acta Met., 16 (1968) pp. 1413-27.

22.   B.E. Sundquist, Acta Met., 17 (1969) pp. 967-78.

23.   M. Hillert, in The Mechanism of Phase Transformations in Crystalline
      Solids, pp. 231-47, Institute of Metals, London, 1969.

24.   J.D. Hunt and K.A. Jackson, Trans.TMS-AIME, 236 (1966) pp. 843-856.

25.   F.M.A. Carpay and J. van den Boomgaard, Acta Met., 19 (1971)
      pp. 1279-86.

26.    J.D.Livingston, in In-Situ Composites, Vol. I, pp. 99-106, National
       Materials Advisory Board, NMAB-308-I, Lakeville, Conn., 1973.

27.    B.L.Bramfitt and A.R. Marder, Internat.Metallog.Soc.Proc., 1968,
       pp. 43-45.

28.    G.F. Bolling and R.H. Richman, Metall.Trans., 1 (1970), pp. 2095-
       2104.

29.    F.M.A. Carpay, Acta Met., 18 (1970), pp. 747-52.

30.    J.D. Livingston, J.Mat.Sci., 5 (1970) pp. 951-54.

31.    J.W. Cahn and W.C. Hagel, Acta Met., 11 (1963) pp. 561-74.

32.    D. Brown and N. Ridley, J.Iron Steel Inst., 204 (1966) pp. 811-16.

33.    B.L. Bramfitt and A.R. Marder, Metall.Trans., 4 (1973) pp. 2291-301.

34.    A.R. Marder and B.L. Bramfitt, Metall.Trans.A, 6A (1975) pp. 2009-14.

35.    A.R. Marder and B.L. Bramfitt, Metall.Trans.A, 7A (1976) pp. 902-905.

36.    G.E. Pellisier, M.F. Hawkes, W.A. Johnson and R.F. Mehl, Trans.ASM,
       30 (1942) pp. 983-1019.

37.    E.E. Underwood, Quantitative Stereology, p. 73; Adison-Wesley,
       New York, N.Y., 1970.

38.    J. Fridberg and M. Hillert, Acta Met., 18 (1970) pp. 1253-60.

39.    M. Mannerkoski, Acta Polytech.Scand., Chap. 26 (1964) pp. 1-63.

40.    A. Hultgren, K. Sven, Skapakad.Handl., 4 (1953) pp. 1-47.

41.    M.L. Picklesimer, D.L. McElroy, T.M. Kegley, E.E. Stansbury and
       J.H. Frye, Trans.TMS-AIME, 218 (1960) pp. 473-80.

42.    F.E. Bowman and R.M. Parke, Trans.ASM, 33 (1944) pp.481-493.

43.    N. Ridley and G.W. Lorimer, in Quantitative Microanalysis with High
       Spatial Resolution, pp. 80-84, The Metals Society, London, 1981.

44.    Al-Salman, G.W. Lorimer and N. Ridley, Metall.Trans.A., 10A (1979)
       pp. 1703-9.

45.    P.A. Beaven, M.K. Miller, and G.D.W. Smith, in Phase Transformations,
       Vol. 2, pp. 1.12-1.14, The Institution of Metallurgists, London, 1979

46.    M.K. Miller and G.D.W. Smith, Metal Sci., 11 (1977) pp. 249-53.

47.    P.R. Williams, M.K. Miller, P.A. Beavan, G.D.W. Smith, in Phase
       Transformations, Vol. 2, pp. 11.98-11.100, The Institution of
       Metallurgists, London, 1979.

48.    P.R. Williams, M.K. Miller and G.D.W. Smith, in Solid-Solid Phase
       Transformations, pp. 813-17, H.I. Aaronson, D.E. Laughlin, R.F.
       Sekerka and C.M. Wayman, eds.; TMS-AIME, Warrendale, PA, 1982.

49.    G. Cliff, University of Manchester, private communication.

50.    R.J. Dippenaar and R.W.K. Honeycombe, Proc.Roy.Soc.A, 333 (1973)
       455-67.

51.    C.S. Smith, Trans.ASM, 45(1953) pp. 533-75.

52.    L.S. Darken and R.M. Fisher, in Decomposition of Austenite by
       Diffusional Processes, pp. 249-88, V.F. Zackay and H.I. Aaronson,
       eds.; Interscience Publishers, N.Y., 1962.

53.    W. Pitsch, Acta Met., 10, (1962) pp. 79-80.

54.    K.W. Andrews and D.J. Dyson, Iron and Steel, 40 (1967) pp. 93-98.

55.    H.G. Bowden and P.M. Kelly, Acta Met., 15 (1967) pp. 105-111.

56.    D.N. Shackleton and P.M. Kelly, J.Iron Steel Inst., 207 (1969)
       pp. 1253-54.

57.    R. Morgan and B. Ralph, J.Iron Steel Inst., 206 (1968) pp. 1138-45.

58.    W. Pitsch, Arch.Eisenhutten, 34 (1963) pp. 381-90.

59.    F.C. Frank and K.E. Puttick, Acta Met., 4 (1956) pp. 206-210.

60.    D. Brown and N. Ridley, J.Iron Steel Inst., 207 (1969) pp. 1232-40.

61.    D. Cheetham and N. Ridley, J.Iron Steel Inst., 211 (1973) pp. 648-52.

62.    J.D. Verhoeven and D.D. Pearson, in Symposium on "Establishment of
       Microstructural Spacing during Dendritic and Cooperative Growth",
       paper presented at TMS-AIME Annual Meeting, Atlanta, Ga., March 1983.
       To be published in Metall.Trans., 1984.

63.    J. Williams, Ph.D. Thesis, University of Birmingham, England, 1969.

64.    J.J. Kramer, G.M. Pound and R.F. Mehl, Acta Met., 6 (1958) pp.763-71.

65.    H.O.K. Kirchner, B.G. Mellor and G.A. Chadwick, Acta Met., 26 (1978)
       pp. 1023-31.

66.    H.I. Aaronson, H.A. Domian and G.M. Pound, Trans.TMS-AIME, 236
       (1966) pp. 753-67.

67.    L. Kaufman, S.V. Radcliffe and M. Cohen, in Decomposition of
       Austenite by Diffusional Processes, pp. 313-52, V.F. Zackay and H.I.
       Aaronson, eds.; Interscience Publishers, N.Y. 1962.

68.    C. Wells, W. Batz and R.F. Mehl, Trans.AIME, 188 (1950) pp. 553-60.

69.    G.A. Chadwick and D.V. Edmonds, in Chemical Metallurgy of Iron and
       Steel, pp. 264-67, Iron and Steel Institute, London, 1973.

70.    M. Hillert, Acta Met., 19 (1971) pp. 769-778.

71.    G. Bolze, M.P. Puls and J.S. Kirkaldy, Acta Met.,20(1972) pp. 73-85.

72.    J.S. Kirkaldy, Can.J.Phys., 36 (1958) pp.907-16.

73.    R.C. Sharma, G.R. Purdy and J.S. Kirkaldy, Metall.Trans.A, 10A
       (1976) pp. 1119-27.

74.    R.C. Sharma, G.R. Purdy and J.S. Kirkaldy, Metall.Trans.A, 10A
       (1976) pp. 1129-39.

75.    N.A. Razik, G.W.Lorimer and N. Ridley, Acta Met., 22 (1974) pp.
       1249-58.

76.    K.W. Andrews, J.Iron Steel Inst., 203 (1965) pp. 721-27.

77.    J. Flugge, W. Heller, E. Stolte and W. Dahl, Arch.Eisenhuttenwes,
       47 (1976) pp. 635-40.

78.    M. Hillert, in Phase Transformations, pp. 181-228, ASM, Metals
       Park, 1970.

79.    N.A. Razik, G.W. Lorimer and N. Ridley, Metall.Trans.A, 7A (1976)
       pp. 209-14.

80.    John Chance and N. Ridley, Metall.Trans.A, 12A (1981) pp. 1205-13.

81.    J. Fridberg, L.E. Torndahl and M. Hillert, Jerkonterets Ann., 153
       (1969) pp. 263-74.

82.    R.W. Parcel and R.F. Mehl, Trans.AIME, 194 (1952) 771-80.

83.    J.R. Blanchard, R.M. Parke and A.J.Herzig, Trans.ASM, 31 (1943)
       pp. 849-68.

84.    C.R. Brooks and E.E. Stansbury, J.Iron Steel Inst., 203 (1965)
       pp. 514-523.

85.    S.A. Al-Salman, G.W.Lorimer and N. Ridley, Acta Met., 27 (1979)
       pp. 1391-1400.

86.    B.G. Mellor and D.V. Edmonds, Metall.Trans.A, 8A (1979) pp. 763-71.

87.    B.G. Mellor and D.V. Edmonds, Metall.Trans.A, 8A (1977) pp. 773-82.

88.    N. Ridley and D. Burgess, Metal Sci., 18 (1984) pp. 7-11.

89.    M.F. Hawkes and R.F.Mehl, Trans.AIME, 172 (1947) pp. 467-72.

90.    D. Cheetham and N. Ridley, Metall.Trans., 4 (1974) pp. 2549-56.

91.    J.W. Cahn and R.L. Fullman, Trans.AIME, 206 (1956) pp. 610-617.

92.    A. Roosz, Z. Gacsi and M.K. Baan, Metallography, 13 (1980)
       pp. 299-306.

93.    A. Roosz and Z. Gacsi, Metallography, 14 (1981) pp. 129-39.

94.    J.M. Hyzack and I.M. Bernstein, Metall.Trans.A., 7A (1976)
       pp. 1217-1224.

95.    A.R. Marder and B.L. Bramfitt,Metall.Trans.A., 7A (1976) pp. 365-372.

96.    G.W. Rathenau and G. Baas, Acta Met. 2 (1954) pp. 875-83.

97.  J.H. Frye, Jr., E.E. Stansbury and D.L. McElroy, Trans.AIME, 150 (1942) pp.185-207.

98.  F.C. Hull, R.A. Colton and R.F. Mehl, Trans.AIME, 150 (1942) pp.185-207.

99.  D.E. Coates, Metall.Trans., 4 (1973) 2313-25.

100. K.R. Kinsman and H.I. Aaronson in Transformation and Hardenability in Steels, pp. 39-53, Climax Molybdenum Co., Ann Arbor, 1967.

101. J.R. Bradley, G.J. Shiflet and H.I. Aaronson, in Solid-Solid Phase Transformations, pp. 819-24, H.I. Aaronson, D.E. Laughlin, R.F. Sekerka and C.M. Wayman, eds., TMS-AIME, Warrendale, PA., 1982.

102. R.C. Sharma and G.R. Purdy, Metall.Trans. 4 (1973)pp. 2302-11.

# OBSERVATIONS ON CURVATURE OF THE
# PEARLITIC FERRITE:CEMENTITE INTERFACE IN AN
# Fe-0.80%C STEEL

S.A. Hackney and G. J. Shiflet

Department of Materials Science

University of Virginia, Charlottesville, VA    22901

Curvature of cementite lamellae within a pearlite colony is examined with transmission electron microscopy in a eutectoid steel.  It is shown that although curvature may _appear_ to be associated with a change in habit plane or even a disordered-like ferrite:cementite interface, in truth it accomplishes this by discrete steps, the broad faces of which maintain the habit plane.

## Introduction

Curvature of lamellae within a pearlite colony is a common observation with optical and electron microscopy in both high purity Fe-C steels and commercial alloys. Although strain effects following transformation have been invoked (1), Mehl's (2) suggestion that curvature occurs during the growth process is now generally accepted. Even though some type of low energy ferrite:cementite (α:c) lamellae interface is often reported in pearlite (3-11) with orientation relationships of either the Bagaryatskii (3), Isaichev (7), or second Petch (5), the presence of curved lamellae has been interpreted by Hillert (12) and Puls and Kirkaldy (13) as an indication that the α:c interface is crystallographically insensitive. Other investigators, however, have suggested that lamellae curvature may occur by a mechanism deferential to a α:c crystallographic relationship. Ohmari, Davenport, and Honeycombe (11) have suggested that the α:c orientation relationship could alternate from the Bagaryatskii to the Isaichev within the same pearlite colony (They differ by an angle of $3.58°$ about [010]c (7)). This would allow for the presence of a series of good fit habit planes parallel to [010]c. This situation could then give rise to a "corrugated habit plane that consisted of small alternate facets comprising these planes . . ." (11). This would allow a change in the habit plane while maintaining a relatively low energy interface.

In observing a surface replica with scanning electron microscopy, Bramfitt and Marder (14, 15) interpreted striated cementite lamellae as being associated with a direction change during growth. They termed these striations "growth steps" and suggested that they could allow the growth direction to change without a change in the crystal orientation.

In these previous studies, however, no attempt to investigate the actual α:c interface of lamellae which had changed habit planes during the growth process was made. When describing the α:c interface, the important point which needs attention was clearly stated by Hillert in his classic 1962 paper on the pearlite transformation (12): ". . . the interesting question is not really whether any orientation relationship exists (or usually exists) between ferrite and cementite, but rather whether such a relationship is of any importance for the development or growth of pearlite." From the experimental evidence available at that time, Hillert concluded ". . . such a relationship is relatively unimportant." However, with the advent of higher resolution TEM techniques, it may be seen that the development of the morphology of the individual lamella is highly sensitive to the crystallographic relationship between the α:c lamellae.

## Experimental Procedure

High purity Fe-0.80%C specimens (kindly provided by Prof. H.I. Aaronson) 10 x 20 x .7 mm were austenitized for 30 min at $1100°C$ in a dynamic Ar atmosphere. They were then quickly transferred to a salt bath at $645°C$ and isothermally heat treated for 12 sec followed by quenching in iced brine. Discs for transmission electron microscopy (TEM) were chemically thinned (16) and polished in a solution of 100 g $NaCrO_4$ dissolved in 500 ml of glacial acetic acid at 60 V and $20°C$ using a twin jet polisher. Examination of thinned discs was performed with a Philips EM 400T electron microscope.

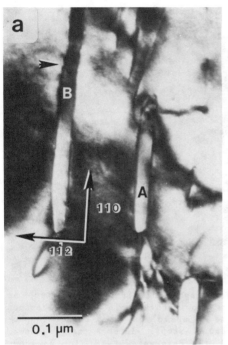

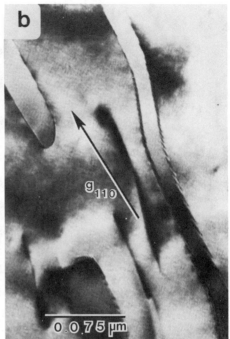

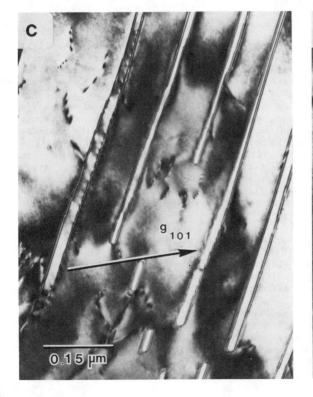

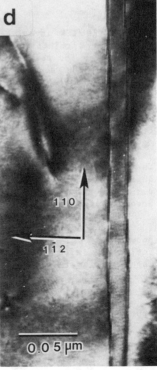

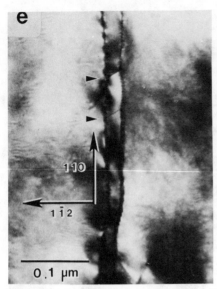
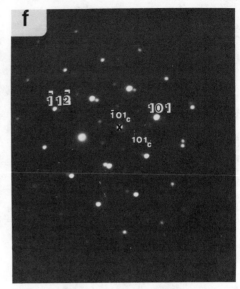

Figure 1 – Transmission electron micrographs illustrating curvature in pearlite lamellae. What might appear as smooth changes in growth direction (a, b, c, and e) and lamellae thickness (c and d) is, upon closer inspection, instead comprised of small interfacial steps. In each example shown, the beam direction is near [$\bar{1}$11]$_\alpha$ with the Bagaryatskii (3) (or Isaichev (7)) orientation relationship predominating (f).

## Results and Discussion

In the TEM investigation of pearlite colonies, it was observed that the abrupt termination (or beginning) of an individual cementite lamella often forces the neighboring lamellae to rapidly readjust the interlamellar spacing. This usually results in a significant change in the <u>apparent</u> habit plane and the appearance of curvature at the $\alpha$:<u>c</u> interface. Examples of this are shown in Fig. 1. In Fig. 1a the termination of cementite lamella A results in the shifting over of lamella B. Note that the change in <u>apparent</u> habit plane does not occur smoothly, but rather by the presence of interfacial steps. From the micrograph, it is clear that lamella B deviates substantially from the ($1\bar{1}2$)$_\alpha$ plane; however, closer inspection reveals that the change in direction$_\alpha$ is accomplished by means of interfacial steps approximately 25 Å in height with a spacing dependent on the angle by which the apparent habit plane varies from ($1\bar{1}2$)$_\alpha$. As may be discerned from Fig. 1b, considerable variation in apparent habit plane may be accomodated by this mechanism.

Variations in the thickness of cementite lamellae which would necessarily cause a change in the apparent habit plane can also occur in conjunction with interfacial steps, as shown in Fig. 1c and at a higher magnification in Fig. 1d. Readjustment of spacing and lamellae thickness seemingly is a continuous process during growth.

Although the cementite lamellae are often faulted (17), there does not appear to be any crystallographic incongruities associated with the steps in Figs. 1a, b, or c. This would imply that the steps are purely structural, thereby ruling out the mechanism outlined by Ohmori et al. (13)

in these instances. However as shown in the micrograph in Fig. 1e, there are faults in some of the cementite lamellae which may be associated with changes in the α:c habit plane.

## Conclusions

The manner by which pearlite maintains the α:c habit plane during perturbations in lamellar spacing and thickness strongly implies that the development and growth of individual lamellae within the pearlite colony is highly sensitive to the crystallographic relationship which exists between the ferrite and cementite lamellae.

## Acknowledgements

This research is based upon work supported by the National Science Foundation under Grant DMR-83-00888. The authors also wish to thank the Department of Materials Science and Dr. Kenneth R. Lawless for supporting this project in its early stages.

## References

1. N.T. Belaiew, JISI, No. 1 (1922) 201.

2. F.C. Hull and R.F. Mehl, Trans. ASM, 30 (1942) 381.

3. Y.A. Bagaryatskii, Dokl. Akad. Nauk. S.S.S.R., 73 (1950) 1161.

4. W. Pitsch, Acta Met., 10 (1962) 79, errata p. 906.

5. N.J. Petch, Acta Crystallog., 6 (1953) 96.

6. K.W. Andrews and D.J. Dyson, Iron Steel Lond., 40 (1967) 93.

7. I.V. Isaichev, Zhur. Tekhn. Fiziki, 17 (1947) 835.

8. D.F. Lupton and D.H. Warrington, Acta Met., 20 (1972) 1325.

9. H.G. Bowden and P.M. Kelly, Acta Met., 15 (1967) 105.

10. R.J. Dippenaar, and R.W.K. Honeycombe, Proc. Roy. Soc. Lond. A, 333 (1973) 455.

11. Y. Ohmori, A.T. Davenport, and R.W.K. Honeycombe, Trans. Iron and Steel Inst. Japan, 12 (1972) 128.

12. M. Hillert, Decomposition of Austenite by Diffusional Processes, Interscience, New York, (1962) 197.

13. M.P. Puls and J.S. Kirkaldy, Met. Trans., 3 (1972) 2777.

14. B.L. Bramfitt and A.R. Marder, Metallog. 6 (1973) 483.

15. B.L. Bramfitt and A.R. Marder, IMS Proc. (1968) 43.

16. M.R. Plichta, H.I. Aaronson, and W.F. Lange, Metallog. 9, (1976) 455.

17. L.S. Darken and R.M. Fisher, Decomposition of Austenite by Diffusional Processes, Interscience, New York (1962) 249.

# MICROSTRUCTURAL CONTROL OF FLOW AND FRACTURE IN PEARLITIC STEELS

D.J. Alexander and I.M. Bernstein

Department of Metallurgical Engineering
and Materials Science
Carnegie-Mellon University
Pittsburgh, PA 15213

## Introduction

The mechanical properties of pearlite have been studied for many years by numerous investigators. Although the effects of pearlite on the mechanical properties of low alloy steels have been recently reviewed (1), much of the information about pearlite is widely scattered in the literature, and there is still considerable debate about the character and details of microstructural influences on flow and fracture. It is intended that this review will clarify some of these issues, and areas where research is needed can be identified. Some of the confusion about microstructural effects has arisen from attempts to extend the behavior of ferrite-pearlite aggregates to fully pearlitic steels; therefore, this article is essentially limited to a discussion of the monotonic room temperature properties, at small to medium strains, of fully eutectoid pearlitic steels, with the goal of clarifying and integrating how microstructure controls mechanical properties.

## Yield Strength

While it is well established that the yield strength ($\sigma_y$) of pearlite increases as the pearlite interlamellar spacing (S) decreases (2-21), the exact form of the relationship between $\sigma_y$ and S is uncertain. Earlier investigators (2-5) found $\sigma_y$ proportional to the inverse logarithm of the mean free path, approximately twice the ferrite lamellae thickness. More recent studies have suggested that the dependence follows a modified Hall-Petch type relation, with $\sigma_y = \sigma_o + kS^x$, where x can be graphically fitted to either $-\frac{1}{2}$ (6-14) or $-1^y$ (8-11), over the range of the interlamellar spacings experimentally achievable (70 to 600 nm).

The $S^{-\frac{1}{2}}$ dependence results in a negative value of $\sigma_o$, the "friction stress"(see Fig. 1), which evidently cannot be justified physically (8-13) and, thus, is considered the less likely dependence. The $S^{-1}$ dependence yields a positive "friction stress", with values in reasonable agreement with the friction stress for pure iron (8-11). Possible models for $S^{-1}$ type behavior have been developed by Gil Sevillano (11) and by Langford and Cohen (22,27). However, the microstructural path defining long-range

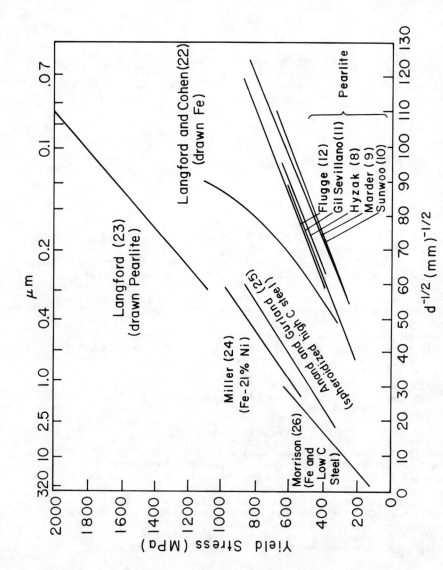

Fig. 1.  Hall-Petch plot for several iron-based alloys (8-12, 22-26).

dislocation movement in the ferrite is not apparent, and while it is
appealing to believe that similar processes control dislocation motion
in pearlite and in pure iron (28), this neglects any contribution from
the cementite except as a barrier which appears unlikely.  Further as
Fig. 1 illustrates, the yield strength and the Hall-Petch slope
for pearlite are considerably less than other iron-based alloys with similar
microstructural dimensions implying weaker dislocation barriers.  It is
clear that yielding in pearlite is far from understood.

## Work Hardening

The flow stress of pearlite greatly increases with strain suggesting
a rapid companion increase in the mobile dislocation density.  It has
been considered that the ferrite-cementite interface acts as a source of
dislocations (14,29,30), or that dislocations are generated within the
ferrite lamella (10,11); however, such sources have not yet been observed.

There is considerable uncertainty as to how microstructure affects
the work hardening behavior of pearlite.  Generally, work hardening is
evaluated by fitting tensile data to an equation of the type $\sigma_y = K\epsilon^n$,
where n is the work hardening exponent.  While it has been reported that
n increases as S increases (10,31,32), this result is somewhat artificial,
since it is a graphical consequence of the lower $\sigma_y$ values, which result
in higher values of n.  If work hardening is instead measured as $d\sigma/d\epsilon$,
then it appears to be independent of S (2,3,10,16,31) although it has
been reported that $d\sigma/d\epsilon$ increases as S decreases (7,14).

The apparent lack of a strong effect of S on the work hardening rate
is quite intriguing, particularly since, as we discuss below, the slip
distribution can be quite different for coarse and fine pearlite spacings.
Most work hardening models, which are based on the increasing dislocation
density with strain (33), predict parabolic stress-strain curves, with
$\sigma$  proportional to  $\epsilon^{\frac{1}{2}}$, and $d\sigma/d\epsilon$ proportional to $S^{-\frac{1}{2}}$, neither of which is
observed.  Ashby's model (34,35) for plate-like particles predicts
similar dependences.  Gil Sevillano (11) suggests that the flow stress
depends on dislocation multiplication within the ferrite plus load
carrying by the cementite plates.  However, this model appears to be more
applicable to large strains more typical of wire drawing.  Tanaka and
Matsuoka (36) have developed a model which is insensitive to the pearlite
spacing, but should apply for very small strains only.

There are several other possible means for rationalizing an increasing
flow stress of pearlite.  While the cementite may remain elastic during
the initial yielding of pearlite, it must become plastic at higher strains.
It is likely that dislocations in the ferrite will either enter into the
cementite, or into the cementite-ferrite interface.  This will change
the nature of this boundary, and perhaps create a slip offset in the
cementite.  The creation of new interfacial area could increase the
flow stress (37), as would changes in the interfacial boundary structure.
A misfit dislocation (38,39) would be produced due to the different
Burgers vectors of cementite and ferrite if slip proceeds into the
cementite, and if the slip planes are not continuous, as is probably the
case, a sessile jog would be created at the interface (39).  These and
other possible mechanisms should be considered.

The increase in flow stress following large strains, as in wire
drawing (23,40-42) is quite substantial (Fig. 1).  It is interesting to
note that although the microstructural dimensions are not altered
markedly, the flow stress rises very dramatically, and the microstructural dependence is changed also, which suggests that the dislocation

barriers have become much more effective obstacles as a result of the prior strain.

It seems clear that our understanding of work hardening in pearlite is incomplete. While a simple work hardening - dislocation density model is insufficient, there are at present no clearly superior alternatives. Further study is certainly warranted.

## Ductility

The reduction of area (RA) during a tensile test is a common measure of ductility, as is the elongation to fracture; for pearlite, both these parameters display similar trends with microstructure, but since changes in elongation are invariably small, we will mainly consider effects of microstructure on RA.

The room temperature ductility depends on both the prior austenite grain size ($\gamma$ g.s.) and the pearlite spacing, increasing as both are refined (2-4,8,9,15,17,21) (Fig. 2). While it may appear surprising at first that the $\gamma$g.s. has an effect, since the austenite is no longer present, this arises from the austenite controlling the cleavage fracture facet size through an orientation dependence, and hence, the effective grain size for pearlite. This will be discussed further in the section on toughness.

The room temperature RA depends on competition between the ductile microvoid coalescence (MVC) process, controlled by the pearlite spacing, and brittle cleavage, related to the austenite grain size. Fracture invariably begins in the center of the specimen by MVC (17,32,43-45), with final failure occurring by cleavage. In general, the RA is found to be primarily controlled by the $\gamma$g.s., increasing as the grain size decreases (4,8,9,17,21). The smaller effective grain size delays the final cleavage fracture, giving larger RA values, as does a finer pearlite spacing (2-4, 8,9,15,17,21).

We can focus on the local stress and strain response to characterize the way the pearlite structure fails (Fig. 3). For example, many investigators have shown that coarse and fine pearlites behave differently (10,17,30,31,46), the former deforming in a less homogeneous fashion, leading in some cases to localized deformation and intense shear bands (Fig. 3a), which are associated with growth faults in the pearlite (10,14,46-48). A truncated cementite lamella can be an effective stress concentration. The resulting slip can cause an offset in the cementite (49,50) and thus propagate across the colony, with the shear largest near the center, and decreasing toward the colony edge (46,50). Eventually, the cementite cracks, due to the combination of fiber loading (36,51,52), the impinging slip band (46,52,53), and the stress concentration due to the slip offset. The resulting voids which form allow further slip in the ferrite, which also eventually fails - the voids then link up. This process is a modification of that proposed by Miller and Smith (51), called shear cracking.

On the other hand, fine pearlite tends to deform in a homogeneous manner with little tendency for shear cracking (7,46). Instead, the cementite thins, and necks down into small fragments (Fig. 3b). Few voids form and failure occurs by void growth in the ferrite through gaps in the fragmented cementite. The resultant dimples will be much smaller than in coarse pearlite. Thus, fine pearlite, by forming fine dimples and avoiding shear cracking, can attain higher levels of ductility.

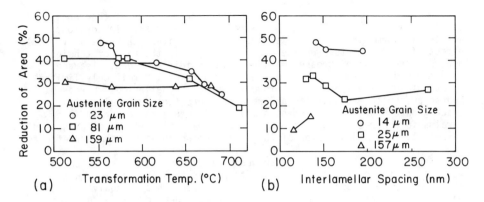

Fig. 2.    Reduction of area at room temperature.    (a) High purity Fe-.81C [9].
            (b) Commercial 1080 [8].

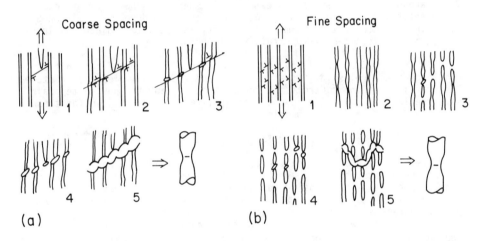

Fig. 3.    Microvoid coalescence in pearlite. (a) Coarse interlamellar spacing.
            (b) Fine interlamellar spacing.

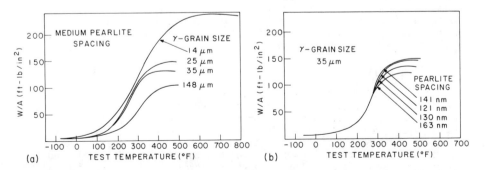

Fig. 4.    Effect of microstructure on precracked Charpy transition curves [8].
            (a) Austenite grain size effect.   (b) Pearlite spacing effect.

Once a central void has formed in the necked region of the tensile specimen, stresses are raised in the remaining material. In torsion, where void formation and tensile stresses are limited, there is very little effect of microstructure on the ductility (5). In the case of tension, however, the tensile stresses will eventually initiate local cleavage. As discussed previously, small austenite grains will permit only limited growth of the cleavage crack, before the crack is deflected, whereas coarse austenite grains, with a larger fracture facet will allow easy crack growth over longer distances. Thus, fracture can occur at lower strains for coarse grained material. This grain size effect is discussed more fully in the discussion of toughness for which it has similar effects.

Several investigators have suggested that ductility is related more strongly to the cementite thickness (t) rather than the spacing, with RA increasing as t decreases (7,12,13). However, this is equivalent to a decrease in S, since the two cannot be varied independently at a given composition. In this regard, the investigation of Houin, et al. (18-20) is quite interesting. They varied the carbon content and cooling rate during transformation to achieve fully pearlitic structures from 0.6 to 0.8% carbon. Their results suggest the RA depends on the volume fraction of cementite, rather than S or t, but the elongation depends mainly on S. Other investigators report a maxima in ductility at intermediate spacings (2-4), but it is not clear why this should occur.

It appears that ductility can be understood in terms of the initial growth of a central void in the specimen, and then final cleavage fracture. The pearlite spacing controls the former process, and the austenite grain size the latter. While we have a good qualitative understanding of these processes, no quantitative models yet exist.

## Toughness

There is also dispute in the literature concerning microstructural effects on the toughness of pearlite. While some of the confusion arises from attempts to extend the trends observed for low-carbon steels to pearlite, even studies of eutectoid steels show little agreement. Two primary parameters control the toughness of pearlite. These are the pearlite spacing, which controls $\sigma_y$, the cleavage fracture stress ($\sigma_{fc}$), and the ductile dimple spacing; and the prior austenite grain size, which controls the cleavage fracture facet size. The cleavage fracture stress is the tensile stress which will cause cleavage fracture. The competition between the ductile and brittle fracture modes can determine the resultant toughness for pearlite. In the following sections we will consider how microstructure exercises a varying influence on different measures of toughness. It is worth noting that intergranular failure is not usually observed in pearlitic steels, so we shall consider transgranular modes of fracture only.

## Transition Temperature

The traditional method for measuring toughness is to determine the ductile to brittle transition temperature (DBTT) using the Charpy impact test. The two most complete sets of data available, those of Gross and Stout (4,21), and of Rinebolt (15), reach almost exactly opposite conclusions. Rinebolt found that as the pearlite spacing increased, the DBTT increased or was constant (depending on how the DBTT was defined), and the upper shelf energy and the absolute energy levels in the transition range decreased. On the other hand, Gross & Stout show a decrease in the DBTT, and an increase of the upper shelf and the transition range energy levels

as the spacing increased.  Both investigations showed little effect of
S on lower shelf levels.  Gross and Stout also show an increase in the
DBTT as the  γg.s. increased, a parameter which Rinebolt did not investigate.  Flugge, et al.(12,13), using a different definition for the DBTT,
showed that it increased with increasing cementite thickness (i.e.,
pearlite spacing), and γg.s.

A recent investigation (8) on the role of interlamellar spacing,
austenite grain size, and pearlite colony size on the transition temperature used precracked instrumented Charpy impact testing.  The data
was analyzed as W/A, the energy absorbed per unit fracture area, to
allow for varying precrack depths.  This investigation showed that the
DBTT (as measured by a particular energy absorbed per unit area) increased
as the  γg.s. increased, largely independent of S and colony size (Fig. 4).
A similar study with hypoeutectoid steels (54) showed a more pronounced
grain size effect, although the effect of the proeutectoid ferrite must
be considered, and the pearlite spacing was varied only slightly.

The most persuasive evidence suggests that the DBTT is controlled by
the  γg.s., increasing as the g.s. increases.  This dependence likely
occurs because the  γg.s. determines the effective grain size for pearlite
which is the fundamental fracture unit.  Fractographic evidence (55-59)
shows cleavage facets readily cross pearlite colony and austenite grain
boundaries, which led to the suggestion that adjacent colonies could
have similar ferrite orientations.  This was later confirmed by TEM
studies (44,57).  This unit of common ferrite orientation would be determined by a common prior austenite parent grain, from which the pearlite
transformed.  This common parentage would result in regions through
which a cleavage crack could easily pass with little change of direction
and hence, little loss of energy.  Fine austenite grains would produce
smaller regions of common orientation and thus reduce the effective grain
and fracture facet sizes, increasing the amount of crack deflection,
leading to more energy absorption during impact tests.  The measured
fracture facet size was indeed found to correlate well with the austenite
grain size (54,57,58) although it was slightly smaller.

We can thus rationalize the role of microstructure on impact energy
and DBTT as follows:  At low temperatures, and high strain rates, the yield
stress is elevated, and thus the cleavage fracture stress is easily reached,
resulting in brittle cleavage fracture with an energy determined by the
resistance to unimpeded crack growth.  As the temperature is increased
into the transition range, the yield stress drops, and eventually fracture
will initiate by a ductile mechanism, until work hardening and crack
tearing can elevate the stresses to high enough levels for cleavage.  When
cleavage is then initiated locally, a coarse austenite grain size would
allow the crack to grow over larger distances, increasing the chance of
failure, while a finer grain size would result in more deflections of the
crack, making it more likely to be arrested and delaying failure until
more energy contributes to crack growth.  This sequence of events would
predict an increase in the transition temperature with increasing γg.s.,
as is observed.

To a certain extent, the discussion of microstructural effects on
DBTT is purely academic.  First, the changes in DBTT through variations
in microstructure are relatively small (less than $50^{\circ}C$ (4,8,15,21)).
Indeed, changing the definition of the DBTT may even change the trends
observed (15).  Also, the difference between the upper and lower shelf
is small, typically less than 30 ft.-lb. (4,15,21).  Finally the different

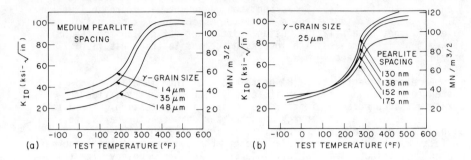

Fig. 5.   Effect of microstructure on dynamic fracture toughness.   (a) Austenite grain size effect (8).   (b) Pearlite spacing effect (61).

test techniques used (blunt notch versus sharp crack) may perhaps show
different material responses as has been observed for other steels (60).

Fracture Toughness

There is little systematic information available concerning the
effect of microstructre on the fracture toughness of pearlite.  A
fairly complete study (8,61) used instrumented precracked Charpy impact
testing to determine the dynamic fracture toughness $K_{ID}$, over a range of
microstructures and temperatures (Fig. 5).  At low temperatures, below
the DBTT, $K_{ID}$ was found to primarily depend on the  γg.s., increasing
as the grain size decreases, while at higher temperatures, it depends
on the pearlite spacing only, increasing as S decreases.  There is also
some evidence that the toughness drops as S increases at lower temperatures,
but is not clear if this is a general trend.

This dynamic behavior can be understood as follows:  at low temperatures, fracture occurs by cleavage, and the effective grain size (i.e.,
the austenite grain size) should control the process.  The high strain
rate and low temperatures elevate the yield stress, so the cleavage
fracture stress is readily exceeded.  The toughness value attained
depends on the amount of crack deflection, so $K_{ID}$ increases as the
effective grain size is reduced.  At higher temperatures, the fracture
now occurs primarily by a ductile MVC process.  Smaller pearlite spacings
result in smaller voids while coarse pearlite tends to develop shear
cracks and large voids, thus lowering its toughness.  No grain size
effect is observed, as expected, since the fracture process is ductile.

A study presently in progress (32) is extending the previous work
to measure the fracture toughness $K_{IC}$ at room temperature, for experimentally obtainable extremes in austenite grain size and pearlite spacing.
One primary purpose is to establish if the aforementioned microstructural
control under high strain rate conditions can be extended to "static"
conditions.  The results to date are not supportive of an extrapolation
of the high strain rate results; now only the combination of coarse
grain size and coarse spacing appears significantly tougher than the
other microstructures (Table I).  Frequent pop-ins which occur during
testing (also observed by Tetelman et al. (62) for a pearlitic steel)
complicate the analysis of the data, but there is little doubt that $K_{IC}$

TABLE I.  Effects of Microstructure on the Room Temperature
Fracture Toughness $K_{1C}$  (ksi $\sqrt{in}$)

| Austenite Grain Size ($\mu$m) / Pearlite Spacing ($\mu$m) | .13 | .32 |
|---|---|---|
| 30 | 46 | 43 |
| 190 | 42 | 56 |

and $K_{ID}$ show different microstructural sensitivities.

It appears that the toughness level depends on, and can be modelled as a complex interaction between the cleavage facet size, determined by the prior austenite grain size, and the plastic zone size, determined by the yield strength, and hence pearlite spacing.  It is believed that the events leading up to the attainment of $K_{IC}$ are as follows: crack front splitting occurs during the initial portion of localized crack growth in differently oriented fracture facets, creating ligaments between adjacent regions of the crack front which are no longer coplanar. It is the subsequent fracture of these ligaments that determines the fracture toughness, now controlled by a strong link rather than a weakest link fracture process.  The coarse grain size tends to allow large deflections of the crack as cleavage occurs, and if the accompanying yield strength is low, a large plastic zone is formed, so regions far from the crack plane may be sampled.  Microcracking can thus develop well away from the crack plane, resulting in large ligaments.  Conversely, microstructures with smaller grains will allow less crack deflection, as the facet size will be smaller, and at higher yield strengths smaller plastic zones will limit the distance from the crack plane over which ligaments can develop.  This will reduce the toughness, as is observed. Although this description is self-consistent much more work is needed to determine the applicability and generality of this model which is a significant departure from the classical approach of fracture initiation and growth from the most sensitive (most brittle) microstructural centers.

There are thus clear differences between microstructural effects observed in the dynamic $K_{ID}$ tests and the slow strain rate $K_{IC}$ tests. These are due both to differences in the test conditions, which for pearlite produce seemingly different responses, and to different roles of microstructure in attaining the conditions for unstable fracture under the different test conditons.  For example, the high strain rate in the impact test raises $\sigma_y$ and thus reduces the ratio $\sigma_{fc}/\sigma_y$.  This should make microcracking easier.  Indeed, Henry (63) has observed increased microcracking in tensile specimens as the strain rate is increased.  Lowering the temperature will also increase $\sigma_y$, and has been observed to result in increased microcracking as well (43).  The higher dynamic yield stress will also result in a greatly reduced plastic zone size, so crack deviation will be limited.

In addition, the pop-ins observed in slow strain rate tests are not seen in impact testing. The high strain rate apparently does not allow the pop-ins, and the subsequent relaxation which pop-ins would produce, which will also affect the resultant toughness. The different test specimens (Charpy and compact) may also have some effect, through geometric size, constraint, or loading conditions.

A parallel but slightly different approach is to consider brittle fracture to be controlled by the peak tensile stress at a particular location in the microstructure. This is usually modelled using the Ritchie-Knott-Rice (RKR) model (64), which suggests that fracture will occur when the maximum tensile stress exceeds the material's cleavage fracture stress over some critical distance $X_o$. Studies show that the cleavage fracture stress for pearlite (31,32,65,66) increases as S decreases, in a manner similar to the yield strength behavior. It appears to be independent of temperature, although it may decrease as the temperature rises (31,66).

The critical distance is usually assumed to be related in some way to the microstructure. While there is no information for this parameter available in the literature for pearlite, it would be reasonable to expect that the critical distance is related to the effective grain size, or the austenite grain size. Recent work (32) shows that for coarse grained austenite, the critical distance is equal to one or two austenite grains; however, as the grain size is reduced, the critical distance becomes equal to about eight austenite grains. Similar behavior has been observed by Curry & Knott (67) who found the critical distance decreased and then became constant as the grain size was reduced in a low carbon steel.

Using this model, it is possible to predict the general trend one might expect as a result of changing the microstructure. Fracture depends on the ratio $\sigma_{fc}/\sigma_y$, which is controlled by the pearlite spacing, and the critical distance $X_0/(K/\sigma_y)^2$, which could depend on the austenite grain size, and the pearlite spacing also. For a fixed pearlite spacing, and thus $\sigma_y$ and $\sigma_{fc}/\sigma_y$, an increase in the grain size should increase the critical distance $X_0$, and thus should increase K, since $X_0/(K/\sigma_y)^2$ is constant. Conversely, for a fixed grain size, and thus constant $X_0$, the toughness will depend on $\sigma_y$ and $\sigma_{fc}/\sigma_y$. For the particular ratio of $\sigma_{fc}/\sigma_y$ found in pearlite (32), the resultant changes in expected toughness are quite small, only a few percent. Thus, using this model, one would expect toughness to increase with increasing grain size, and to be essentially independent of pearlite spacing. While this is observed to some extent in the $K_{IC}$ results mentioned earlier, this approach is clearly incapable of rationalizing all the observed microstructural dependences of $K_{IC}$ and certainly $K_{ID}$.

The toughness of pearlite appears to depend on both the plastic zone size (as determined by the yield strength), and the fracture facet size, which are controlled by the pearlite spacing and the prior austenite grain size, respectively. The relative importance of these microstructural parameters differ for different toughness parameters. However, it is clear that much further effort is needed to determine the full role of microstructure on the toughness of pearlitic steel.

## Summary

Table II shows how the microstructure affects the mechanical properties of pearlite. The trends shown are fairly well established for the more basic properties such as yield strength, ductility, and transition temperature, but the fracture process is less clear. In all cases, successful quantitative models of the microstructural effects remain to be established.

TABLE II.   Effects of Microstructural Refinement on the
            Mechanical Properties*

| | $\sigma_y$ | n | $\dfrac{d\sigma}{d\varepsilon}$ | RA | DBTT | $K_{1D}$ Low T | High T | $K_{1C}$ |
|---|---|---|---|---|---|---|---|---|
| Inter-lamellar Spacing | ↑ | ↓ | ↔ | ↑ | ↔ | ↔ | ↑ | ↔ ↓ |
| Austenite Grain Size | ↔ | ↔ | ↔ | ↑ | ↓ | ↑ | ↔ | ↔ ↓ |

*Room Temperature (unless noted otherwise)
↑ Increase    ↓ Decrease    ↔ No effect

By way of a summary, let us consider the behavior of different test specimens as the load is increased until fracture occurs (Fig. 6).  The resultant properties will depend on the specimen type and testing mode, but clear microstructural effects can nonetheless be seen.

A smooth tensile specimen will begin to yield when the stress is high enough to move dislocations long distances or many dislocations short distances through the pearlitic ferrite.  The dislocation density will then rapidly increase, either through dislocation multiplication from sources within the ferrite or by emission from sources in the ferrite-cementite interface, and most of these dislocations pile up at the ferrite-cementite interfaces.  For coarse pearlite, a truncated cementite lamella at a growth fault triggers localized deformation, with the resulting slip band impinging on adjacent cementite lamellae, producing an offset.  It thus propagates across the colony, and eventually the cementite will fracture, producing a void and allowing increased shear in the ferrite.  Eventually, the ferrite's ductility is exhausted leading to link-up of adjoining voids, producing a shear crack.  Fine pearlite continues to

**Smooth Specimen**

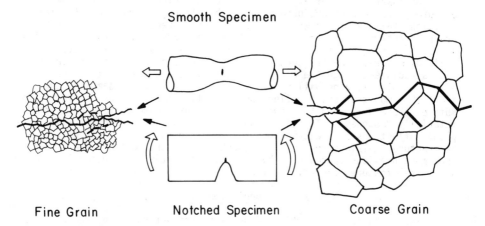

Fine Grain          Notched Specimen          Coarse Grain

Fig. 6.  Fracture processes in different microstructures.

deform in a more homogeneous manner to greater strains with much less
tendency to develop shear bands. Instead, the cementite thins down,
and fragments into small pieces. Eventually, voids form in the ferrite,
and link up through gaps in the fragmented cementite lamellae.

The result of these processes is a central void in the neck of a
tensile specimen. The stresses in the remaining material are then raised
until cleavage is triggered. Fine-grained material will result in more
limited cleavage crack growth, as the crack will encounter differently
oriented ferrite and be arrested. Coarse-grained material will allow a
larger increment of crack growth, further increasing the stresses on the
remaining material, and final cleavage fracture will occur at lower strains.

For a specimen containing either a blunt notch or a sharp crack
(Fig. 6), yielding will begin at the notch tip. For the blunt notch, a
broad gradient of stress develops, while the sharp crack will develop
a steep stress gradient ahead of it, with the peak stresses dependent
on the yield stress and the work hardening exponent. The ease and type
of fracture will depend on the ratio of the cleavage fracture stress to the
yield stress, which is influenced by the temperature and the strain rate.

If the ratio of $\sigma_{fc}/\sigma_y$ is large, as for a low yield stress case
(coarse spacing, slow strain rate, high temperature), failure will occur
by microvoid coalescence. The smaller dimples associated with a finer
pearlite spacing will absorb more energy, and the pearlite spacing will
be the controlling microstructural feature.

At intermediate ranges of the ratio, the ductile and brittle modes
of fracture will compete. A sharper crack will raise the local stresses,
favoring cleavage; a blunt crack will limit the stress elevation, favoring
ductile failure. If MVC is initiated, the crack is sharpened and with
work-hardening the stresses are increased, promoting brittle cleavage.
When cleavage is initiated, a fine austenite grain size will result in
more frequent crack deflections, making crack arrest more likely. A
coarse austenite grain size will allow larger crack advances, allowing the
crack to accelerate, and promote brittle fracture. Thus, the DBTT is
increased by an increase in the austenite grain size.

At lower values of $\sigma_{fc}/\sigma_y$ (fine spacing, high strain rate, low temperature), fracture occurs predominantly by cleavage. For a blunt notched
specimen, failure occurs when the peak stresses reach $\sigma_{fc}$. For a sharp
crack, fracture occurs when the peak stresses exceed the cleavage fracture
stress over some critical distance, which appears to be a few austenite
grain diameters. The fracture toughness in this case will depend on the
microstructure through the cleavage facet size, and the plastic zone size.
A smaller plastic zone will result in less material being sampled, and
limit the amount of crack deflection, as will a smaller fracture facet size.
Both of these will tend to reduce the resultant ligament size, reducing the
material's toughness. Thus, toughness will be greatest when large, out-of-
plane deflections of the crack occur, producing large ligaments which must
be fractured. This strong link approach to fracture control is a departure
from most previous approaches which suppose that the most susceptible microstructural features control fracture.

## Conclusions

The mechanical properties of pearlite are determined by the pearlite
interlamellar spacing, which controls the yield and cleavage fracture
stresses, and by the prior austenite grain size, which determines the

effective grain size for cleavage. Predictive models are still needed for both yielding and the work-hardening behavior of pearlite. Fracture of this material can be qualitatively understood in terms of the competition between ductile MVC and brittle cleavage modes. It is clear, however, that in spite of the apparent simple nature of the microstructure of pearlite and the many studies of its properties, much more work remains to be done to complete our understanding.

## Acknowledgment

We wish to thank Dr. A.W. Thompson for many helpful discussions and suggestions. This work was sponsored by the Association of American Railroads.

## References

1.  A.R. Rosenfield, G.T. Hahn, and J.D. Embury, Met. Trans., 3 (1972) pp. 2797-2804.

2.  M. Gensamer, E.B. Pearsall, and G.V. Smith, Trans. ASM, 28 (1940) pp. 380-395.

3.  M. Gensamer, E.B. Pearsall, W.S. Pellini, and J.R. Low, Jr. Trans. ASM, 30 (1942) pp. 983-1019.

4.  J.H. Gross and R.D. Stout, Weld. Res. Supp., 30 (1951) pp. 481s-485s.

5.  D.C. Lemmon and O.D. Sherby, J. Materials, 4 (1969) pp. 444-456.

6.  T. Gladman, I.D. McIvor, and F.B. Pickering, JISI, 210 (1972) pp. 916-930.

7.  F.B. Pickering, pp. 9-31 in Towards Improved Ductility and Toughness, Climax Molybdenum Co., 1971.

8.  J.M. Hyzak and I.M. Bernstein, Met. Trans., 7A (1976) pp. 1217-1224.

9.  A.R. Marder and B.L. Bramfitt, Met. Trans., 7A (1976) pp. 365-372.

10. H. Sunwoo, M.E. Fine, M. Meshii, and D.H. Stone, Met. Trans., 13A (1982) pp. 2035-2047.

11. J. Gil Sevillano, pp. 819-824 in ICSMA 5, Aachen, W. Germany, P. Haasen, ed.; Pergamon Press, New York, 1979.

12. J. Flugge, W. Heller, E. Stolte, and W. Dahl, Arch. Eisen., 47 (1976) pp. 635-640.

13. J. Flugge, W. Heller, and R. Schweitzer, Stahl u. Eisen., 99 (1979) pp. 841-845.

14. T. Takahashi and M. Nagumo, Trans. JIM, 11 (1970) pp. 113-119.

15. J.A. Rinebolt, Trans. ASM, 46 (1954) pp. 1527-1543.

16. B. Karlsson and G. Linden, Mat. Sci. Eng., 17 (1975) pp. 153-164.

17. Y. Yamada, Trans. ISIJ, 17 (1977) pp. 516-522.

18. J.P. Houin, A. Simon, and G. Beck, Trans. ISIJ, 21 (1981) pp. 726-731.

19.  J.P. Houin, A. Simon and G. Beck, Mem. Sci. Rev. Met., (1978) pp. 149-159.

20.  J.P. Houin, A. Simon and G. Beck, Mem. Sci. Rev. Met., (1978) pp. 227-235.

21.  J.H. Gross and R.D. Stout, Weld. Res. Supp., 34 (1955) pp. 117s-122s.

22.  G. Langford and M. Cohen, Trans. ASM, 62 (1969) pp. 623-638.

23.  G. Langford, Met. Trans., 8A (1971) pp. 861-875.

24.  R.L. Miller, Met. Trans., 3 (1972) pp. 905-912.

25.  L. Anand and J. Gurland, Met. Trans., 7A (1976) pp. 191-197.

26.  W.B. Morrison, Trans. ASM, 59 (1966) pp. 824-846.

27.  G. Langford and M. Cohen, Met. Trans., 1 (1970) pp. 1478-1480.

28.  S. Karashima and T. Sakuma, Trans. JIM, 9 (1968) pp. 63-66.

29.  K.W. Burns and F.B. Pickering, JISI, 202 (1964) pp. 899-906.

30.  F.B. Pickering, Iron and Steel, 38 (1965) pp. 110-117.

31.  J.J. Lewandowski and A.W. Thompson, unpublished research, Carnegie-Mellon University, Pittsburgh, PA 15213, 1983.

32.  D.J. Alexander and I.M. Bernstein, unpublished research, Carnegie-Mellon University, Pittsburgh, PA 15213, 1983.

33.  J.C.M. Li and Y.T. Chou, Met. Trans., 1 (1970) pp. 1145-1159.

34.  M.F. Ashby, Phil Mag., 21 (1970) pp. 399-424.

35.  M.F. Ashby, pp. 137-192 in Strengthening Methods in Crystals, A. Kelly and R.B. Nicholson, ed.; Wiley, New York, 1971.

36.  K. Tanaka and S. Matsuoka, Acta Met., 22 (1974) pp. 153-163.

37.  J. Gil Sevillano, Prog. Mat. Sci., 25 (1981) pp. 69-412.

38.  R.L. Fleischer, Acta Met., 8 (1960) pp. 598-604.

39.  A. Kelly and R.B. Nicholson, Prog. Mat. Sci., 10 (1963) pp. 149-391.

40.  G. Langford, Met. Trans., 1 (1970) pp. 465-477.

41.  J.D. Embury and R.M. Fisher, Acta Met., 14 (1966) pp. 147-159.

42.  V.K. Chandhok, A. Kasak, and J.P. Hirth, Trans. ASM, 59 (1966) pp. 288-301.

43.  U. Lindborg, Trans. ASM, 61 (1968) pp. 500-504.

44.  Y.-J. Park and I.M. Bernstein, Met. Trans., 10A (1979) pp. 1653-1664.

45.  K.E. Puttick, JISI, 185 (1957) pp. 161-176.

46.   D.A. Porter, K.E. Easterling, and G.D.W. Smith, Acta Met., 26 (1978) pp. 1405-1422.

47.   T. Inoue and S. Kinoshita, Trans. ISIJ, 17 (1977) pp. 245-251.

48.   B.R. Butcher and H.R. Pettit, JISI, 204 (1966) pp. 469-477.

49.   J. Gil Sevillano, Mat. Sci. Eng., 21 (1975) pp. 221-225.

50.   A. Inoue, T. Ogura, and T. Masumoto, Trans. JIM, 17 (1976) pp. 149-157.

51.   L.E. Miller and G.C. Smith, JISI, 208 (1970) pp. 998-1005.

52.   A.R. Rosenfield, E. Votava  and G.T. Hahn, Trans. ASM, 61 (1968) pp. 807-815.

53.   J.T. Barnby and M.R. Johnson, Met. Sci. J., 3 (1965) pp. 155-159.

54.   G.K. Bouse, I.M. Bernstein, and D.H. Stone, pp. 145-166 in Rail Steels-Developments, Processing and Use, ASTM STP 644, D.H. Stone and G.C. Knupps, ed.; 1978.

55.   A.M. Turkalo, Trans. TMS-AIME, 218 (1960) pp. 24-30.

56.   J.R. Low, Jr., pp. 68-90 in Fracture, B.L. Averbach, ed.; Technology Press, MIT, Cambridge, Mass., 1959.

57.   Y.-J. Park and I.M. Bernstein, pp. 33-44 in Fracture 1977, ICF4, D.M.R. Taplin, ed.; University of Waterloo Press, Waterloo, Canada, 1977.

58.   Y.-J. Park and I.M. Bernstein, pp. 287-302 in Rail Steels - Developments, Processing and Use, ASTM STP 644, D.H. Stone and G.C. Knupp, ed.; 1978.

59.   A.S. Tetelman and A.J. McEvily, Jr., Fracture of Structural Materials, pp. 518-520; Wiley, New York, 1967.

60.   R.O. Ritchie, B. Francis, W.L. Server, Met. Trans., 7A (1976) pp. 831-838.

61.   J.M. Hyzak and I.M. Bernstein, unpublished research, Carnegie-Mellon University, Pittsburgh, PA 15213, 1976.

62.   A.S. Tetelman, S. Ensha, and J.N. Robinson, Report to Association of American Railroads, Chicago, Illinois, September 28, 1972.

63.   R.J. Henry, Met. Trans., 1 (1970) pp. 1073-1075.

64.   R.O. Ritchie, J.F. Knott, and J.R. Rice, J. Mech. Phys. Sol., 21 (1973) pp. 395-410.

65.   S. Ensha, Ph.D. Thesis, UCLA, 1974; DAHC-04-69-C-0008, U.S. Army Research Office, Durham.

66.   K. Kuhne, D.-Ing. Thesis, Rheinisch-Westfalischen Technischen Hoehschule, Aachen, West Germany, 1982.

67.   D.A. Curry and J.F. Knott, Met. Sci., 10 (1976) pp. 1-6.

# CARBIDE PRECIPITATION IN FERRITE

R W K Honeycombe

Department of Metallurgy and Materials Science
University of Cambridge, UK

The formation of ferrite from austenite in alloy steels is frequently accompanied by precipitation of alloy carbides in several different morphologies. The precipitation can occur progressively at the $\gamma/\alpha$ interphase boundaries during transformation, either at coherent or semi-coherent planar interfaces along which ledges migrate, or at less regular incoherent boundaries, leading to periodic bands of precipitate (interphase precipitation) which mark the movement of the $\gamma/\alpha$ interfaces during the transformation. The likely mechanisms in each case are described. Alternatively, fine fibrous carbide growth takes place, often in the same grain. This is explained in terms of a change in the nature of the $\gamma/\alpha$ interface, and the variables which encourage fibrous carbides are considered. A third possibility is that ferrite forms initially as a supersaturated solution which subsequently precipitates carbides at grain boundaries, on dislocations and in the matrix generally. Finally the coarsening behaviour of the carbide dispersions is briefly considered.

## 1.  Introduction

One of the most important basic features of the iron-carbon equilibrium diagram is that the carbon solubility in austenite is very substantially greater than that in ferrite.  Consequently during the $\gamma/\alpha$ phase transformation carbon is precipitated as iron carbide, if equilibrium conditions are approached.  The most familiar manifestation of this is the formation of pearlite as a result of the eutectoid reaction, however at low carbon concentrations, discrete cementite particles can form, while at high carbon levels  hyper-eutectoid cementite laths grow prior to the nucleation of pearlite.

The addition of alloying elements results in profound changes both in the kinetics of the basic reactions, and often in the microstructures resulting from the $\gamma/\alpha$ transformation.  Non-carbide forming elements such as nickel or weak carbide formers such as manganese influence greatly the reaction kinetics, but the basic microstructures which result are similar to those formed in binary iron-carbon alloys.  In contrast, alloying elements which form more stable carbides than iron can not only change the kinetics of the transformation, but also the microstructure by replacing iron carbide either in part or in whole by an alloy carbide.  For example in micro-alloyed steels concentrations of substantially less than 0.1 wt.%[†] of niobium, vanadium or titanium added to a low carbon steel result in the replacement of iron carbide by NbC, VC or TiC which form much finer dispersions in the ferrite.  With the growth in importance of these micro-alloyed steels, it has become essential to understand how these alloy carbide dispersions are formed, and to determine the significant variables influencing the distribution of the carbides in ferrite.

In this review we shall deal with the formation of carbides (and certain other phases) during the $\gamma/\alpha$ transformation, from the point of view of understanding the basic mechanisms involved.  However it is also useful to utilize the dispersions to throw light on the basic modes of transformation of austenite to ferrite.  By control of cooling rates or transformation temperatures, the newly formed ferrite can be supersaturated so that precipitation is delayed until after the transformation, or suppressed altogether so that a subsequent ageing treatment is necessary to result in precipitation.  This review will not deal with the latter situation, which broadly encompasses the field of quench ageing.

## 2.  Survey of Precipitate Morphologies Produced Via The $\gamma/\alpha$ Transformation

### 2.1  Pearlite

This reaction is dealt with elsewhere in this symposium, but several relevant points should be made here.  Firstly in micro-alloyed steels when carbon is present in excess of the stoichiometric concentration required for combination with the carbide-forming micro-additions, pearlite can form during the $\gamma/\alpha$ transformation.  Secondly, in highly alloyed steels, pearlitic structures can form, in which the carbide phase is an alloy carbide e.g. $M_{23}C_6$[1]. For example these have been observed in steels with high concentrations of chromium (Fig. 1)[1,2].  Thirdly, in some steels exhibiting pearlitic structures the presence of micro-additions such as Nb or V leads to precipitation of NbC, VC within the pearlitic ferrite.

---

[†]  Throughout this paper, compositions are stated in wt. percentages.

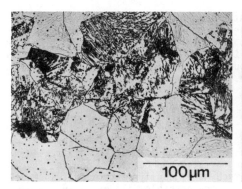

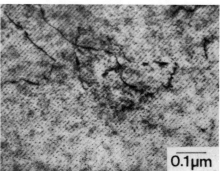

Figure 1 - Fe-12Cr-0.2C 30 min. at 775°C. $Cr_{23}C_6$ "pearlite". Optical micrograph. (Campbell)

Figure 2 - Fe-0.75V-0.15C 5 min. at 725°C. Interphase precipitation of VC. TEM. (Batte)

## 2.2  Interphase Precipitation

This type of precipitation occurs normally in bands which can be planar, or irregular curved surfaces(3,4) and has been shown to form periodically at the γ/α interfaces during the phase transformation (Fig. 2). It is an important mode of precipitation in micro-alloyed steels(5,6,7) in which these banded structures occur on a very fine scale and involve formation of NbC (7,8), VC(9) and TiC(10), but similar, much coarser structures based on $Fe_3C$ have been observed. The phenomenon also occurs frequently in higher alloyed steels, notably those containing chromium, vanadium, molybdenum and tungsten. The precipitating phase is not restricted to carbides, for example, ε-Cu can precipitate in this way in Cu-containing steels(11,12), and Au(13) forms similar dispersions in appropriately contrived alloys.

## 2.3  Fibrous Carbides

This morphology involves colonies of very fine (∿20-100 nm diam.) parallel carbide fibres, the length being determined by the size of the ferrite colony, but can be up to 10μm with little or no branching (Fig. 3); they have been occasionally observed in heat treated low alloy steels (14,15) as well as in micro-alloyed steels. It is tempting to explain these fibres in terms of alloy carbide pearlite structures, but in some of the same

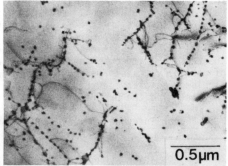

Figure 3 - Fibrous VC precipitation in an Fe-1V-0.4C alloy. C-extraction replica. (Parsons)

Figure 4 - Fe-2-Cu-2Ni 20 min. at 650°C. ε-Cu precipitated mainly on dislocations. TEM. (Ricks)

systems, usually at higher transformation temperatures, much coarser, more irregular alloy carbide/ferrite aggregates can occur which are closer to the pearlite model. All the familiar fcc alloy carbides, NbC, TiC, VC as well as $Mo_2C$, $W_2C$, $Cr_7C_3$ and $Cr_{23}C_6$ can form these fine fibrous aggregates. In micro-alloyed steels they are frequently found adjacent to regions in which inter-phase precipitation has occurred, i.e. within the same original austenite grain.

## 2.4  Precipitation On Dislocations

Dislocations are generated during the $\gamma/\alpha$ transformation as a result of stresses set up when the small volume change takes place. The dislocation density increases with decreasing transformation temperature, and in particular, Widmanstätten ferrite is much more heavily dislocated than equi-axed ferrite(16). This may well reflect the possible role of shear processes in the formation of Widmanstätten ferrite. At lower transformation temper-atures, bainitic ferrite has a still higher dislocation density and the possibility of shear processes occurring during transformation is still more likely. TEM studies have shown that the dislocations are favoured nucleating sites for precipitates (Fig. 4), and also play a dominant role in subsequent coarsening of the dispersion. Ferrite can be supersaturated by rapid cooling to room temperature, and subsequent ageing (quench ageing) leads to precipi-tation on dislocations. $\varepsilon$-iron carbide and cementite nucleate in this way in the range 50-300°C, while alloy carbides will do so at higher temperatures (500-600°C).

## 2.5  Grain Boundary Precipitation

The $\gamma/\alpha$ transformation starts preferentially at the $\gamma$-grain boundaries but carbide precipitates also form at these sites. They may nucleate while the steel is still fully austenitic, but they can also form and grow during and after the transformation. The ferrite/ferrite boundaries formed also become sites for carbide nucleation and growth. In general this type of pre-cipitation is coarser than the other morphologies already referred to.

## 3.  Interphase Precipitation

Steels grain refined by the addition of small concentrations of niobium often exhibit very fine "rows" of NbC in the as-rolled and normalized con-ditions(5,6,7,8). Similar banded structures are found in steels with small amounts (<0.1%) of titanium(10) or vanadium(17), also in higher alloy steels e.g. Cr(3), V(18) and Mo(9,19) steels. One view held was that these precipi-tates nucleated  on  dislocations either in the austenite or in the ferrite (8), however the alternative explanation that nucleation was occurring at regular intervals on the $\gamma/\alpha$ interfaces(7,9) during transformation gradually became accepted after more detailed structural studies had been carried out (2,4). Tilting experiments in the electron microscope clearly revealed that the rows were indeed planar distributions of precipitate, moreover examination of partially transformed specimens showed directly that nucleation of the carbide particles occurred in the $\gamma/\alpha$ interfaces, and the bands of particles thus revealed the progression of the interfaces through the steel(2,4).

## 3.1  Planar Interphase Precipitation

3.1.1  General. We shall first consider the planar arrays of precipi-tates which are now known to form on low energy semi-coherent interfaces. It might be assumed that these are associated mainly with Widmanstätten ferrite and consequently with lower transformation temperatures. However there is ample evidence that planar precipitate arrays can form at the highest trans-formation temperatures e.g. 850°C in an Fe-0.25V-0.05C alloy. Moreover

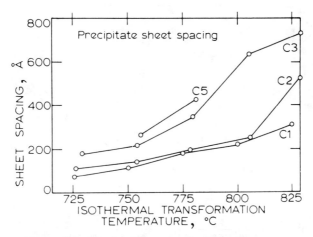

Figure 5 - Spacings between precipitate sheets
for several vanadium steels. (Batte & Honeycombe)
C1: 1.04V 0.20C 0.02Nb; C2: 0.75V 0.15C 0.02Nb
C3: 0.48V 0.09C 0.02Nb; C5: 0.55V 0.04C 0.02Nb

microscopic examination, including in situ photo-emission microscopy of transforming steels(20) has shown that facetted (planar) boundaries are very
common at all transformation temperatures including those at which equi-axed
ferrite exclusively forms.  Interphase precipitation occurs much less
frequently in Widmanstätten ferrite, indeed there is only limited evidence
that it forms at all in these circumstances.

   3.1.2  Sheet Spacing.  The spacing of the precipitate bands has been
measured as a function of transformation temperatures and composition for
several simple vanadium steels(18) (Fig. 5); the spacing decreases with decreasing temperature, typically in a 0.75V 0.15C steel varying from 50nm at
825°C to 10nm at 725°C (Fig. 2).  Similarly, work(21) on a 0.55Ti 0.11C steel
has shown that spacings of bands of TiC precipitate vary from 15nm after
transformation at 600°C to 60nm at 850°C.  Dunlop and Honeycombe(22) made a
comparison between the spacings of the precipitate in three alloys of similar
stoichiometry, but involving vanadium, titanium and an alloy with both these
elements.  Typical results for band spacings and particle sizes within the
bands after transformation at 800°C are shown in Table 1.

Table 1

Interphase Precipitation of VC, TiC and (V,Ti)C

Average Dispersion Parameters after Isothermal Transformation
(10 min. at 800°C)

|  | Mean band spacing | Mean particle size |
|---|---|---|
|  | nm | |
| Fe-0.4V-0.08C | 50-60 | 7-10 |
| Fe-0.18V-0.13Ti-0.08C | 17-27 | 4-5 |
| Fe-0.42Ti-0.09C | 20-25 | 6.5-10.5 |

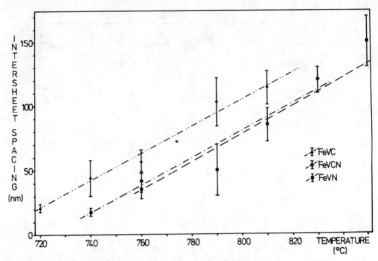

Figure 6 - Intersheet spacing of interphase precipitation as a function of transformation temperature. FeVC: 0.27V 0.05C; FeVCN: 0.26V 0.02C 0.022N; FeVN: 0.29V 0.032N. (Balliger and Honeycombe)

Similar measurements have been made on three iron-0.25V alloys with 0.05C, (0.02C + 0.022N), 0.032N respectively(23). Figure 6 shows the intersheet spacing of the precipitates (which vary from VC through V(CN) to VN) as a function of transformation temperature in the range 720-850°C. The coarsest spacings are achieved by slow cooling through the transformation range, for example the Fe-0.25V-0.05C alloy cooled at 5°C/min. from 1150°C produces bands between 250 and 400nm apart.

The use of much higher alloyed steels, e.g. 12%Cr 0.2%C, slowed down the $\gamma/\alpha$ reaction greatly, and enabled the $Cr_{23}C_6$ precipitation to occur on a scale which permitted the detailed morphology to be determined(2). Firstly and most importantly, this work showed that the interplanar spacing of the precipitate was determined by the passage of steps along the $\gamma/\alpha$ interface (Fig. 7a-c). It is well known that step migration is the main mechanism of growth of coherent low energy interfaces during phase transformations, but the significance of the steps in initiating precipitation reactions is not so widely appreciated. In a static interfacial situation, nucleation would be expected to occur preferentially on the high energy steps, and not on the low energy planar interfaces. However in a dynamic situation prevailing during a phase transformation, the steps are moving too fast to act as nucleating sites, while the gradually extending new planar interface (Fig.7a) represents a viable alternative for nucleation. Occasionally nucleation of precipitate does take place on a step which is then impeded or completely stopped. Fig. 8a illustrates this in a 10Cr 0.2C alloy where particles of $Cr_{23}C_6$ have nucleated on a step. Such events lead to discontinuities in the precipitate distribution(4) as shown in Fig. 8b for an Fe-0.85V-0.23C alloy cooled at 5° min$^{-1}$ from 1000°C, where a discontinuity is visible at A.

3.1.3 Crystallography. The planar $\gamma/\alpha$ interfaces which are associated with step migration occur typically between $\gamma$ and $\alpha$ which have the Kurdjumov-Sachs relationship (K-S):

$$(111)_\gamma \ || \ (110)_\alpha$$
$$[\bar{1}10]_\gamma \ || \ [1\bar{1}1]_\alpha$$

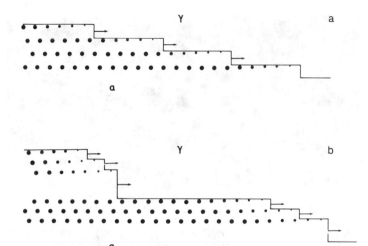

Figure 7a - schematic: planar interphase precipitation
a) uniform steps; b) irregular steps.

Figure 7b - Example of steps associated with $Cr_{23}C_6$ precipitated in an Fe-12Cr-0.2C alloy transformed 30 min. at 650°C. TEM. (Campbell)

Figure 7c - Same alloy and treatment as b. More irregular precipitation associated with steps. TEM. (Campbell)

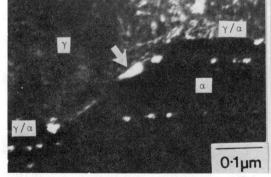

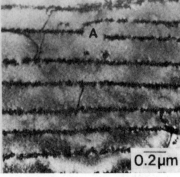

Figure 8a - Fe-10Cr-0.2C. Ledge on a $\gamma/\alpha$ boundary on which $Cr_{23}C_6$ has nucleated. TEM. (Ricks)

Figure 8b - Fe-0.85V-0.23C furnace cooled from 1000°C at 5°C min$^{-1}$. VC precipitation in sheets, discontinuity at A. TEM. (Davenport & Honeycombe)

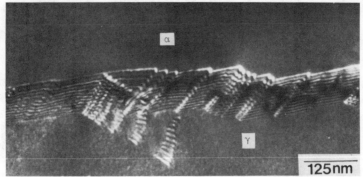

Figure 9 - Steps on a planar austenite-ferrite interface.
Weak beam using $111_\gamma$ reflection. TEM. (Southwick)

The interphase plane is normally $(111)_\gamma \parallel (110)_\alpha$. Recent work(24,30)
on a stainless $\gamma/\alpha$ steel in which both phases co-exist at room temperature
has allowed the $\gamma/\alpha$ interfaces to be examined by high resolution TEM.
Periodic arrays of intrinsic dislocations were not usually observed in the
planar boundaries with the $(111)_\gamma \parallel (110)_\alpha$, and only occasional extrinsic
dislocations were present. These boundaries have low energies and low
mobility, consequently the transformation proceeds by the movement of small,
higher energy steps along the interface (Fig. 9). Weak beam TEM micrographs
did not reveal intrinsic dislocations in this type of boundary. Preferential
sites for carbide nucleation are most probably the extrinsic dislocations;
Fig. 10a illustrates nucleation of $Cr_{23}C_6$ at extrinsic dislocation sites
along a $(111)_\gamma \parallel (110)_\alpha$ interface. On the semi-coherent part of the same
boundary (containing one visible set of intrinsic dislocations) the $Cr_{23}C_6$
precipitation is much more pronounced (Fig. 10b).

Turning from the crystallography between $\gamma$ and $\alpha$ phases to that of the
precipitating phases with the matrix, it is necessary to distinguish between
types of structure which are encountered. The following phases have been
found in association with planar $\gamma/\alpha$ interfaces: VC, NbC, TiC, TaC, $Cr_{23}C_6$,
$Cr_7C_3$, $M_6C$, $W_2C$, $Mo_2C$, $\varepsilon$-Cu and Au. The crystal structures, habit and
orientation relationships with ferrite for some of these phases are given in
Table 2.

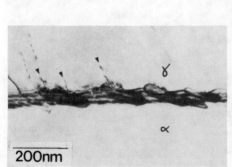
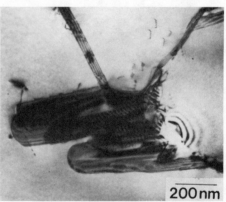

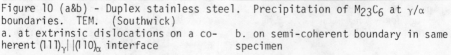

Figure 10 (a&b) - Duplex stainless steel. Precipitation of $M_{23}C_6$ at $\gamma/\alpha$
boundaries. TEM. (Southwick)
a. at extrinsic dislocations on a co-        b. on semi-coherent boundary in same
herent $(111)_\gamma \parallel (110)_\alpha$ interface       specimen

Table 2

Crystallography of Precipitates and Matrix

| Phase | Structure | Habit | Orientation Relationship with $\alpha$ matrix |
|-------|-----------|-------|-----------------------------------------------|
| VC, NbC, TiC, TaC | fcc | plates | $(100)_p \vert\vert (100)_\alpha : [010]_p \vert\vert [\bar{1}10]_\alpha$ (Baker-Nutting) |
| $Cr_{23}C_6$ | complex cubic | rods or laths | $(111)_{Cr_{23}C_6} \vert\vert (101)_\alpha : [\bar{1}10]_{M_{23}C_6} \vert\vert [\bar{1}\bar{1}1]_\alpha$ (K-S) |
| $Mo_2C$ | c-p hex. | rods $[\bar{1}00]_\alpha$ growth dir. | $(0001)_{Mo_2C} \vert\vert (011)_\alpha : [2\bar{1}\bar{1}0]_{Mo_2C} \vert\vert [\bar{1}00]_\alpha$ |
| $\varepsilon$-Cu | fcc | spheres, rods | $(111)_{\varepsilon\text{-Cu}} \vert\vert (101)_\alpha : [\bar{1}10]_{\varepsilon\text{-Cu}} \vert\vert [\bar{1}\bar{1}1]_\alpha$ (K-S) |

An important result from precipitate spot dark field imaging studies is
that the precipitating phase very frequently adopts only one variant of the
orientation relationship with ferrite in a particular region(4,25).  There
appear to be two different ways in which this arises, dependent on the crystal
structure.  The first is best illustrated by the precipitation of $Cr_{23}C_6$ on
$\gamma/\alpha$ interfaces.  Dark field TEM studies(26) on an isothermally transformed
Fe-10Cr-0.2C alloy has shown that the $Cr_{23}C_6$ is related to the ferrite by the
K-S relationship, but also to the austenite by a cube/cube relationship.
Bearing in mind that the $\alpha$ is related to the austenite by the K-S relationship, the following three phase crystallography results.

$$(111)_\gamma \ \vert\vert \ (110)_\alpha \ \vert\vert \ (111)_{Cr_{23}C_6}$$

$$[\bar{1}10]_\gamma \ \vert\vert \ [\bar{1}11]_\alpha \ \vert\vert \ [\bar{1}10]_{Cr_{23}C_6}$$

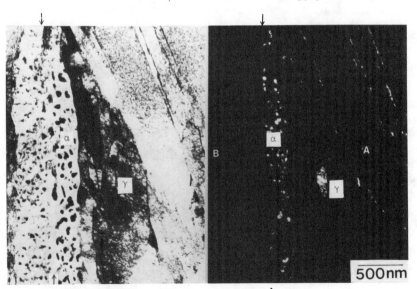

500nm

Figure 11 - Fe-10Cr-0.2C.  Ferrite allotriomorph (L.H.S.) growing into
$\gamma$ grain (R.H.S.) with precipitation of $Cr_{23}C_6$ at the $\gamma/\alpha$ interface.
a. bright field TEM;  b. dark field, showing both retained $\gamma$ and
$Cr_{23}C_6$ with the same orientation.  (Ricks & Howell)

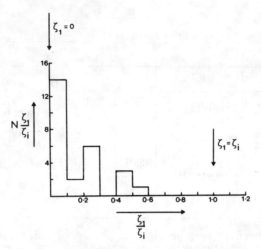

Figure 12 - Histogram of the angle between
the VC habit plane and the local $\gamma/\alpha$ interface normal.  (Law)

The $Cr_{23}C_6$ has adopted that variant of the K-S relationship with ferrite
which allows it also to have a cube-cube relationship with austenite (Fig.
11).  The precipitation of $\varepsilon$-copper in Fe-Cu and Fe-Ni-Cu alloys provides
another good example of this three phase crystallography which results in
single variant particles of $\varepsilon$-Cu being formed(12,26).

    The second situation involving single variant precipitation relates to
the fcc carbides VC, NbC and TiC forming at the $\gamma/\alpha$ interface, which has been
observed particularly in vanadium steels(4,27).  Analyses show that none of
the three possible Baker-Nutting variants adopted by fcc carbides in ferrite,
allows the simultaneous formation of a cube/cube relation between the carbides and austenite(28).  These carbides all normally adopt a plate-like

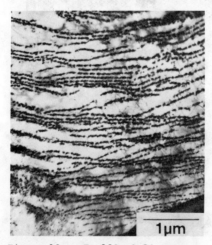

Figure 13a - Fe-12Cr-0.2C transformed at 650°C.  Irregular banded
precipitation of $Cr_{23}C_6$.  TEM.
(Campbell)

Figure 13b - Fe-1V-0.2C transformed
12 sec at 700°C.  Non planar banded
VC dispersion in ferrite.  TEM.
(Ricks & Howell)

morphology, and the orientation variant chosen with respect to the ferrite is that which results in a habit plane of the carbide $((001)_{VC}||(001)_{\alpha})$ most closely parallel to the $\gamma/\alpha$ interface, proposed originally by Davenport and Honeycombe(4) after an examination of the behaviour of a vanadium steel.  A histogram of the angle $\zeta_1$ between the VC habit plane normal and the local $\gamma/\alpha$ interface normal (Fig. 12) shows that $\zeta_1$ is always much smaller than the two other possible angles $\zeta_2$ and $\zeta_3$.  Such a choice lowers the free energy of activation for critical nucleus formation(29).

## 3.2  Non-Planar Interphase Precipitation (Incoherent)

Examination of many isothermally transformed ferrous alloys has indicated that frequently the periodic arrays of precipitate are not planar, but are curved, often in quite an irregular way (Fig. 13) clearly indicating that the $\gamma/\alpha$ boundary on which the precipitation has taken place was of a high energy incoherent type, and not a low energy planar interface of the $(111)_{\gamma}||(110)_{\alpha}$ type.  This, of course, would be expected, as ferrite grains should have both coherent and incoherent boundaries with austenite, and TEM observations on a $\gamma/\alpha$ stainless steel have confirmed that this is normally the case at least for transformations occurring above 600-650°C(30).

3.2.1  Bowing Mechanism.  Early observations of this non-planar inter- phase precipitation in vanadium(31) and chromium steels(2) led to the view that the precipitates pinned the $\gamma/\alpha$ boundaries and that these boundaries were then forced to move around the particles by bowing.  This is analogous to the Orowan mechanism for the looping of dislocations around precipitates, except that the interphase boundary does not appear to leave dislocation loops behind.  Recently Ricks and Howell(32) have confirmed this mechanism in a 10%Cr 0.2%C steel (Fig. 14a) and developed a simple model for the pro- cess (Fig. 14b) involving a free energy balance, the transformation being resisted by the increase in $\gamma/\alpha$ interfacial energy produced by the bulge (c.f. strain induced boundary migration during recrystallization).  The free energy change $\Delta G_T$ with respect to bulge height h (see Fig. 14b) was shown to be:

$$\frac{d\Delta G_T}{dh} = \tfrac{1}{2}\pi(a^2 + h^2)\frac{\Delta G}{V} + 2\pi h\sigma. \qquad (1)$$

where:

a = $\tfrac{1}{2}$ spacing of precipitates      V = the molar volume of austenite
$\Delta G$ = the free energy change per mole of $\gamma$   $\sigma$ = the surface energy/unit area

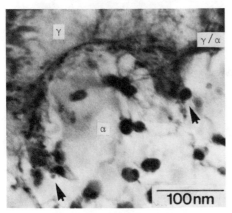

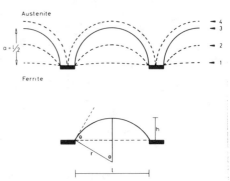

Figure 14a - Bowing of $\gamma/\alpha$ interface   Figure 14b - Model for the bowing between $Cr_{23}C_6$ particles in an Fe-    mechanism.  (Ricks & Howell) 10Cr-0.2C alloy.  (Ricks & Howell)

The change in free energy becomes zero at a = $a_{crit}$ which is determined from (1) by equating to zero:

$$a_{crit} = \frac{2\sigma V}{\Delta G} \qquad (2)$$

There is clearly a minimum value of a, below which it is not energetically feasible for a bowing mechanism to occur. Those alloy systems which produce relatively coarsely spaced precipitates, e.g. Cr steels, Cu-containing steels, readily exhibit this mechanism, whereas steels in which precipitates are finer and closer spaced, e.g. those containing V, Nb and Ti, do not. It is reported(32) that experimentally measured critical precipitate spacings in Cr and V-bearing steels are in good agreement with spacings predicted from the theory but it should be pointed out that the above calculation does not take account of the influence of alloying elements on $\Delta G^{\gamma-\alpha}$. It might be expected that increasing the transformation temperature should increase the probability of non-planar arrays, however at these temperatures $\Delta G$ will be lower, so the driving force for bowing, other things being equal, will be reduced.

   3.2.2 Quasi-Ledge Mechanism. The bowing of the $\gamma/\alpha$ interface between appropriately spaced carbide particles can also lead to a quasi-ledge mechanism for the subsequent movement of the interface as shown schematically in Fig. 15a(32). Abrupt changes have been observed in the $\gamma/\alpha$ boundary curvature which subsequently lead to the movement of a pair of steps moving in opposite directions along the interface. The ferrite and austenite are related by the K-S orientation relationship, but the $\gamma/\alpha$ boundaries are remotely oriented to the low-energy $(111)_\gamma||(110)_\alpha$ interface, the trace of which is indicated on Fig. 15b, taken using an Fe-10Cr-0.2C alloy transformed 60 min. at 655°C. The quasi-ledge mechanism becomes necessary when the $\gamma/\alpha$ boundary is largely immobilized by copious precipitation, and a bulge occurs between two of the more widely spaced precipitates thus forming two steps which "unzip" the interface from the other particles holding it in position. In general this results in more obviously banded structures, than the simple bowing mechanism which can lead to almost random arrays of precipitate, although the individual rods or platelets are all similarly oriented in a particular region.

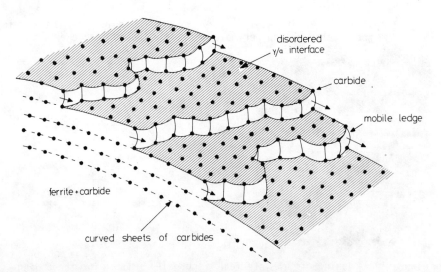

Figure 15a - Formation of non-planar sheets of precipitate by a "quasi-ledge" mechanism. (Ricks & Howell)

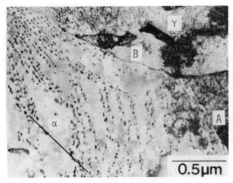 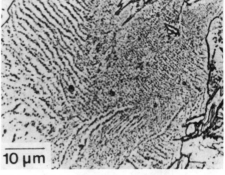

Figure 15b - Fe-10Cr-0.2C quasi-ledge growth on two $\gamma/\alpha$ interfaces (A & B). The trace of the low energy $(111)_\gamma \| (011)_\alpha$ plane is marked.(Ricks & Howell)

Figure 16 - Fe-0.07C. Iced brine quench from 1000°C. Interphase precipitation of Fe₃C. Optical micrograph. (Balliger)

### 3.2.3 Interphase Precipitation of Cementite.

Occasionally, interphase precipitation of $Fe_3C$ has been reported, for example Ohmori(33), Davenport & Becker(34) and more recently Balliger(35). Ohmori, working with an Fe-0.8C steel, found interphase precipitation of $Fe_3C$ in the temperature range 400-450°C, while Davenport and Becker observed similar precipitation in an Fe-0.5Mn-0.1C alloy after continuous cooling. Balliger worked with Fe-0.07C and Fe-2Ni-0.07C, the former alloy exhibiting some interphase precipitation of $Fe_3C$ after brine cooling from 1000°C which was coarse enough to be visible in the optical microscope (Fig. 16); the precipitation followed the formation of clean ferrite at the $\gamma$ grain boundaries. To allow systematic isothermal studies 2%Ni was added to the alloy, and transformations carried out in the range 650-300°C. Above 500°C classical Widmanstätten and allotriomorphic ferrite was formed, but at 450°C and below equi-axed ferrite containing interphase precipitation of $Fe_3C$ was observed. At temperatures of 400°C and below bainite also occurred.

### 3.2.4 Interphase Precipitation in Pearlite.

It has been found that medium carbon ferrite-pearlite steels can be substantially strengthened by small amounts of Nb or V ($\sim$0.1wt.%)(36) which was attributed to strengthening of the ferrite by precipitation of NbC or VC. Dunlop et al(37) have found that VC precipitates not only in the pro-eutectoid ferrite but also in the pearlitic ferrite laths. Moreover the precipitation within the laths occurred in bands, so it was assumed to form by an interphase reaction at the boundaries between the ferrite lamellae and the austenite (Fig. 17). Moreover the VC adopted only one variant of its habit, a well defined characteristic of interphase precipitation of fcc carbides. The spacing of the VC bands in the ferrite was between 15 and 40nm, increasing with decreasing cooling rate as expected. Depending on the cooling rate, vanadium would be retained in super-saturated solution and precipitated after subsequent ageing at 700°C. In contrast, this precipitation occurs in all three variants of the orientation relationship (Baker-Nutting). Recent work by Parsons(38) has shown that if isothermal transformation is used, the spacings of the VC precipitate both within the pro-eutectoid ferrite and the pearlitic ferrite are similar; this is not so if the steel is transformed during cooling. The pearlitic ferrite/austenite interfaces are shown to be of high energy incoherent type, so this would appear to eliminate a conventional ledge mechanism to explain the precipitation reaction. The precipitate was probably too fine to enable a bowing mechanism to operate, so it was proposed that a "quasi-ledge" type of transformation was responsible.

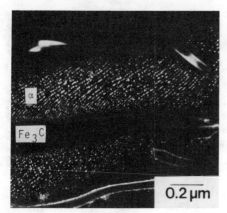

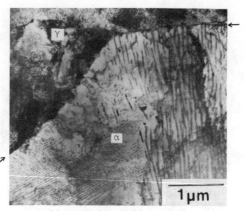

Figure 17 - 0.8Mn 0.3V 0.5C steel
slowly cooled.  Interphase precipitation of VC in pearlitic ferrite.
TEM dark field(Parsons)

Figure 18 - Fe-5Cr-0.2C transformed
for 30 min. at 650°C.  γ/α interface.
TEM.  (Campbell)

## 4.    Fibrous Carbide Growth

### 4.1    General

Fine carbide fibres have been observed from time to time in heat treated
low alloy steels(14), and in particular in normalized steels containing V,
Mo and Cr(15,39,40).  In isothermally transformed steels, they are frequently
observed in close association with interphase precipitation and most of the
carbide phases so far mentioned can exist in this morphology.  For example
the fibrous form has been observed for $Mo_2C$(19), VC(18,23), TiC(41), NbC(42),
$Cr_{23}C_6$(2), $Cr_7C_3$(2), $\varepsilon$-Cu(43) and Au(13), and for Mo-containing steels(44)
are typically 100-300Å diam. with spacings between 200 and 500Å.  It is a
reasonable assumption that these structures in association with ferrite
represent the alloy carbide equivalent of pearlite, however there are some
important differences in morphology, for example the fibres are rarely if
ever branched, and they usually grow nearly normal to the direction of the
transformation front.  In a given region they are identically oriented, but
arise from separate nucleation events.  Some elements, notably Cr, participate
in the formation of an alloy pearlite on a coarser scale at high transformation temperatures (750-850°C)(2).  Thus it is better to approach the study
of this carbide morphology without a pre-conceived model based on pearlite
nucleation and growth characteristics.

### 4.2    Influence Of The γ/α Interface

The intimate association of fibrous carbides and interphase precipitation
is shown in Fig. 18 for an Fe-5Cr-0.2C alloy isothermally transformed 30 min.
at 650°C, in which the precipitating carbide phase $Cr_7C_3$ has adopted both
morphologies, moreover dark field microscopy shows that they possess the
same orientation with respect to the ferrite(2).  Close inspection of the
γ/α interface indicates that the section associated with the interphase
precipitation is planar and contains small steps, whereas the interphase
associated with the fibres is more irregular, showing local curvatures.  It
is thus concluded that in this example, a change in orientation of the interface has altered the interfacial structure, and as a result the basic
mechanism by which it moves.  This, in turn, has influenced the morphology
of the carbide phase.  There is a short transitional region of the γ/α
boundary where both modes of precipitation co-exist.

The importance of the exact nature of the $\gamma/\alpha$ interface is illustrated in an Fe-4Mo-0.2C alloy transformed 5 min. at 700°C where there is extensive fibrous precipitation of Mo$_2$C (L.H.S.) in contrast to the R.H.S. of the field (Fig. 19) where interphase precipitation predominates. Again, dark field TEM examination revealed that the fibrous and interphase Mo$_2$C had the same orientation with respect to the ferrite matrix(19). This relationship tends to vary but two predominate:

$$(011)_\alpha \,\,||\,\, (0001)_{Mo_2C}$$

$$(100)_\alpha \,\,||\,\, (2\bar{1}10)_{Mo_2C}$$

$$[100]_\alpha \,\,||\,\, [\bar{1}010]_{Mo_2C}$$

which is also the relationship for Widmanstätten precipitation of Mo$_2$C in ferrite. Ideally the growth direction of the fibres should be $[100]_\alpha || [\bar{1}010]$, (45) because this leads to the minimum degree of misfit between atoms(46), but it has been found that this is not accurate in many examples. The second orientation frequently observed is:

$$(011)_\alpha \,\,||\,\, (0001)_{Mo_2C}$$

$$(101)_\alpha \,\,||\,\, (10\bar{1}1)_{Mo_2C}$$

$$[11\bar{1}]_\alpha \,\,||\,\, [\bar{1}2\bar{1}0]_{Mo_2C}$$

a familiar one for an hexagonal phase in a bcc matrix, as found for $\varepsilon$-iron carbide in tempered martensite(Jack).

Transformation at higher temperatures gives coarser microstructures which allow further investigation of $\gamma/\alpha$ interfaces, sometimes in the optical microscope. Fig. 20 is taken from a 4%Mo 0.2C steel transformed at 850°C showing two adjacent $\gamma/\alpha$ interfaces, both planar, but the one associated with interphase precipitation of Mo$_2$C is stepped(47). However the fibrous Mo$_2$C is also associated with a planar, but un-stepped interface; this morphology is frequently observed in molybdenum-containing steels. Fig. 19 shows one of these interfaces at high resolution.

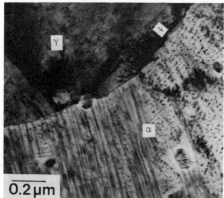 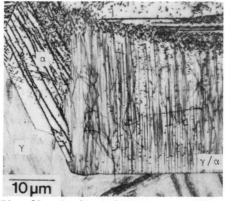

Fig. 19 - Fe-4Mo-0.2C transformed 5 min. at 700°C. $\gamma/\alpha$ interface. TEM. (Berry)

Fig. 20 - Fe-4Mo-0.2C transformed at 850°C. Interphase & fibrous precipitates of Mo$_2$C. Opt.micrograph(Barbacki)

In many cases in Mo-containing(46) and V-containing(48) steels, within the one ferrite allotriomorph formed at an austenite grain boundary the carbide phase ($Mo_2C$ or VC) forms as interphase precipitation on one side and as fibres on the other (Fig. 21). The crystallography of these situations in a V-steel(48) has revealed that on the interphase precipitate side of the boundary, ferrite has a Kurdjumov-Sachs relationship with the austenite, and the bands of precipitate are parallel to the transformation front which is $(110)_\gamma || (111)_\alpha$. In contrast, on the fibrous carbide side, the ferrite bears no relationship to the austenite in which it is growing, and the $\gamma/\alpha$ boundary does not correspond to a rational plane, so it can be described as a high energy incoherent interface. Close examination of such interfaces by TEM reveals that they are locally curved, but in contrast many boundaries associated with interphase precipitation can be closely planar. However in the cases discussed earlier (Figs. 18,19), fibrous carbide is growing adjacent to interphase precipitation in the same ferrite grain and into the same austenite grain, so there are situations where the K-S relationship applies, and the $\gamma/\alpha$ boundary is not planar.

## 4.3 Factors Determining The Relative Proportions Of Fibres & Interphase Precipitation

4.3.1 Alloying. Fibrous $Mo_2C$ predominates in a 4%Mo 0.2C steel after isothermal transformation in the range 600-900°C, whereas fibrous VC is rare in a 1%V 0.2C steel. These two steels possess different reaction kinetics, the C curve of the Mo steel being displaced to substantially longer times. However if 1.5wt.%Ni or Mn is added to the vanadium steel, the amount of fibrous carbide is much increased(18,48). For example, after transformation at 700°C, an Fe-1V-0.2C alloy contains about 5%VC as fibres, however on addition of 1.5%Mn this increases to 20%, while the addition of 2%Cr leads to 40% of the carbide in the fibrous morphology. This type of result suggests that more incoherent interfaces are formed when the transformation is slowed down, so that more fibrous growth is likely to be nucleated at such interfaces. It is interesting to note that when the austenite is rendered completely stable by addition of Mn, and then heat treatments carried out to precipitate VC in an Fe-13Mn-2V-0.8C steel, the precipitation occurs much in the fibrous mode at the austenite grain boundaries(49). An implication of the observed effect of alloying elements on the morphology of

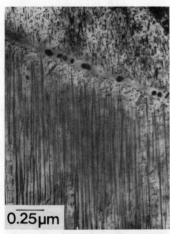

Fig. 21 - Fe-4Mo-0.2C transformed 30 min. at 700°C. TEM. (Berry)

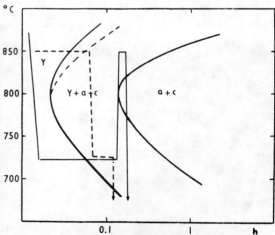

Fig. 22a - Transformation of Fe-4Mo-0.2C at two temperatures (a) TTT diagram with heat treatments superimposed. (Barbacki)

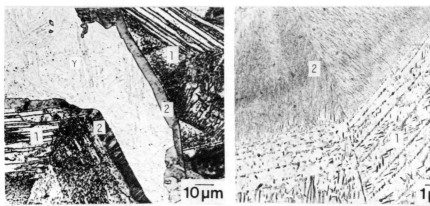

Figure 22b&c - Transformation of Fe-4Mo-0.2C at two temperatures.  (b) partial transformation at 850°C then at 725°C; change from interphase to fibrous precipitation.  Optical micrograph.  (c) same as (b), higher magnification carbon extraction replica.  (Barbacki)

the carbides is that a rapid reaction appears to favour the movement of the γ/α interfaces by step migration, and thus the interphase precipitate morphology.

4.3.2  Effect of Transformation Temperature.  Detailed investigation of an Fe-4Mo-0.2C alloy transformed over the range 600-900°C(19) has shown that fibrous $Mo_2C$ predominates from 850 to 600°C and tends to become more pronounced with falling temperature.  This is in broad agreement with the above results on alloying effects, in so far as the nose of the TTT curve for the ternary alloy is at around 770°C, consequently the transformation occurs at increasingly longer times below this temperature, and this should encourage growth of fibrous carbides.  Berry found an optimum development of $Mo_2C$ fibres at 650°C at which temperature the γ/α reaction takes 3 days to go to completion.

One way of determining more precisely the effect of transformation temperature is to transforn an alloy partially at $T_1$, then rapidly change the temperature to $T_2$ and allow the transformation to continue(47) (Fig. 22a).

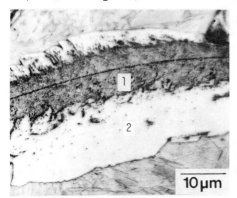

Figure 23 - Fe-4Mo-0.2C upquenched from 725 to 850°C during transformation.  Fibrous reaction at 725°C while at 850°C ferrite only is formed. Optical micrograph.  (Barbacki)

Figure 24 - VC precipitation in an Fe-1V-0.2C alloy.  TEM.  (Batte)

Fig. 22b shows an optical micrograph from a 4Mo 0.2C steel partially transformed at 850°C then quenched to 725°C for further transformation. The high
temperature structure was dominated by interphase precipitation of $Mo_2C$, but
the structure formed at 725°C contained mainly fine fibrous $Mo_2C$ (Fig. 22c).
The transition zone was examined by TEM-electron diffraction which showed
that in a region such as that in Fig. 22c the ferrite matrix did not change
its orientation as a result of the temperature change, despite the marked
change in carbide morphology. However the nature of the $\gamma/\alpha$ interface had
changed from a planar stepped one to another which migrated in the direction
of the carbide fibre axis without the need for step propagation. A simpler
transition from coarse $Mo_2C$ formed at 850°C to fine fibres formed at 750°C
was also observed, but this did not involve structural changes in the $\gamma/\alpha$
interface. Upquenching the same alloy (Fig. 23) from 725°C to 850°C resulted
in a change from a fine fibrous aggregate to a practically dispersion-free
ferrite. The original interface appeared quite irregular, whereas the latter
was coarsely stepped (Fig. 23). These experiments not only throw light on
the importance influence of transformation temperature on structure, but also
demonstrate that often,within a specimen, ferrite of identical orientation
can be associated with widely different carbide morphologies.

However one observation remains unexplained. It is frequently observed
that an interphase precipitation will suddenly change to a fibrous form in a
limited region during isothermal transformation, then often revert quickly to
the original morphology (Fig. 24)(50). This implies that locally the $\gamma/\alpha$
interface has changed character for a short period during transformation.
This would occur if, instead of steps moving along the interface, locally
the interface bulged between precipitate particles. As explained in section
3.2.1 this occurs and leads to non-planar interphase precipitation, however,
with the appropriate reaction kinetics, this could result in fibre growth
behind a section of incoherent interface.

## 5.  Precipitation On Dislocations

So far, the precipitation reactions examined have been initiated at the
$\gamma/\alpha$ interfaces during transformation. There is however much evidence to
support the view that ferrite can form in the supersaturated condition, and
only after some time at the transformation temperature does precipitation
occur, usually on dislocations, within the matrix and at ferrite grain
boundaries. Alternatively the ferrite may remain supersaturated both at
temperature and subsequently at room temperature. Work on Fe-0.25V-0.04(C,N)
(23) and Fe-Cu-Ni alloys(51) has confirmed that V(C,N) and $\varepsilon$-Cu can precipitate within ferrite in this manner.

TTT curves of all these alloys, determined by isothermal dilatometry,
showed well-defined discontinuities (Fig. 25) which marks a change in
mechanism of the reaction. This has been examined in detail for Fe-2Cu,
Fe-2Cu-2Ni and Fe-2Cu-5Ni, and below the discontinuity (low temperature
reaction) only supersaturated ferrite is formed initially, although precipitation on dislocations may occur if the alloy is held for a long time at the
transformation temperature. It has been proposed that the low temperature
reaction takes place by simple transfer of atoms across an interface. In
contrast the reaction above the discontinuity (high temperature reaction) in
the case of Fe-2Cu-2Ni and Fe-2Cu-5Ni leads to three dissimilar ferritic
products:

1.  Supersaturated ferrite, which often ages during subsequent transformation.

2.  Ferrite with interphase precipitation of $\varepsilon$-Cu occurring in planar sheets
    and also in more irregular curved sheets.

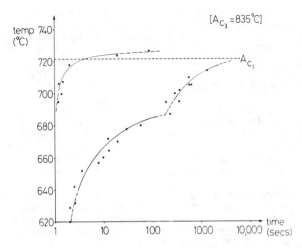

Figure 25 - Dilatometrically determined TTT curve for Fe-2Cu-2Ni. (Ricks)

3.  Ferrite with precipitation of $\varepsilon$-Cu on dislocations.

The presence of the second mode implies that interphase boundary diffusion of copper takes place, leading to the observed $\varepsilon$-Cu precipitation.

As the degree of undercooling and hence the supersaturation of the ferrite increases, the driving force for nucleation of precipitates will increase, however this is more than balanced by the increase in velocity of

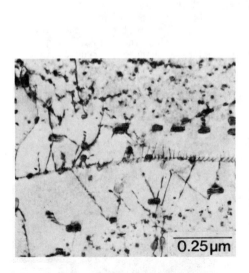

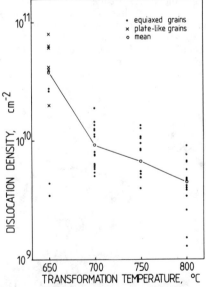

Figure 26 - Fe-2Cu-5Ni isothermally transformed 40 min. at 600°C.  $\varepsilon$-Cu precipitation on dislocations, boundaries and within the matrix.  TEM. (Ricks)

Figure 27 - Fe-0.21Ti-0.08C: dislocation densities as a function of transformation temperature. (G M Smith)

the $\gamma/\alpha$ interfaces at lower temperatures which reduces the time available for nucleation on these interfaces. Consequently while interphase precipitation will predominate at higher transformation temperatures, precipitation on dislocations and within the matrix generally will take over as the transformation temperature is lowered, because this occurs in the longer time available after the $\gamma/\alpha$ has passed by (Fig. 26). Finally, at still lower temperatures, the ferrite remains supersaturated.

A further important point is that the dislocation density within the ferrite increases with decreasing transformation temperature. This is illustrated in Fig. 27 for an Fe-0.21Ti-0.08C alloy transformed in the range 650-800°C. This means that at low transformation temperatures there is a much higher concentration of suitable nucleation sites on dislocations, moreover the dislocations play a significant part in the coarsening process.

Similar observations on Fe-0.25V-0.04(C,N) alloys have been made, confirming that vanadium carbide forms both on $\gamma/\alpha$ interfaces and on dislocations. The dislocation precipitation occurs as platelets on all three variants of the Baker-Nutting relationship (Fig. 28) whereas the interphase precipitation of VC exhibits only one variant in a particular region, namely that which allows the plates to be most closely aligned to the plane of the $\gamma/\alpha$ interface. In addition to dislocation precipitation, matrix precipitation is also frequently observed in specimens transformed at 740°C and lower.

## 6. Coarsening of Precipitate Dispersions

The banded carbide dispersions described above, coarsen on prolonged holding at the transformation temperature, and in doing so their banded structure rapidly disappears(52-54). This is illustrated for an Fe-0.26V-0.044(C,N) alloy transformed and aged at 790°C (Fig. 29) where the V(C,N) precipitate is imaged in dark field, the specimen being tilted to show up the banded structure. The bimodal nature of the precipitate is clearly revealed. It has been shown by weak beam TEM that the coarse particles lie on dislocation networks, and eventually it is these particles which almost completely comprise the dispersion. The coarsening kinetics in the early stages of the reaction obey a (time)$^{1/2}$ law, but soon there is a change to a (time)$^{1/5}$ regime, which has been interpreted as a changeover from an interface reaction controlled coarsening to dislocation coarsening controlled by

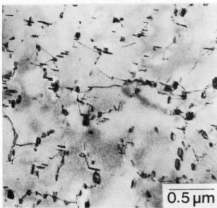

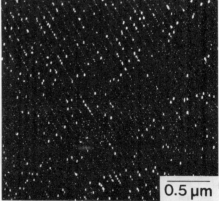

0.5 µm

0.5 µm

Figure 28 - Fe-0.27V-0.05C isothermally transformed and held at 740°C for 2.5hr. Precipitation of VC on dislocations. TEM. (Balliger)

Figure 29 - Fe-0.26V-0.02C-0.22N transformed 135s. at 790°C. Bimodal distribution of V(C,N). TEM dark field. (Balliger)

pipe diffusion along the dislocations.  The two coarsening equations which apply are, firstly, one arising from the Lipshitz-Wagner theory of diffusion controlled particle coarsening, and secondly that used by Ardell for particles on dislocations(56).

$$(\bar{r}_t^2 - \bar{r}_0^2) = (K/RT)V^2 Ck\gamma . t \tag{3}$$

$$(\bar{r}_t^5 - \bar{r}_0^5) = (K'/RT)V^2 CD_d Nq\gamma . t \tag{4}$$

where $\bar{r}_0$ and $\bar{r}_t$ are mean particle radii at zero time and time t, V is the molar volume of precipitate, C is the concentration of solute in equilibrium with a particle of infinite radius, $D_d$ is the solute diffusion coefficient for pipe diffusion, q is the effective cross sectional area of the dislocation diffusion path and $\gamma$ is the particle/matrix interfacial energy.  K and K' are constants.

Work on three vanadium steels, Fe-0.24V-0.023C, Fe-0.28V-0.028(C,N) and Fe-0.29V-0.032N showed that the use of nitrogen to form VN and V(CN) led to a marked reduction in coarsening compared with the Fe-V-C alloy, there being a factor of 50 difference between the coarsening rate of the latter alloy and Fe-V-N, at least in the early stages.

Similarly modification of the metal in the carbide phase has been examined in a series of alloys precipitating VC, TiC and (V,Ti)C(52,53), these carbides being isomorphous and mutually soluble in each other.  The coarsening rate increased in the order (V,Ti)C-VC-TiC, demonstrating that the use of a mixed carbide in these circumstances leads to a finer micro-structure, other things being equal.  The reason for the low coarsening rate of the mixed carbide is probably related to the greater interatomic binding energy compared with that of the straight VC or TiC(55).

## Acknowledgements

I am grateful to my colleagues Dr H K D H Bhadeshia and Dr R A Ricks for helpful comments during the preparation of this paper, also Miss Rosemary Leach for her careful typing of the manuscript.

## References

1.  R F Mehl and W C Hagel, Prog. in Metal Physics, 6 p. 74, 1956.
2.  K Campbell and R W K Honeycombe, Met.Sci., 8 (1974) p. 197.
3.  M Mannerkoski, Acta Polytech.Scand., Ch. 26 (1964); Met.Sci., 3 (1969) p. 54.
4.  A T Davenport and R W K Honeycombe, Proc.Roy.Soc., A332 (1971) p. 191.
5.  W B Morrison and J H Woodhead, J.Iron Steel Inst., 201 (1963) p. 43.
6.  W C Leslie, "The Relation Between Structure & Mechanical Properties of Metals", paper presented at NPL Conference, 1963 p. 337, HMSO London.
7.  J M Gray and R B S Yeo, Trans.ASM, 61 (1968) p. 255.
8.  M Tanino and K Aoki, Trans.Iron Steel Inst. Japan, 8 (1968) p. 337.
9.  A T Davenport, F G Berry and R W K Honeycombe, Met.Sci., 2 (1968) p. 104.
10.  S Freeman, in: The Effect of Second Phase Particles on the Mechanical Properties of Steels, p. 152, Iron Steel Inst. London, 1971.
11.  I D McIvor, J.Iron Steel Inst., 207 (1969) p. 106.
12.  R A Ricks, P R Howell and R W K Honeycombe, Met.Trans., 10A (1979) p. 1049.
13.  R A Ricks, J.Mat.Sci., 16 (1981) p. 3006.
14.  J McCann and J A Ridal, J.Iron Steel Inst., 202 (1964) p. 441.

15.  R G Baker and J Nutting, J.Iron Steel Inst., 192 (1959) p. 257.
16.  G M Smith and R W K Honeycombe, Proc. 6th Int.Conf. on Strength of Metals and Alloys, Melbourne, 1982 p. 407.
17.  T Greday, Proc. 3rd European Conf. on Electron Microscopy, Prague, 1964 p. 85.
18.  A D Batte and R W K Honeycombe, J.Iron Steel Inst., 211 (1973) p. 284.
19.  F G Berry and R W K Honeycombe, Met.Trans., 1 (1970) p. 3279.
20.  D V Edmonds and R W K Honeycombe, Met.Science, 12 (1978) p. 399
21.  S Freeman and R W K Honeycombe, Met.Science, 11 (1977) p. 59.
22.  G L Dunlop and R W K Honeycombe, Met.Science, 12 (1978) p. 367.
23.  N K Balliger and R W K Honeycombe, Met.Trans., 11A (1980) p. 421.
24.  P D Southwick, PhD Dissertation, University of Cambridge, 1978.
25.  P R Howell, J V Bee and R W K Honeycombe, Met.Trans., 10A (1979) p. 1213.
26.  P R Howell et alia, Phil.Mag., 41 (2) (1980) p. 165.
27.  N C Law et al, unpublished work.
28.  P R Howell and R W K Honeycombe in "Solid→Solid Phase Transformations" edited by H I Aaronson, D E Laughlin, R F Sekerka and C M Wayman, Met. Soc. AIME  p. 399, 1982.
29.  J K Lee and H I Aaronson, Acta Met., 23 (1975) p. 799.
30.  P D Southwick and R W K Honeycombe, Met. Science, 14 (1980) p. 253.
31.  V K Heikinnen, Acta Met., 21 (1973) p. 709.
32.  R A Ricks and P R Howell, Acta Met., 31 (1983) p. 853.
33.  Y Ohmori, PhD Dissertation, University of Cambridge, 1969.
34.  A T Davenport and P C Becker, Met.Trans., 2 (1971) p. 2962.
35.  N K Balliger, PhD Dissertation, University of Cambridge, 1977.
36.  A Von den Steinen et al, Stahl u. Eisen, 95 (1975) p. 209.
37.  G L Dunlop, C J Carlsson and G Frimodig, Met.Trans., 9A (1978) p. 261.
38.  S A Parsons, PhD Dissertation, University of Cambridge, 1981.
39.  R M Hobbs et alia, J.Iron Steel Inst., 205 (1967) p. 207.
40.  J D Newton and H G Brinkies, J.Iron Steel Inst., 208 (1970) p. 507.
41.  S Freeman, PhD Dissertation, University of Cambridge, 1971.
42.  P Rios,unpublished work, Cambridge.
43.  R A Ricks, PhD Dissertation, University of Cambridge, 1979.
44.  D V Edmonds and R W K Honeycombe, J.Iron Steel Inst., 211 (1973) p. 209.
45.  D J Dyson, S R Keown, D Raynor and J A Whiteman, Acta Met., 14 (1966) p. 867.
46.  F G Berry, PhD Dissertation, University of Sheffield, 1968.
47.  A Barbacki and R W K Honeycombe, Metallography, 9 (1976) p. 277.
48.  N C Law, PhD Dissertation, University of Cambridge, 1977.
49.  M H Ainsley, G J Cocks and D R Miller, Met.Science, 13 (1979) p. 20.
50.  D V Edmonds, J.Iron Steel Inst., 210 (1972) p. 363.
51.  R A Ricks, P R Howell and R W K Honeycombe, Met.Science, 14 (1980) p. 562.
52.  G L Dunlop and R W K Honeycombe, Met.Science, 12 (1978) p. 367.
53.  G L Dunlop and R W K Honeycombe, Phil. Mag., 32 (1975) p. 61.
54.  N K Balliger and R W K Honeycombe, Met.Science, 14 (1980) p. 121.
55.  R H J Hannink and M J Murray, J.Mat.Sci., 9 (1974) p. 223.
56.  A J Ardell, Acta Met., 20 (1972) p. 601.

# ISOTHERMAL DECOMPOSITION OF RECRYSTALLIZED AUSTENITE

## IN STEELS MICROALLOYED WITH NIOBIUM AND VANADIUM

K.M. Tiitto †, L.E. Estrada †/††, A.J. DeArdo †

† Department of Metallurgical and Materials Engineering
848 Benedum Hall
University of Pittsburgh
Pittsburgh. PA. 15261

†† Instituto Mexicano de Investigaciones
Siderurgicas
Apartado Postal 491
Saltillo Coahuilo
Mexico

Extensive optical and electron microscopic studies were conducted on Nb and/or V microalloyed steels in order to investigate the isothermal decomposition of recrystallized austenite at temperatures ranging from 710°C to 500°C. Volume fraction of ferrite, either polygonal or nonpolygonal, was determined by quantitative optical microscopy. It was observed that the transformation kinetics for these steels are extremely rapid during the early stages of isothermal treatment so that the time-temperature-transformation (TTT) diagrams could not be constructed in a conventional fashion. Instead, the TTT diagrams representing 90% transformation were compiled. They exhibited single-or double-nose behavior depending on the austenitization temperature. To explain this behavior, both optical and electron microscopy were employed to correlate the appearance of the bay in the TTT diagrams to the changes in ferrite morphology and the occurrence of the interphase precipitation of the microalloy particles. The results of these experiments appear to favor the model in which the bay is caused by the pinning of the alpha-gamma boundary by the interphase precipitation.

## INTRODUCTION

The microstructure and properties of hot rolled HSLA steels are controlled by their chemical composition and processing. For example, the final ferrite microstructure is controlled by the austenite microstructure, the hardenability and the cooling rate. An important tool in understanding the evolution of microstructure is the isothermal transformation or TTT diagram. Whereas much work has been conducted on plain carbon [1,2] and simple alloy steels [3,4], relatively little systematic research has been done on steels of approximate HSLA commercial composition. The purpose of this current study, therefore, is to examine the kinetics and morphology of the evolution of the ferrite microstructure in Nb-V microalloyed steels. Of special interest in this work is the interplay which exists among transformation kinetics, ferrite morphology and the precipitation of microalloying (MA) elements which attends the transformation.

## EXPERIMENTAL PROCEDURE

### Materials

The study was conducted on four steels whose compositions are shown in Table I:

### Table I

### Steel Composition, Wt%

| Steel | C | Mn | Si | Nb | V | N |
|-------|-----|------|-----|-----|-----|------|
| 1 | .08 | 1.25 | .38 | – | – | .024 |
| 2 | .08 | 1.23 | .40 | – | .14 | .025 |
| 3 | .08 | 1.22 | .39 | .04 | .14 | .029 |
| 4 | .08 | 1.23 | .42 | .07 | .14 | .027 |

The steels were air-induction melted, silicon killed and teemed as 25kg ingots. The ingots were then homogenized and hot rolled to plates of thickness 12mm and air cooled to room temperature. Specimens of size 4 x 4 x 12mm were cut from these hot rolled plates for use in subsequent heat treatment.

### Heat Treatment

Austenitizing treatments were carried out at either 950 or 1200°C for 60 minutes under a purified Ar atmosphere. After reheating, the specimens were quenched into a lead bath for isothermal transformation. The transformation temperatures varied from 550 to 740°C and the transformation times from 5s to 4h. Since the specimens could not reach thermal equilibrium in less than 5s, no observations were made at times shorter than this. [5] Following the isothermal treatment, the specimens were quenched into iced brine.

### Metallography

Optical metallography was accomplished using standard laboratory techniques. The volume fraction transformed was measured by using a point counting method. [6,7] These data were used to construct TTT diagrams. Since the kinetics of transformation were very rapid, conventional TTT

diagrams indicating the start of transformation could not be constructed. [5]
However, it was possible to measure the kinetics at the later stages of
transformation; hence, TTT diagrams denoting 90% transformation were constructed. For the case of the acicular transformation product, an effort was
made to insure that the product observed had, in fact, formed isothermally
as opposed to during the subsequent quench. This was accomplished by observing no alteration in the isothermal product with changes in cooling rate
from the isothermal transformation temperature. The precipitation of MA
elements was studied using conventional TEM employing a JEM-100U instrument
operated at 120 KeV.

### RESULTS AND DISCUSSION

#### Microstructure and Reheated Austenite

The austenite microstructure which resulted from one-hour reheat at
950°C consisted of an equiaxed structure of average grain size 48μm for the
plain carbon (steel 1) and approximately 10μm for the three microalloyed
steels. Since 950°C is a rather low reheat temperature for steel 3, the
major portion of the microalloying elements would be expected to be in undissolved and coarsened carbonitrides. This was confirmed by TEM, Fig. 1.

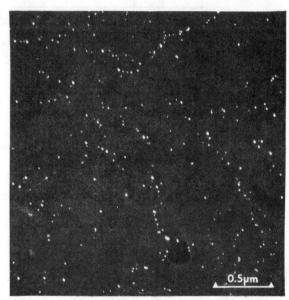

The undissolved carbon-
itrides appear to be located
on the substructure of the
austenite. They probably
formed during the earlier
processing on the sub-
structure of the aust-
enite and coarsened during
the reheating. These carbonitrides were identified by
electron diffraction as
Nb rich (Nb, V) CN with a
NaCl structure and a lat-
tice parameter of 0.429 nm.

As expected, larger
austenite grain sizes were
observed after the 1200°C
reheating. For example,
steel 1 had a grain size
of 330μm. According to
solubility data [8], no
precipitates would be
expected in steels 2 and
3 after reheating at 1200°C,
but some would be
expected in steel 4.
Limited TEM work
confirmed these
predictions.

Figure 1.  Dark-field electron micrograph on
steel 3 after reheat at 950°C for 1 h and quench
to room temperature.

#### Transformation Kinetics

The kinetics of the decomposition of austenite, based on optical
metallography, were summarized as TTT diagrams for 90% of transformation
and are shown in Figures 2 and 3, depending upon the reheat temperature.
When the steels were reheated at 950°C, they all exhibited a single-nose
TTT diagram, Figure 2. A single-nose diagram was also observed for steel 1
after reheating at 1200°C. However  when the microalloyed steels were re-

heated at 1200ºC, the resulting TTT diagrams showed a double-nose behavior, with a bay at 600ºC, Figure 3.

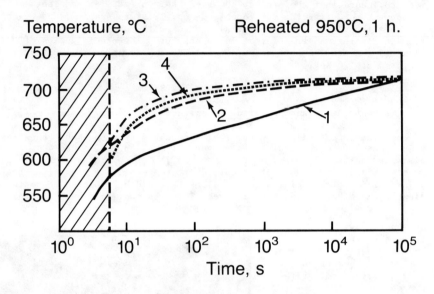

Figure 2. Isothermal transformation diagrams at 90% transformation in steels reheated at 950ºC.

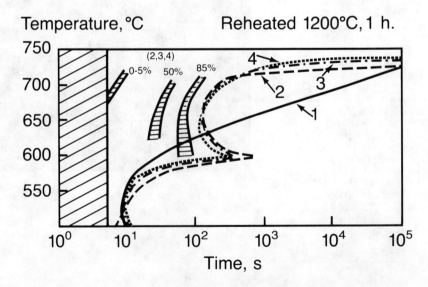

Figure 3. Isothermal transformation diagrams at 0 to 5, 85 and 90% transformation in steels reheated at 1200ºC.

After 950ºC reheating, the transformation kinetics of the three MA steels were remarkably similar. However, as is shown in Figure 2, the kinetics exhibited by the carbon steel were more sluggish than those of the MA steels at temperatures above 600ºC.

The kinetics exhibited after 1200ºC reheating were more complex. Once again, the three MA steels showed remarkably similar behavior, i.e., a double-nose curve with a bay at 600ºC. Steel 1, on the other hand, showed the same basic behavior as it did after reheating at 950ºC. That is, the kinetics for this steel above approximately 625ºC were nearly identical for the two reheat temperatures. Below 625ºC, however, the kinetics exhibited by steel 1 after the 950ºC reheat were more rapid than after the 1200ºC reheat. A comparison of the times corresponding to 90% transformation revealed that at temperatures above approximately 625ºC the MA steels transformed more rapidly than did the carbon steel. Between 625ºC and 550ºC, the opposite behavior was observed. The transformation kinetics were nearly identical near 500ºC.

The slower overall transformation kinetics observed after the 1200ºC reheating permitted TTT diagrams to be constructed at less than the 90% transformed used in Figure 2. Observations of the TTT diagrams at various amounts of transformation indicate that there is an ever increasing discontinuity in the TTT curve as the transformation proceeds near 600ºC. This discontinuity changes to a clearly observable bay in the TTT diagram at 85% transformation.

### Ferrite Microstructures

Optical metallography revealed that the nature of the ferrite formed depends on the reheating temperature, transformation temperature and steel composition, Table II. Under certain conditions, the ferrite morphology was observed to change from being essentially polygonal to essentially acicular with decreasing temperature.

TABLE II

Morphology of Transformation Products *

| Steel | $T_{Aust.}$, ºC | | | | $T_{Transf.}$, ºC | | |
|-------|-----------------|-----|-----|-----|------|-----|-----|
|       |                 | 740 | 710 | 680 | 650  | 600 | 500 |
| 1 | 950  | P | P | P   | A | A | A |
| 2 | 950  | P | P | P   | P | P | P |
| 3 | 950  | P | P | P   | P | P | P |
| 4 | 950  | P | P | P   | P | P | P |
| 1 | 1200 | P | P | P   | A | A | A |
| 2 | 1200 | P | P | P+A | A | A | A |
| 3 | 1200 | P | P | P   | P | A | A |
| 4 | 1200 | P | P | P   | P | A | A |

P= Polygonal    *= Predominant Feature
A= Acicular

However, this change was not observed in every case, since the MA steels
showed polygonal ferrite, with grain sizes from 2 to 7μm, over the entire
range of transformation temperatures after reheating at 950°C. In addition,
the final microstructures observed at 90% transformation were rather homogeneous, with the one exception of steel 2 reheated at 1200°C and transformed at 680°C. This homogeneity in structure was taken as evidence that there
was no appreciable change in transformation product as the gamma-alpha
transformation proceeded. Hence, the nature of the final product was assumed
to be indicative of that of the product present at earlier stages of
transformation.

In an attempt to relate the temperature of the bay with the temperature
associated with the change in ferrite morphology, a comparison was made
between the ferrite morphology observations, Table II, with the TTT diagrams,
Figures 2 and 3. The data indicate that whereas these two temperatures are
very close in steels 3 and 4, this appears not to be the case in steel 2.
It further appears that there is a change in ferrite morphology in the base
steel near 675°C for both reheating temperatures despite the fact that no
bay is observed in the TTT diagrams.

Grain size considerations would indicate that small starting austenite
grain sizes (950°C reheat temp-steels 2,3 and 4) contribute to the development of polygonal ferrite grains. It also appears that the base steel transforms more rapidly at low temperatures ($\leq$ 600°C) after 950°C reheating than
after 1200°C reheating. This may be at least partly attributed to the
small starting austenite grain size after 950°C reheating (48μm) when
compared to the grain size after 1200°C reheating (330μm).

### Precipitation Observations

The precipitation study conducted on steel 3 reheated at 950°C revealed
that very little precipitation accompanied the austenite-to-ferrite transformation which took place during the isothermal treatment. Some evidence
of a limited amount of precipitation on dislocations in ferrite was found,
but this precipitation was only rarely observed. The lack of precipitation
during the isothermal transformation is a direct result of the very limited
amount of MA elements being dissolved in the austenite during the reheating
at 950°C.

Since most of the MA elements in steel 3 were taken into solution
in the austenite upon reheating to 1200°C, the resulting precipitation
upon isothermal transformation was relatively extensive. The precipitation
behavior observed can be best described by considering three transformation
temperature regimes. At temperatures clearly above the bay, e.g. $>$ 625°C,
interphase precipitation was observed in the polygonal ferrite in all MA
steels. An example is given in Figure 4. The precipitates were identified
by selected area diffraction as being essentially V-rich (V, Nb) CN,
and they were crystallographically related to the ferrite by the Baker-
Nutting orientation relationship [9].

When the transformation took place below the bay, e.g. $\leq$ 600°C, the
resulting ferrite was essentially acicular in nature, Figure 5. No evidence
of either interphase or general dislocation-related precipitation was found
at the completion of the ferrite transformation. In fact, no precipitation
was observed in these structures unless they had undergone extensive aging
after the completion of transformation at these low temperatures. This lack
of concurrent precipitation with ferrite formation can be rationalized by
assuming that the ferrite formed below the bay is supersaturated with respect
to the MA elements.

The third temperature range is defined by the bay i.e., 600°C. Specimens representing various volume fractions of ferrite trans- ' formed at the temperature of the bay were examined to determine the progress of precipitation with the evolution of the ferrite structure. No precipitation was observed at 600°C until 120s had passed. At this time, interphase precipitation was observed in certain isolated regions of the TEM foils. An example of this precipitation at 600°C shown in Figure 6. Because of this low temperature, the precipitate size (3 nm) and row spacing (7 nm) were very small. With increasing transformation time, the extent of interphase precipitation also increased. It was further noted that no interphase precipitation was observed below this temperature.

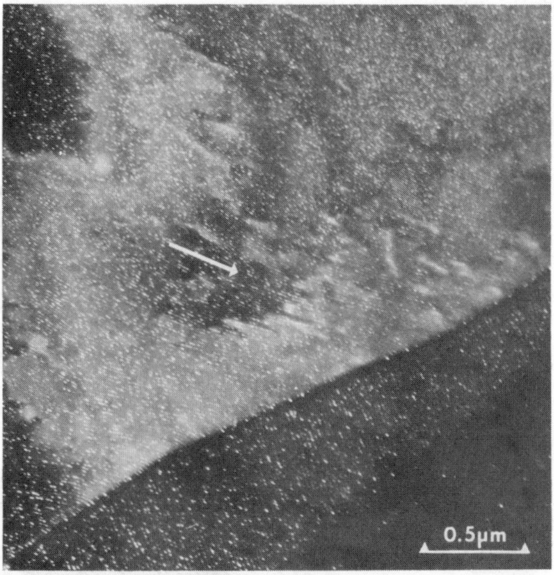

As shown in Table II, the evolution of the ferrite structure with time at 600°C starts and proceeds predominatly as acicular ferrite. As discussed above, the transformation kinetics became sluggish and the bay became evident at about 80% transformation. The precipitation studies discussed above clearly indicate that the point at which the bay becomes observable represents the same time at which interphase precipitation was first observed at this temperature.

Figure 4. Dark-field electron micrograph on steel 3 after reheat at 1200°C for 1 h and transformation at 700°C for 500s. The arrow points the direction of rows of interphase precipitates.

Figure 5. Bright-field electron micrograph on steel 3 after reheat at 1200°C for 1 h and transformation at 550°C for 10 s.

It should be noted that the ferrite which contained this interphase precipitation had a rather high dislocation density, indicative of its low temperature of formation.

## Summary of Evolution of Transformation Products

It is clear from the results given above that there is a strong relationship between transformation kinetics, ferrite morphology and precipitation behavior in microalloyed steels which have been reheated at 1200°C.

When the transformation takes place at high temperatures, the resulting transformation products are polygonal ferrite of low dislocation density plus MA carbonitrides. Since the precipitation is of the interphase type, the ferrite and the precipitates form at the same time, and,

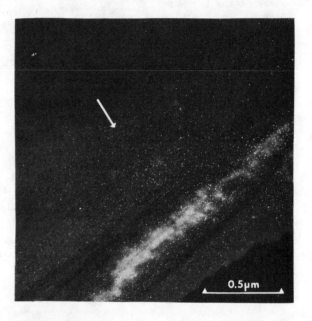

Figure 6. Dark-field electron micrograph on steel 3 after reheat at 1200°C for 1 h and transformation at 600°C for 120 s. The arrow points the direction of rows of interphase precipitates.

in fact, the precipitates are nucleated at the moving austenite-ferrite boundary. [3,4] The high transformation temperature explains both the low dislocation density and the presence of the interphase precipitates in the ferrite. It is well-known that the austenite-to-ferrite transformation is accompanied by a volume expansion which leads to the formation of dislocations. When the transformation takes place at a high temperature, these dislocations can be readily annihilated. This would result in a rather dislocation free polygonal ferrite. If, however, the polygonal ferrite were to form at lower temperatures, fewer of these dislocations would be eliminated, and the resulting polygonal ferrite would have a higher dislocation density. Since the supersaturation of MA elements has been relieved by the interphase precipitation, further aging of fully transformed polygonal ferrite does not result in additional precipitation. In fact, the precipitation hardening increment associated with interphase precipitation is actually reduced by aging subsequent to complete transformation[3,4]. This is caused by the coarsening of precipitates during the aging treatment.

When the transformation temperature is low, below the bay, the resulting transformation product is a highly dislocated acicular ferrite. It is believed that acicular ferrite forms by a shear mechanism [10,14]. This mechanism is responsible for both rapid transformation kinetics and a transformation product with a high dislocation density. The rapid ferrite reaction, when coupled with the low transformation temperature, eliminates the possibility for interphase precipitation. The resulting acicular ferrite is, therefore, supersaturated with respect to the MA elements. This super-saturation can be relieved through precipitation if the acicular ferrite is aged once the ferrite has formed. This is consistent with the observation that aging treatments are often used to increase the precipitation hardening increment in MA steels with acicular ferrite microstructures. [15]

The transformation products formed at the bay exhibit characteristics of both the high and low temperature products described above. The product formed at the initial stage of transformation was a highly dislocated, acicular ferrite, supersaturated with MA elements. As the transformation progressed, the formation of polygonal ferrite was observed. Similar to the high temperature polygonal ferrite, the polygonal ferrite formed in the later stages of transformation at 600°C also contained interphase precipita-tion. However, unlike the high temperature polygonal ferrite, the polygonal ferrite formed at 600°C contained a high dislocation density.

The simultaneous appearance of the polygonal ferrite with interphase precipitation and the bay can furnish important information concerning the cause of the bay in the TTT diagram. The occurrence of a bay has been variously attributed to factors such as (i) partitioning of elements between $\gamma$ and $\alpha$ [16], (ii) change in ferrite morphology [17], (iii) solute drag effects [18,19], and (iv) pinning effect of interphase precipitation. [3,4] The results of the present investigation appear to favor the importance of the interphase precipitation model while, at the same time, appear to discount the model based on a change of ferrite morphology.

## CONCLUSIONS

The results of this investigation have led to the following conclusions:

1. The C-Mn-Si base steel exhibited a single-nose TTT diagram after both 950 and 1200°C reheating. This observation is quite significant since the ferrite morphology changed from polygonal to acicular with decreasing transformation temperature.

2. The MA steels exhibited a single-nose TTT diagram after reheating at 950°C, and a double-nose diagram after reheating at 1200°C. The ferrite morphology remained polygonal at all transformation temperatures after 950°C reheating, whereas it changed from polygonal to acicular with decreasing temperature after 1200°C reheating.

3. No MA precipitation was observed during transformation after 950°C reheating.

4.  When the MA steels were reheated at $1200^{\circ}C$, two distinct types of precipitation were observed depending on the transformation temperature. At temperatures of $600^{\circ}C$ and higher, interphase precipitation which accompanied the transformation was observed. At temperatures below the bay, supersaturated acicular ferrite was found at the completion of transformation. This supersaturation led to general precipitation in the ferrite upon aging after the transformation.

5.  The lowest temperature at which interphase precipitation was observed corresponded to the temperature of the bay.

6.  The occurrence of the bay in the TTT diagram after $1200^{\circ}C$ reheating appears to be caused by the pinning of the alpha/gamma boundary by interphase precipitates, rather than by a change in the ferrite morphology.

## Acknowledgement

The authors wish to thank the Metals Division of Union Carbide Corporation and Niobium Products Company, Ltd, a subsidiary of CBMM, for supporting this research. Thanks are also due to the Graham Research Laboratory of the Jones and Laughlin Steel Corporation for providing the steels.

REFERENCES

1.  E.S. Davenport and E.C. Bain, Trans. AIME, 1930, V. 90, 117.

2.  The Making, Shaping and Treating of Steel, H.E. McGannon, Editor, Ninth Edition, United States Steel Corporation, 1971, Pittsburgh, PA, 1075.

3.  R.W.K. Honeycombe, Metall. Trans., 1976, V. 7A, 915.

4.  R.W.K. Honeycombe, Met. Sci., 1980, V14, 201.

5.  L.R. Estrada, Transformation Studies of V-Nb-N Microalloyed Steels, M.S. Thesis, Dept. of Met. and Matls. Eng., University of Pittsburgh, 1981.

6.  E.E. Underwood, Metals Handbook, 8th Ed., 1973, ASM Metals Park, OH, 37.

7.  C.I. Garcia and A.J. DeArdo, Metall. Trans., 1981, V. 12A, 521.

8.  K.J. Irvine, F.B. Pickering and T. Gladman, J. Iron Steel Inst., 1967, V. 205, 161.

9.  R.G. Baker and J. Nutting, Precipitation Processes in Steels, 1959, Special Rept. No. 64, ISI, London, 1.

10. L.J. Habraken and M. Economopoulos, Tranformation and Hardenability in Steel, 1967 Climax Molybdenum Co., Ann Arbor, MI, 69.

11. A.J. McEvily, R.G. Davies and C.L. Magee, Ibid, 179.

12. F.B. Pickering, Ibid, 192.

13. R.W.K. Honeycombe and F.B. Pickering, Metall. Trans, 1972, V.3,1099

14. A.P. Coldren, R.L. Cryderman and M. Semchyshen, Steel-Strengthening Mechanisms, 1969, Climax Molybdenum Co., Ann Arbor, MI, 17.

15. Y.E. Smith, A.P. Coldren and R.L. Cryderman, Towards Improved Ductility and Toughness, 1971, Climax Molybdenum Co., Ann Arbor, MI, 119.

16. A. Hultgren, Trans. ASM, 1947, V39, 915.

17. A.J. McEvily, R.G. Davies, C.L. Magee and T.L. Johnson, Ref. 10, 179.

18. G.R. Purdy, D.H. Weichert and J.S. Kirkaldy, Trans. TMS-AIME, 1964, V230, 1025.

19. K.R. Kinsman and H.I. Aaronson, Ref. 10, 33.

# THE BAINITE TRANSFORMATION

J. W. Christian and D. V. Edmonds

Department of Metallurgy and Science of Materials,
University of Oxford,
Oxford OX1 3PH, U.K.

Studies of the bainite transformation in steels are reviewed with the objective both of summarising recent research and attempting to evaluate strongly opposing views on the transformation mechanism based either on a diffusionless martensitic model or on the diffusional migration of interfacial growth ledges. Thus, the main microstructural features of bainite are first presented, and the various crystallographic relationships that have been observed between parent austenite and the ferrite and carbide product phases are documented. The question of the initial carbon content of bainitic ferrite is then addressed, firstly through a formal description of the thermodynamics of ferrite formation with a full supersaturation, partial supersaturation or equilibrium content of carbon, and then by a discussion of recent experiments employing modern analytical techniques which have been made in an attempt to deduce this carbon content. The importance of surface relief effects associated with bainite formation is emphasised and an attempt is made to classify the possible growth processes for bainite by using a set of virtual operations, with the result that a martensitic-type process emerges as the most likely candidate to result in the observed systematic shape change. The results of several investigations into the overall reaction kinetics in the bainitic temperature range, and the growth rates of individual bainitic ferrite plates, which have been variously quoted in support of particular transformation mechanisms are also considered. Finally, the proposed mechanisms of formation of bainite are considered in the light of the accumulated results, and it is concluded that the totality of available evidence favours a diffusionless martensitic mechanism.

## Introduction

Bainite forms from austenite in steels in a temperature range below that in which pearlite is produced and above that of the martensitic transformation. In pure iron-carbon alloys and in plain carbon steels, the pearlitic and bainitic temperature ranges overlap each other to a considerable extent, and this makes the interpretation of microstructure and kinetics very difficult. Aaronson (1,2) has emphasised that microstructural and kinetic definitions of bainite are not always consistent with one another, and this is one question which we shall attempt to resolve. We begin with a kinetic distinction.

In many alloy steels, bainite forms at temperatures below a pronounced bay in the isothermal transformation (T-T-T) diagram so that it may be studied in isolation without direct interference from competing reactions. In such circumstances, isothermal transformation to bainite exhibits the classical kinetic features of nucleation and growth reactions with one important exception; the transformation ceases before complete decomposition of the austenite (or establishment of a metastable ferrite-austenite equilibrium) has occurred, and the extent of transformation is a function of the reaction temperature, increasing as the temperature is lowered. A curve relating the limiting percentage of austenite transformed to the reaction temperature may thus be plotted, and extrapolation back to 0% transformation defines the so-called kinetic $B_S$ (bainitic start) temperature, above which bainite cannot form (3). A similar procedure may also be used to define a kinetic $B_S$ in some low alloy steels for which, despite a partial overlap of the bainitic and pearlitic temperature ranges, an isothermal curve of fraction transformed vs. time shows a prominent inflection at an intermediate stage because of the cessation of the bainite reaction in advance of the initiation of the pearlite reaction (4). The temperature of the bay in a T-T-T curve clearly should correspond to the kinetic $B_S$ temperature in those alloys in which the bainitic and pearlitic ranges are well separated, and this provides a more usual (though possibly less accurate) means of defining $B_S$.

Bainite usually consists of a non-lamellar aggregate of roughly lath- or plate-shaped ferrite grains with carbide precipitates either within the ferrite or formed directly from austenite in the inter-lath regions. The widest microstructural definition of bainite (1) would encompass all non-lamellar aggregates of ferrite and carbide (irrespective of the ferrite morphology), and structures conforming to this description certainly form well above the kinetic $B_S$, as, for example, in 'divorced pearlite'.Indeed,it has been proposed that a microstructural $B_S$ temperature should,in principle, be the eutectoid temperature for all hypoeutectoid alloys, and should follow the extrapolated $\gamma/\gamma + \alpha$ curve for hypereutectoid alloys. A narrower microstructural definition is based on non-lamellar carbide precipitation in association with Widmanstatten ferrite, but this also leads to discrepancies. However, when proper attention is paid to the substructure and composition of the ferrite, most workers believe that a clear distinction emerges between high temperature pro-eutectoid Widmanstätten ferrite and bainitic ferrite, and that the formation of the latter takes place in hypoeutectoid steels only below a microstructural $B_S$ which is equivalent to the kinetic $B_S$. According to this view, there is a clear distinction in growth mechanisms sub-structure and kinetics of ferrite regions formed above and below $B_S$. An opposing view (5) is that the bainitic bay and the apparent change in kinetics arise from drag caused by substitutional solute segregated to the ferrite-austenite interface, and that apart from this special effect, the formation of bainite is part of a process which is continuous through the kinetic $B_S$.

A strict application of the microstructural definition would imply that carbide-free structures should not be classed as bainite. However, it is possible for entirely ferritic products to form in a manner which is closely similar in morphology, growth mechanism and kinetics to the formation of classical bainite; this is observed, for example, in steels containing silicon. The extent to which the formation of ferrite precedes the carbide precipitation, and the extent of the initial carbon supersaturation in the first-formed ferrite plates in normal (duplex) bainitic structures and in silicon steels, are controversial matters which are difficult to investigate experimentally, and recent attempts with modern techniques have been only partly successful.

Early replica (6) and transmission (7-9) electron microscopy revealed what are now well-recognised differences in the distribution of the carbides formed in the upper and lower parts of the bainitic temperature range, and thus justified the original division into upper and lower bainite (10), for which some kinetic evidence has also been presented (11,12). Another microstructural form has been called 'granular bainite' (13); it consists of relatively coarse ferrite plates or regions of other shapes separated by retained austenite or by high carbon martensite. This structure is formed only in certain alloy steels and appears more readily after transformation by continuous cooling rather than isothermal treatment. Microstructures apparently equivalent to those of lower bainite are also formed in hypereutectoid alloys after appropriate isothermal heat treatment, but at higher temperatures, the carbide phase apparently precipitates first, leading to a structure which has been termed 'inverse bainite' (14).

The remaining characteristic feature of bainite formation is that the ferritic plates or laths have macroscopic shapes which are markedly different from those of the austenitic regions from which they are formed. This shape change is usually considered to be an invariant plane strain in which the broad (habit plane) face of the ferrite lath corresponds to the invariant plane. Thus the shape change is similar to that found in martensitic plates, and, in particular, there is a large shear component. Analogous changes of shape are also found in pro-eutectoid ferrite plates (formed at appreciably higher temperatures) and in many non-ferrous alloys which decompose isothermally at a finite rate; such reactions are also called 'bainitic' by some authors, although the implication of a common mechanism is not universally accepted. This third definition of bainite as the product of a transformation combining 'surface relief' with composition change (1,2) will not be adopted here since we regard it as essential in steels to distinguish between bainite and high temperature pro-eutectoid ferrite. This is not to deny the importance of the shape change, the interpretation of which in terms of growth mechanism has been one of the most controversial aspects of the continuing debate on bainite.

Very many excellent reviews of bainite formation have been published (1-3, 15-18) and the authors are under no illusion that what they write will constitute any kind of breakthrough in interpretation or that it is likely to command general acceptance among scientists who hold strongly opposed views. Nevertheless, progress is being made in understanding the reaction through the application of modern high resolution structural and analytical techniques, and it seems useful occasionally to attempt to survey the current position. As in all such reviews, the authors' prejudices will, no doubt, be embarrassingly revealed; we can only promise to attempt to be objective, and especially to distinguish clearly between hard experimental information and soft theoretical inference.

## Morphology and Crystallography of Ferrous Bainite

The classical structure of upper bainite consists of a 'feathery'
aggregate of laths or needle-like ferrite particles with cementite precipitated in the inter-lath regions parallel to the long axes of the laths.
The laths appear to nucleate on the austenite grain boundaries, and at
temperatures just below $B_s$, where individual laths are observed, the amount
of carbide precipitated is very small. As the temperature is lowered, laths
increasingly form adjacent to one another in parallel groups, a process
which has been described as 'sympathetic nucleation' (19), and subsequently
amalgamate into irregular, roughly plate-shaped regions, often described as
'sheaves' of upper bainite. The widths of the individual laths decrease
with decreasing temperature, whilst the axial lengths are largely unchanged
and according to Pickering (20) are governed mainly by the austenitic grain
size or its sub-division by previously formed ferritic regions. The volume
fraction of ferrite and the amount of carbide precipitated both increase
with decreasing temperature, and at sufficiently low temperatures, carbide
precipitates within the larger ferrite plates are also observed, although
this may only represent engulfment by the ferrite of carbides originally
precipitated in the austenite. Very similar ferritic structures without any
visible carbides are formed in silicon-containing steels, and close
examination shows that the austenite remaining after cessation of transformation is then enriched in carbon; as already emphasised, such structures
are generally regarded as bainitic.

In general, the effect of increasing the carbon content of the steel is
similar to that of decreasing the reaction temperature (20); that is, the
laths become thinner and amalgamate more readily and more carbide is precipitated. Eventually, the interlath carbides form nearly complete
cementite films between the ferrite grains so that the resulting structure
has an almost lamellar appearance. In hypereutectoid steels, there is
evidence that cementite is nucleated first to give the structure which
Hillert (14) described as inverse bainite. The initial cementite crystal
develops as a long lath or plate and according to Pickering (20) rapid
side-by-side nucleation then leads to a spear-like structure in which laths
of cementite and ferrite appear to grow alternately. In a more detailed
study, however, Kinsman and Aaronson (21) have shown that although inverse
bainite begins with the formation of individual single spines of cementite,
the region around any one such spine usually develops into a structure
resembling normal bainite with large ferrite laths and small carbide precipitates. They ascribe this limited ability of the initial inverse
bainite structure to reproduce itself to the greater volume fraction and
higher growth velocity of the ferrite regions.

Transmission electron microscopy by many authors has shown that an
upper bainite sheaf consists of a parallel arrangement of ferrite laths or
structural sub-units with a high dislocation density. A typical structure
is shown in Figure 1 and as emphasised by Oblak and Hehemann (22), there is
a marked contrast between this internal structure and that of proeutectoid
Widmanstätten ferrite which has no lath or cellular substructure and has a
rather low dislocation density. The laths are in immediate contact with
each other, but occasional carbides are precipitated from austenite trapped
between them. More recently, Bhadeshia and Edmonds (23) have shown that
sub-units can also be distinguished in a silicon-containing steel where no
carbides form but where the separately nucleated ferrite regions are separated by films of carbon-enriched retained austenite (see Figure 2).
These workers also found that the sub-units have a lenticular shape, sharply
pointed tips being particularly apparent near the edge of the sheaf, that the

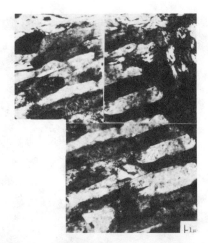

Figure 1 - Micrograph (TEM) of upper bainite in a
0.4%C-3.34%Ni steel isothermally transformed at
540°C for 15 hours. After Oblak and Hehemann (22).

width of a sub-unit is approximately the same at the tip as it was at the
original nucleation site, and that sympathetic nucleation occurs near the
tips of existing sub-units.

Several workers have found by diffraction techniques and other methods
that the various lath sub-units in a sheaf either all have the same
orientation or have a small spread of orientations so that where they
impinge on one another only low angle boundaries are formed.  Direct
measurement of the ferrite-austenite orientation relationship is difficult
since the remaining austenite after partial reaction normally transforms
to martensite during cooling to room temperature.  Smith and Mehl (24) in
early work used an X-ray pole figure method to deduce that upper bainitic
ferrite has a Nishiyama-Wasserman (25,26) orientation relative to austenite
and that lower bainitic ferrite has a Kurdjumov-Sachs (27) orientation.
More recent direct measurements have often given approximately Kurdjumov-
Sachs relations for upper bainite, although Sandvik (28) reports an
irrational relation which is rather closer to Nishiyama-Wasserman.  The
relative orientations of the cementite and austenite do not appear to have
been measured; the Pitsch (29) relationship is generally assumed.

Many measurements have been made of the relative orientations of the
ferrite and cementite phases in a bainitic product.  Shackleton and Kelly
(9,30) found that in two-thirds of the 36 plates of bainite which they
examined, the ferrite and cementite were relatively orientated in a variant
of the Bagaryatskii (31) relationship commonly observed for the precipitation of cementite in tempered martensite or quench-aged ferrite.
However, various (irrational) ferrite-cementite relationships found in the
remaining cases differed from the Bagaryatskii relation but could all be
obtained by combining variants of the usual (Kurdjumov-Sachs and Pitsch)
ferrite-austenite and cementite-austenite relations.  Since appropriate
combinations of variants will also give the Bagaryatskii relation, all the
data are consistent with the hypothesis that the cementite regions in upper
bainite form directly from the austenite.  Pickering (20) also states that
in no case did he find an orientation indicative of direct precipitation of
cementite from ferrite or martensite.  Slightly different results have been

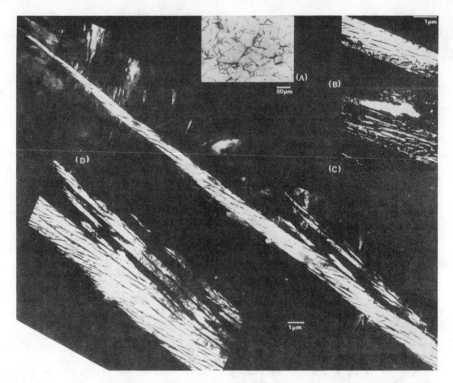

Figure 2 - Microstructural appearance of upper bainite in a silicon steel isothermally transformed at 286°C for 30 min. After Bhadeshia and Edmonds (23,54). [(A) Micrograph (optical) showing upper bainite sheaves in a martensitic matrix. (B) Micrograph (TEM) of a sheaf of upper bainite. (C) Corresponding retained austenite dark-field image. (D) Montage of micrographs (TEM) illustrating morphology and substructure of a sheaf of upper bainite].

reported by Ohmori and Honeycombe (32) who found the Isaichev (33) rather than the Bagaryatskii orientation relation, but they do not challenge the conclusion that carbide precipitates from the austenite.

This hypothesis is strengthened by determination of the habit plane of the cementite which is reported (9,30) to be $(101)_{cem}$ in terms of its own lattice with the long direction $<010>_{cem}$ for all orientations investigated. In some cases, this habit plane when referred to the ferrite lattice does not correspond to the $\{211\}_\gamma$ habit which is expected for cementite precipitation from ferrite.

The true habit of bainitic ferrite is difficult to determine because of the morphological complexities. For upper bainite, Smith and Mehl (24) found a $\{111\}_\gamma$ habit, and this is equivalent to the $\{110\}_\alpha$ habit later reported by Ohmori (34) if the habit plane corresponds to the parallel close-packed planes of either of the usually reported rational orientation relations. However, Davenport (35) found the habit plane to be close to a $\{223\}_\gamma$ plane which is 12° away from this pair of close-packed planes, and his result appears to be consistent with the recent work of Hoekstra (36) who determined a number of individual habit planes in a commercial nickel-chromium-carbon steel transformed at 365°C and found appreciable scatter with a mean habit plane close to {569}$\gamma$. In another recent investigation, Sandvik (28) has measured the ferritic habit planes in four commercial iron-silicon-carbon alloys at temperatures from 290° to 380°C; these would normally be considered to straddle the transition from lower to upper bainite, but no carbide precipitates were observed. An irrational mean habit of approximate indices {0.37, 0.66, 0.65}$\gamma$ was obtained. Finally, we should note that there is fairly general agreement that the long direction of the ferrite laths is parallel to the close-packed direction <111>$_\alpha$.

The division of the bainitic range into upper and lower regions was first suggested by Mehl (10) on the basis of microstructural differences, and this implies that in most steels there should be a fairly sharp change in morphology. The transition is usually found to occur at ~350°C and not to be very dependent on composition, but Pickering (20) found appreciably higher transition temperatures in lower carbon steels (containing 0.5% molybdenum), rising to a maximum of ~550°C at ~0.5 wt% carbon and then decreasing again (see Figure 3). However, Ohmori and Honeycombe (32) in an

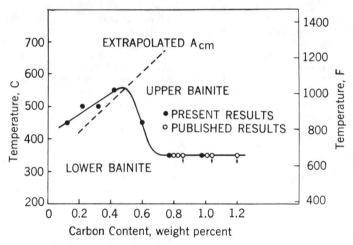

Figure 3 - Effect of carbon content on the transition temperature between upper and lower bainite. After Pickering (20).

investigation of plain carbon steels of similar composition except for molybdenum found a constant transition temperature of 350°C down to 0.4% carbon and no lower bainite below this composition. In most steels, there is an intermediate transformation range in which both upper and lower bainite may form.

Although the division was first made by optical metallography, the most striking differences between the structures of upper and lower bainite arise from the distribution of the carbides, and are hence most clearly revealed by electron metallography. In lower bainite, the carbide particles are contained within the ferrite regions and are usually oriented at a characteristic angle of about 60° to what appears in section as the long axis of the ferrite (see Figure 4). Actually, many authors report that lower bainite has a plate rather than a lath morphology, as was verified by Srinivasan and Wayman (37) by careful two-surface examination. The plates

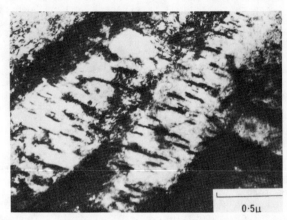

Figure 4 - Micrograph (TEM) of lower bainite in a 0.6%C
steel transformed at 300°C.   After Honeycombe and
Pickering (15).

are reported by Oblak and Hehemann to be an amalgam of smaller sub-units, as
in  upper bainite, but this was not observed by Srinivasan and Wayman and
sub-units do not appear to be obvious in many published micrographs.   The
precipitated carbides have been found to have either the ε-carbide or the
cementite structure, the latter being perhaps more usual; it is not
entirely clear whether in some conditions, ε-carbide forms first and is then
succeeded by cementite as in the low-temperature tempering of martensite, or
whether direct formation of the two carbides occurs competitively.
Bhadeshia (38) has attempted to rationalise this in terms of the arguments
put forward by Kalish and Cohen (39), who have estimated that a dislocation
density $\sim 2 \times 10^{12}$ cm$^{-2}$ will prevent ε-carbide precipitation in martensite
for steels up to 0.2% carbon.   The hypothesis is that if the dislocation
density is high, sufficient carbon atoms are bound to the dislocations to
prevent ε-carbide formation, and cementite is formed directly.   Bhadeshia
estimates that ε-carbide will not be formed in bainitic ferrite for steels
with carbon contents <0.55%, which agrees fairly well with published
observations.

     Determinations of the orientation relations between lower bainitic
ferrite and austenite give results close to either the Kurdjumov-Sachs or
the Nishiyama-Wasserman relations, which may indicate that the true
relation is irrational.   The relation between cementite and ferrite has
often been reported to be the Bagaryatskii relation found in tempered
martensite, but Shackleton and Kelly (9) additionally found a few rare
examples of another (irrational) orientation relation also encountered in
tempered martensite.   Although the Bagaryatskii relation is consistent
with both ferrite and cementite forming directly from the austenite, the
absence of the other ferrite-cementite relations found in upper bainite led
both Pickering and Shackleton and Kelly to conclude that cementite in lower
bainite is probably precipitated directly from supersaturated ferrite.
However, later work by Ohmori (34) and Huang and Thomas (40) has found the
ferrite-cementite relation to be that of Isaichev (33) which differs from
the Bagaryatskii relation by a rotation of about 4° around <111> .   This
small difference probably does not affect the argument about whether the
cementite forms from the austenite or the ferrite, but the observation of a
unique variant of the cementite in each lower bainitic ferrite plate and
the alignment of the particles within the plate (see Figure 5) has led to
the suggestion that it forms by interphase precipitation (32,41) at the

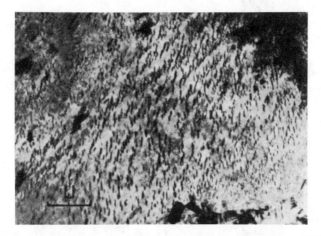

Figure 5 - Micrograph (TEM) of lower bainite in a 0.69%C
steel transformed at 300°C. After Ohmori and Honeycombe
(32).

ferrite-austenite interface.  On the other hand, up to four different
variants of the cementite habit plane were observed recently by Bhadeshia
and Edmonds (23) in a single lower bainitic ferrite crystal of a silicon-
manganese steel.  These workers found an unusual rational orientation
relationship in which $(011)_{cem}$ is parallel to $\{011\}_\alpha$ and $[1\bar{2}2]_{cem}$ is
parallel to $<100>\alpha$, but state that this cannot be combined with the
Kurdjumov-Sachs relation to give the expected three-phase $\alpha$-$\gamma$-cementite
relationship.  A rigorous crystallographic analysis in an iron-chromium-
carbon alloy has also confirmed that the expected $\alpha$-$\gamma$-cementite relationship
does not exist, and this is interpreted as evidence that an interphase
precipitation mechanism does not operate (38).  Bhadeshia and Edmonds also
determined the habit plane as $\{011\}_\alpha$ , in contrast to the $(201)_{cem}$ or
$\{213\}_\alpha$ found by Ohmori with the Isaichev orientation relation.

When $\varepsilon$-carbide has been detected in lower bainite, it has been found to
have an orientation relation with the ferrite close to that given by Jack
(42) which is again a relation found in martensite tempered at low
temperatures.  As in the case of cementite, this orientation relation can be
interpreted either in terms of direct precipitation from the ferrite or as
arising from precipitation of both ferrite and $\varepsilon$-carbide from the austenite
with suitable variants of the usual rational approximation to the
orientation relations.

It remains to discuss the crystallography of the austenite-ferrite
transformation in lower bainite.  The habit plane, like the orientation
relation, is generally found to be irrational; in a pure iron-carbon alloy
(32) and a commercial silicon-manganese containing plain carbon steel (34),
the habit is close to $\{496\}\gamma$ whereas in an iron-chromium-carbon alloy (37)
it is near $\{254\}\gamma$.  An early suggestion (16), based partly on Smith and
Mehl's results, was that the crystallography of upper bainite should be
similar to that of martensite in a steel of appreciably lower carbon content
than the specimen under investigation, whilst the crystallography of lower
bainite should be very similar to that of martensite in the same steel; it
has, in fact, often been doubted whether there is any difference between

lower bainite and auto-tempered martensite. However, in the first application of the theory of martensite crystallography to bainite, Bowles and Kennon (43) showed that the observed habit is not predicted by the hypothesis of transformation twinning on $\{112\}\alpha$, but is consistent with a different lattice invariant shear. No evidence of transformation twinning has ever been found in bainite despite the attractiveness of internal twinning as an explanation of the aligned carbides in lower bainite, and Srinivasan and Wayman, moreover, verified directly that in their 7.9% chromium, 1.1% carbon steel, the orientation relations, habit planes and shape deformation are all different from those found for martensite in the same alloy. These workers and Ohmori (34) were able successfully to apply the crystallographic theory of martensite to explain their results on lower bainite by choosing a mode of lattice invariant deformation which is not equivalent to fine twinning of the product.

## Thermodynamic Relations

In conventional martensite theory, the temperature at which the free energy per atom (or mol) of unconstrained austenite is equal to that of martensite (i.e. supersaturated, tetragonal 'ferrite') of the same compositon is denoted by $T_0$; this is a slightly different usage from the original definition of Zener (44) in which $T_0$ is the temperature of equilibrium with supersaturated cubic ferrite. If interstitial solutes are present, the two temperatures differ by an amount which represents the free energy of Zener ordering (45). Diffusionless transformation of austenite to martensite without Zener relaxation of the carbon atoms can take place only below the first of these temperatures, which we shall now call $T_{0m}$, whereas diffusionless transformation to cubic ferrite is thermodynamically possible below the second, $T_0$, temperature. In both cases, however, diffusionless transformation immediately below the appropriate $T_0$ is not possible except in unusual experimental configurations because of the additional free energy which is stored in the martensitic product and the surrounding matrix by the transformation process itself. This additional strain and surface free energy depends on the details of the morphology, crystallography and substructure of the product phase, and on the extent of the transformation, but a typical figure for the initial stages of transformation in carbon containing steels is of the order 400-600 J $mol^{-1}$ (46,47). Thus, two further temperatures $T'_{0m}$ and $T'_0$ may be defined for any alloy composition by the condition that the actual first formed regions of martensite or supersaturated ferrite are in constrained equilibrium with the austenite. Finally, the observed $M_S$ temperature represents a further cooling below $T'_{0m}$ because of the required driving force for nucleation of martensite; estimates of the total chemical driving force at $M_S$ vary for various ferrous alloys from $\sim$1,000 to $\sim$2,400 J $mol^{-1}$ (47). The question whether the $B_S$ temperature similarly represents a necessary further supercooling below $T'_0$ to allow nucleation of supersaturated ferrite remains to be discussed.

Curves of $T_0$ as a function of composition may be calculated for many steels using procedures first given by Aaronson et al (48) and subsequently improved by later workers (49,50). In an early discussion of such calculations, Kinsman and Aaronson (51) showed that for a number of steels investigated by Oblak and Hehemann (22), the measured $B_S$ temperatures lay on or below the calculated $T_0$ curve, so that these results appear to endorse the thermodynamic feasibility of proposals that the initial ferrite regions to form in upper bainite have the composition of the parent austenite. Kinsman and Aaronson further calculated that with typical parameters, the excess carbon could diffuse out of a ferrite sub-unit in a time of $\sim$0.003 s whereas measured growth rates indicate an interval of $\sim$10 s between the

nucleation of successive sub-units.  However, the excess carbon concentration at the interface dissipates relatively slowly, so that sub-units
forming at fixed 10 s intervals have gradually increasing carbon contents.
Kinsman and Aaronson refer to this kinetic effect as a 'thermodynamic
barrier' to growth, and point out that growth must eventually slow down to
prevent the composition of the new sub-units falling above the $T_O$ curve.

For hypereutectoid steels, the $B_S$ temperature often lies well above the
calculated $T_O$ temperature (1, 48, 51) but the situation is complicated by
the formation of inverse bainite in which carbide precipitates first
(14, 21).  However, Bhadeshia (46) plotting the critical free energy for
nucleation at $B_S$ from the extensive data of Steven and Haynes (52) for the
two rival assumptions of a critical nucleus of unchanged carbon composition
and of equilibrium composition, concluded that the latter assumption is much
more likely to be valid; in a few of the steels examined, the former assumption apparently requires nucleation with a negative driving force, or in
other words $B_S$ lies above the calculated $T_O$.

Consider next the possibility that ferrite forms by partitioning
carbon between the product regions and the austenite which remains, but that
there is no partitioning of any substitutional elements nor any precipitation of carbides.  The metastable $\alpha + \gamma/\gamma$ equilibrium line under these
conditions may be denoted $Ae'_3$, and if the mechanism of formation
necessarily involves non-chemical stored energy, there will similarly be a
lower temperature $Ae''_3$ representing the actual constrained equilibrium
between ferrite regions and austenite.  Further supercooling will then lead
to a temperature, $W_S$, at which the nucleation rate of low-carbon ferritic
regions of a particular type becomes appreciable.  (For reasons which will
become apparent, W stands for Widmanstätten ferrite).

We now consider the thermodynamics of possible growth processes above
and below $T_O$.  Schematic free energy vs. composition curves are shown in
Figure 6 for an alloy of mean carbon content $\bar{x}$ at a temperature $T_1$ above $T_O$.

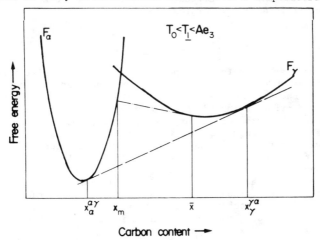

Figure 6 - Schematic austenite and ferrite free energy
curves ($T_1 > T_O$).  After Bhadeshia (46).

Growth of a ferrite region will decrease the free energy for all compositions between the equilibrium $x_\alpha$ and the composition $x_m$ defined by the
tangent to the $\gamma$ free energy curve at $\bar{x}$.  Bhadeshia (46) suggests that

growth with any partial supersaturation between $\bar{x}$ and $x_m$ will be unstable
to fluctuations in the interface composition, and hence will rapidly cascade
towards the equilibrium segregation, as is normally assumed in diffusional
theories of growth. The corresponding free energy curves at a temperature
below $T_O$ are shown in Figure 7 (53), and the same conclusion is valid
provided fluctuations in composition are feasible. However, if there is

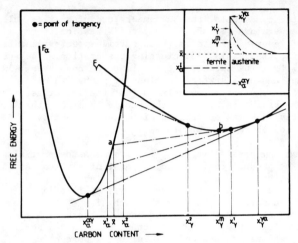

Figure 7 - Schematic austenite and ferrite free energy
curves ($T_1 < T_o$). After Bhadeshia and Waugh (53).

virtually no diffusion in the growth interval, growth at constant composition is now an alternative. The general conclusion that growth is not
likely to produce partially supersaturated regions of ferrite might be
challenged by supposing that particular interface configurations confer
stability, but it is a useful working hypothesis.

## Initial Composition of Bainitic Ferrite

The initial carbon content of upper bainitic ferrite plates is a
difficult problem which has given rise to appreciable controversy. Micrographs of upper bainite are all consistent with partitioning of the carbon
to the austenite prior to carbide precipitation; those upper bainite
structures where the carbides appear inside the ferrite can probably be
ascribed to subsequent engulfment by the ferrite. What is not clear,
however, is whether segregation accompanies growth, and indeed controls the
growth rate, or whether it follows the rapid diffusionless growth of
individual sub-units of the ferrite structure. Direct evidence about the
initial composition is almost impossible to obtain since the calculations
(51) indicate that the carbon could diffuse out of a sub-structural lath in
a few milliseconds. In recent years, attempts have been made to answer
this question by utilising the phenomenon of incomplete reaction on isothermal holding at any fixed temperature.

The basic assumption made in the interpretation of the cessation of
reaction at a fixed temperature is that this is a consequence of the carbon
enrichment of the austenite. There are then two possibilities, either
two-phase (metastable) equilibrium has been attained between the ferrite
and the austenite or the carbon content has risen to a level at which
diffusionless growth of ferrite is no longer possible. In the first case,

the mean carbon content of the austenite after growth has ceased would be expected to correspond approximately to the $Ae_3'$ or $Ae_3''$ curve whereas in the second case, this content should be close to the $T_0$ or $T_0'$ curve. The experiment thus simply consists in determining the carbon content of the austenite, either directly with high resolution techniques such as atom-probe analysis or lattice imaging, or indirectly from the extent of transformation.

This method was introduced by Bhadeshia and Edmonds (54) and a number of different investigations have now been made (53-57). Some typical results are shown in Figure 8. It will be seen that there is an appreciable

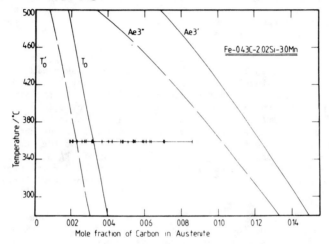

Figure 8 - Calculated phase boundaries and experimental carbon determinations. After Bhadeshia and Waugh (53).

spread in the individual carbon determinations, but that all values fall well short of the $Ae_3''$ line. On the other hand, the mean carbon content is appreciably in excess of the $T_0'$ value, so that at first sight neither theory is supported and the results seem to indicate that upper bainitic ferrite grows with a partial supersaturation. Bhadeshia and co-workers, however, have emphasised that once a local region of austenite has acquired sufficient carbon to prevent diffusionless transformation at the reaction temperature, it may nevertheless continue to accumulate carbon up to a maximum level given by the $Ae_3''$ curve. In other words, the only true criterion is whether or not reaction has ceased because of the attainment of a lever-rule equilibrium.

Bhadeshia and Waugh (53) have argued that growth involving partial supersaturation, whilst not requiring the austenite composition to attain $Ae_3''$, nevertheless does imply some minimum terminal carbon level of the austenite which is represented in Figure 7 as $x_\gamma^m$. The majority of the experimental data from Figure 8 represent carbon contents lower than the calculated value of $x_\gamma^m$ and hence are in apparent disagreement with the hypothesis of partial segregation.

As indicated above, measurement of the carbon content of the ferrite is likely to prove inconclusive as an indicator of the initial carbon content because of the rapidity of carbon diffusion at the temperatures of bainite formation. Thus in electron microprobe studies of various forms of ferrite produced in the early stages of transformation above and below the bay in

the T-T-T diagram, Kinsman et al (58) found the carbon content always to be
below 0.03%. No difference in carbon content was found for high temperature
grain boundary allotriomorphs, Widmanstätten sideplates and plates of
bainitic ferrite. Slightly different results have, however, been obtained
in the recent atom-probe studies (53). The post-transformation ferrite
contained less than 1 at% carbon compared with the overall level of 1.93at%
but the measured levels were nonetheless appreciably higher than the
equilibrium carbon concentrations. This result is tentatively attributed to
the high dislocation density of the bainitic ferrite, since binding of
carbon atoms to dislocations might prevent or hinder their diffusion out of
the ferrite; there is, in fact, some evidence from the imaging atom-probe of
inhomogeneous distribution of carbon in ferrite. In any event, the
existence of residual supersaturation strengthens the view that the initial
supersaturation was much greater.

Whilst the atom-probe studies may certainly be interpreted as
supporting the belief that initial growth of ferrite takes place without
carbon diffusion, the interpretation requires that individual sub-units
cease growth for some other reason, possibly arising from the strain
energy - shape relation or from accumulating defects or a loss of
coherency at the interface. It might then be argued that the cessation of
the whole reaction is a result of a similar interaction, and cannot be
attributed simply to the increase in the carbon level of the austenite.
The strain energy which is used in the calculation of $T_0'$ and $Ae_3'$, for
example, is strictly not a constant per unit volume of ferrite produced but
varies with the aspect ratio of a plate and hence probably with the total
volume fraction of ferrite.

The initial carbon content of lower bainite presents equally
difficult problems. The fact that the carbides are all precipitated within
the ferrite and the crystallographic results already described have been
used as arguments for believing that the ferrite forms in a  diffusionless
manner and subsequently relieves the supersaturation by internal precipitation. In support of this it has also been reported (23, 59) that
occasionally carbides could not be detected within lower bainite plates
at an initial stage in their formation. This theory has to face the
difficulty that although several variants of the precipitate
crystallography are sometimes encountered within a single ferrite plate, the
usual morphology (Figure 4) indicates the predominance of a single variant
and some tendency towards alignment of the carbide particles along certain
planes. One interpretation of this is in terms of preferential formation
of precipitates along some internal defects left by the ferrite-austenite
interface; since twins have never been observed (and also do not seem to be
consistent with the observed crystallography when the usual martensitic
theory is  applied (37, 43)), tentative suggestions have been made that the
interface dislocations of the lattice invariant shear may leave behind some
kind of fault. A different view (38) is that the particular carbide
variant which forms is the one that will allow most relief of the
transformation strain. As mentioned above, the alternative explanation for
the observed morphology is that precipitation of carbide takes place at
the interphase interface, and thus is similar to the high temperature
interphase precipitation of many alloy carbides which have been extensively
studied in recent years (41). There is, however, an obvious conceptual
difficulty in accepting that upper bainitic ferrite may form rapidly and
without diffusion whereas lower bainitic ferrite has a growth rate
controlled by the interphase precipitation.

## The Shape Change in Bainite Formation

The observation of a systematic change of shape in the form of an invariant plane strain on the habit plane of a plate-shaped particle was originally proposed (60, 61) as the definitive experimental observation characterising a martensitic transformation. However, since the observations of Ko and Cottrell (61) on bainite and Ko (63) on pro-eutectoid Widmanstätten ferrite, evidence has steadily accumulated that such shape changes may also be produced in the course of phase transformations which involve long-range diffusion. The significance of shape change observations has thus been questioned. Several excellent analyses of aspects of this problem have been published (64-68) but there is also much confused discussion in the literature, and we therefore address it again.

At the risk of labouring the obvious, we first emphasise that a macroscopically homogeneous deformation can only arise during growth if the net atomic displacements include common non-random components which accumulate to give the observed effects. If we focus attention on those equivalent lattice points which define unit cells (not necessarily primitive cells) of the two structures containing the same number of atoms, we see that a shape deformation will result only if the new set of such lattice points are related to the original set either by a homogeneous deformation or by a combination of such a lattice deformation with a lattice invariant deformation. As in the theory of martensite crystallography, the lattice invariant deformation modifies the macroscopic shape change produced by the lattice deformation but leaves the unit cells unchanged; we do not restrict the lattice invariant deformation to simple shears, however, but allow any deformation which would be produced by the glide or climb motion of lattice dislocations right through the transformed region.

The above description is in terms of lattice points rather than atoms in order to permit atomic displacements which are merely interchanges between equivalent sites, and hence are 'redundant' as far as macroscopic effects are concerned. The assertions are equivalent to the previous claim that a systematic shape change implies a lattice correspondence (69, 70) even if accompanied by some diffusion of the atom species; the important factor is that the possible atom sites are fixed by the growth mechanism and form only a sub-set of those sites which are permitted by the translational symmetry of the product lattice. The argument does not rest on any model of the interface or of the growth process, although it is an immediate further deduction that the required correlation in lattice sites could not be enforced unless the interface is at least semi-coherent.

Since the observed shape change is always an invariant plane strain (IPS) (and indeed the material could not remain coherent with any other systematic shape change), and since the shear component is usually appreciable, it follows that it should be an acceptable simplification to use such phrases as 'growth involving shear' or 'shear mechanisms of growth'. However, despite the almost tautological nature of these descriptions, they unfortunately seem to arouse passions on both sides of an imaginary fence dividing proponents of diffusional mechanisms from those of diffusionless mechanisms. In our opinion, the question of whether growth is associated with a shear depends solely on the measured shape deformation and is not dependent on detailed models of the interface. There is, of course, a further question, namely whether growth accompanied by shear can be thermally activated, and in particular diffusion-controlled, but it confuses the issue to suppose that if this is possible, the assumption of diffusion control somehow makes the shear less real.

When particles are formed within a constraining matrix, straightforward deductions about growth processes from measured shape deformations
are complicated by secondary factors such as accommodating plastic
deformation or the formation of product regions in mutually self-accommodating pairs or groups. In the case of bainite additional difficulties
might be caused by the precipitation of carbide. Nevertheless, we shall now
attempt to classify the various possibilities for the growth mechanism in
bainites by using a slightly novel approach based on the well-known set of
virtual operations used by Eshelby (71) to represent the formation of a
coherent particle of unchanged composition. We include additional
operations to allow the particle to acquire a composition different from
that of the matrix, or further to change its size or shape, and we then
examine the growth processes which are consistent with these virtual
operations. Thus we begin by assuming a full lattice correspondence
between the sites of the parent and product crystals and then examine
various ways in which this correspondence may be modified or destroyed.
This procedure was previously used for the particular case of martensite
(72) in order to demonstrate the equivalence of the basic assumption of the
standard crystallographic theories and the hypothesis that a confined
martensite plate forms with minimum elastic strain energy.

Figure 9 illustrates the modifications envisaged. We can allow the

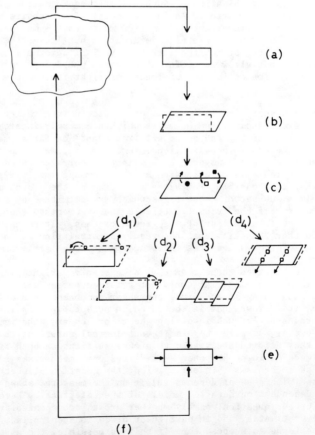

Figure 9 - Schematic diagram illustrating the virtual operations required
to form a particle in a constraining matrix.

particle to have any required composition by transferring suitable numbers of solute atoms between interstitial sites in the particle and the matrix, and/or by interchanging atoms of substitutional species in the particle with atoms in the matrix (operation (c) in Figure 9). We suppose the particle composition remains homogeneous; the solute atoms removed from or added to the matrix may also be distributed homogeneously, or they may be concentrated in the region around the hole, giving composition gradients and elastic strains. The accompanying changes of lattice parameters must now be included in revised 'stress-free strains' as defined by Eshelby in order correctly to calculate the strain energy of the (coherent) particle when forced back into the hole. However, in order to reduce the strain energy, additional operations on the particle may be carried out before it is reinserted into the matrix, and we classify these as follows:

[1] The volume of the particle may be made equal to that of the hole in the matrix by transferring the appropriate number of atoms of each species to or from the particle to the outer surface of the matrix or to and from internal sinks within the matrix, and the shape of the particle can be similarly adjusted by transferring atoms within the particle from existing sites to new, crystallographically equivalent sites. The strain energy then vanishes (Operation $(d_1)$ ).

[2] The total number of atoms in the particle may be conserved but its shape may nevertheless be adjusted by creation and removal of atom sites. In the case of a plate-shaped particle, such a process will minimise the elastic energy of the final assembly if the net shape change after this stage consists solely of an expansion or contraction in the direction of the small dimension of the plate; in other words, if it is an IPS with no shear component (Operation $(d_2)$).

[3] The shape of the particle may be changed by plastic deformation, i.e. by dislocation glide in which dislocations are not retained within the particle but pass out of its free surfaces, leaving atomic height steps. In this case, the lowest strain energy for a plate-shaped particle will be obtained if this plastic deformation converts the lattice deformation into a shape deformation which is an IPS on the habit plane (72) as in the standard theory of martensite crystallography (Operation $(d_3)$).

[4] The shape of the particle may be changed by the removal or addition of particular planes of atoms, e.g. by dislocation climb from one surface to another, again giving lowest strain energy for an IPS on the habit plane of a plate-shaped region. This is a non-conservative process, but as previously discussed (65, 73), if there is effectively no reconstruction of the atom sites, the shape change may retain an appreciable shear component (Operation $(d_4)$).

Now consider how these virtual operations correspond to physical growth processes. For particles of type [1], growth is necessarily accompanied by long range transport of matter; this will only be possible at reasonably high temperatures where the lattice diffusion coefficient is relatively large, and it follows that any required chemical diffusion of substitutional or interstitial species should be accomplished simultaneously. The growth rate may be controlled by chemical composition gradients established around the particle, or by the rate of matter transport. There is no shape change. 'Epitaxially-coherent' particles of type [4] (73) also have their growth rates controlled by the appropriate lattice diffusion coefficient but grow only at finite values of the driving force because of the stored strain energy. There is some difficulty in accepting this possibility, however, since it has often been emphasised (65) that further atomic fluxes could destroy the correspondence and hence the shear component of the shape change.

Particles of type [2] like those of type [1] are undergoing the kind of transformation which Buerger (74) classed as 'reconstructive' and which is alternatively described as 'civilian' (69). There is no systematic shape change and the volume change manifests itself as 'surface rumpling'. This leaves only the martensitic type of change (type [3]) as a likely candidate for an IPS macroscopic shape change. This need not be diffusionless, especially if there are rapidly diffusing interstitial species present in a virtually immobile lattice framework. This type of growth is also consistent with an ordering process in which individual solute atoms migrate only a very few interatomic distances during the reaction period, but it is difficult to reconcile it with long range substitutional diffusion.

In the above classification, the overall shape of the particle is material only in so far as the shear component in the stress-free strains remaining after operations [3] or [4] give an elastic energy which is linear in the aspect ratio of a plate, and so will be minimised by a thin plate shape. The effect of the volume change in [2], [3] or [4] is similarly minimised if it is achieved by a uniaxial expansion or contraction in the smallest dimension of the particle. It is thus an interesting question whether observed plate morphologies should be attributable mainly to this strain energy condition (equivalent to the IPS condition of the formal theories (72)), or to the relative immobility of the faces of the plate, as is generally assumed in 'ledge' descriptions of growth (67, 68, 75).

Turning now to the interfaces between the particle and the matrix, we emphasise first that these may be fully coherent, partly coherent or incoherent with different operations, but that the type of coherence is a property of the particle as a whole, and it is not meaningful to regard some interfaces as coherent whilst others are incoherent. Thus all interfaces were fully coherent in the original Eshelby operations, whilst the additional operations [3] and [4] lead to partly coherent interfaces. In both cases, the interfaces might be irrational or macroscopically curved, thus requiring structural steps or ledges when described on an atomic scale, but the important result is that any displacement of any element of the interface will produce a shape change which is the resultant of the lattice and lattice invariant deformation. In cases [1] and [2], the interface is incoherent and there is no required correlation of atomic positions.

Aaronson and co-workers have laid considerable emphasis on a growth model which involves an immobile (or sessile) planar interface which is traversed by mobile ledges which are supposed to have a relatively disordered structure. They have suggested that the existence of the semi-coherent interface will somehow ensure that the movement of the ledge will produce a shape change, but as pointed out by Watson and McDougall (76), this concept is not self-consistent. If there is a shape change, a lattice correspondence must be maintained during growth across any portion of the interface surrounding the particle, so that the ledge cannot be regarded as disordered in this sense. Rigsbee and Aaronson (77) have made an electron microscope study of the dislocation structure of the broad faces of ferrite plates embedded in retained austenite, and have compared their results with a computer simulation of the same interface (78). The model begins with an interface parallel to both $\{111\}_\gamma$ and $\{110\}_\alpha$ with various relative rotations of the lattices about the normal to these planes to give orientations ranging from Nishiyama-Wasserman to beyond Kurdjumov-Sachs. Within the expected range of lattice parameter ratios, this unrelaxed interface has poor atomic fit (defined by near coincidence of atom positions) and the modelling is essentially a geometrical procedure to increase these near coincidences by the introduction of 'structural ledges' and misfit dislocations. The structural ledges (atomic height steps) have the effect

of rotating the macroscopic interface plane away from the close-packed
orientation, and the misfit dislocations concentrate the effects of the
different planar densities of atoms into linear discontinuities.  The procedure is thus first to define the orientation relationship and (within
narrow limits) the habit plane and then to determine the dislocation
structure (this can be done mathematically by application of the Frank-
Bilby or Bollmann equations (see 72)).  Quite a different procedure is used
in the theory of martensite crystallography where (in effect) the dislocations of the interface are specified and the habit plane and
orientation relation are then determined (79-81).

In their experimental investigation, Rigsbee and Aaronson (77) show
micrographs which are claimed to support the computer models.  If this
identification is valid, the implication is certainly that such a plate
cannot grow rapidly since the motion of the semi-coherent interface is
necessarily non-conservative.  We have, in fact, the equivalent of
operation [4 ] above, in which the net shape change is the combination of
the lattice deformation and the lattice invariant deformations caused by
climb of the misfit dislocations.  Although we concluded above that
particles of type [4] are unlikely to be formed, this may still be considered an open question.  However, we consider it important to emphasise that
any difficulty in moving the so-called sessile interface is a consequence
of the non-conservation of atoms in the swept volume, and this difficulty
can in no way be obviated by a description in terms of the migration of
growth ledges across the interface.

Figure 10  shows schematically two sections of a sessile interface
separated by a (growth) ledge.  If section A moves forward to position B as

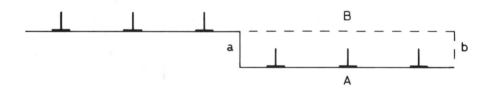

Figure 10 - Schematic diagram illustrating the displacement of a partially
coherent misfit-dislocation interface.

indicated by the broken lines and the interface retains its structure,
excess atoms represented by the climb of the misfit dislocations must have
diffused out of the swept region.  However, this operation is clearly
equivalent in final result to migration of the ledge from a to b; this must
similarly transfer the misfit dislocation structure from its original site
at A to its new position B.  Thus whatever is the detailed atomic structure
at the ledge, problems associated with the migration of misfit dislocations
in the broad faces cannot be avoided by supposing that the migration is
accomplished by moving the ledge.

In concluding this section, we have attempted to show that a growth
process involving an atomic correspondence is characteristic of a particle
as a whole and should not be associated with particular interfaces.  There
is a basic difference between growth processes which preserve a lattice

correspondence, are associated with a shape change and require coherent interfaces and those which allow independent atom migration over large distances, produce no shape change and may involve incoherent interfaces. Ledge mechanisms are possible in both cases and so are irrelevant to the distinction.

## Kinetics of the Reaction

There has been much discussion of whether bainite formation should be regarded as a separate reaction with its own C-curve on a T-T-T diagram or as continuous with the proeutectoid ferrite and pearlite reactions which produce the same phases. In some alloy steels, a continuous T-T-T diagram contains a distinct bay below which pearlite is not observed and above which the characteristic structure and sub-structure of upper bainite is absent, so that it seems a natural interpretation to regard this as a mani-festation of independent C-curves for the two reactions. The interpret-ation is strengthened by the agreement between the temperature of the bay and the kinetic $B_s$ temperature as plotted from the extent of incomplete transformation to bainite, but is weakened by the absence of a bay in many plain carbon steels where a single C-curve seems to represent the ini-tiation of transformation over the whole pearlite-bainite range. Recently, however, Kennon and Kaye (82) have used optical and electron metallography to plot separately the curves representing 1% volume fraction of pearlite and bainite in a eutectoidal 0.77% manganese steel isothermally reacted in the range 350-600°C. In this way, they were able to effect a separation of the normally drawn single C-curve (shown by the broken line in Figure 11) into two closely overlapping but distinct C-curves (solid lines in Figure 11) for the beginning of the two reactions.

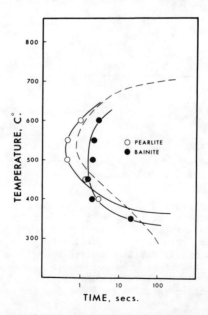

Figure 11 - Partial isothermal transformation diagram showing C-curves representing the beginning of austenite decomposition to pearlite and bainite. After Kennon and Kaye (82).

A different interpretation has been used by Kinsman and Aaronson (1, 2, 5) who ascribed the bay in the T-T-T diagram to an effect on the mobility of the ferrite-austenite interface caused by solute drag. According to this view, which is linked to the so-called microstructural definition of bainite, the formation of bainitic ferrite is continuous with that of higher temperature pro-eutectoid ferrite. Although it is agreed that substitutional alloying elements do not partition during bainitic growth, it is proposed that they may nevertheless segregate to the transformation interface and then exert a dragging force which restricts its mobility. Elements such as molybdenum which are strong carbide formers are assumed to be strongly bound to the interface, and as the temperature is reduced, the amount of segregation initially increases, thus progressively lowering the mobility. Eventually, the segregation approaches saturation and the effect of the increased driving force for growth begins to overcome the drag, so that the mobility is again increased and the bay effect is **thus** produced. In support of this proposal, the absence of a bay in steels containing only carbon and low manganese is claimed to be consistent with the relatively small effect of manganese in reducing the carbon activity in austenite. It is also claimed that the progressively smaller reductions in the activity of carbon produced by given amounts of chromium, manganese and nickel are consistent with their relative effectiveness in producing a bay in the T-T-T diagram.

Only qualitative descriptions of the solute drag theory have been given, and in our opinion the evidence in favour of a separate C-curve for bainite below the kinetic (and properly-defined microstructural) $B_s$ is now quite strong. Both Hehemann (3) and Bhadeshia and Edmonds (23) have pointed out that it is possible to obtain bays in steels containing alloying elements which increase the activity of carbon in austenite, and the recent atom-probe investigations have failed to reveal any substitutional solute segregation at the bainitic ferrite-retained austenite interface in a manganese-silicon steel.

Early research documented the overall reaction kinetics in the bainitic range, and some evidence for separate transformation mechanisms for upper and lower bainite was obtained by the measurement of different activation energies in their respective temperature ranges (11, 83). Barford (12) plotted T-T-T curves for high purity iron-carbon-manganese alloys and found a lower 'nose' on the curves in addition to a change in activation energy; the existence of an additional bay between the upper and lower bainite ranges has recently received further experimental support (23, 84).

Extensive investigations on upper bainite formation in alloy steels, generally chosen to avoid the complications of simultaneous formation of pearlite and pro-eutectoid ferrite, were made by Troiano and his collaborators (4, 85-88) and others (89-91); this work which established the incomplete reaction phenomenon, is summarised by Hehemann (3). In the incomplete reaction range, each ferrite lath is limited in volume; following edge impingement on grain boundaries or other obstacles, only a limited amount of thickening occurs, and unless pearlite is nucleated, the untransformed austenite may remain stable for weeks or months at the same temperature. Transformation is resumed by the formation of new laths on further lowering of the temperature. Although the reasons for the cessation of growth are not known, as we have already discussed, the similarity to the athermal kinetics of martensite formation is striking.

Stepped transformation studies by Hehemann and his collaborators (3) have shown that the rate of transformation to upper bainite at a given temperature is increased by prior partial transformation at a **lower**

temperature. The lower temperature may be in either the upper or lower
bainite range, but in the latter case, formation of lower bainite ceases
after the upquench. The acceleration seems to be achieved by providing more
ferritic nuclei which can grow at the upper reaction temperature; moreover,
it can be eliminated by giving a treatment at a temperature above $B_S$
between the two transformation treatments. If a transformation is allowed
to continue to its maximum extent in the incomplete transformation range
(between $B_S$ and $B_f$), a brief amount  of additional transformation at a lower
temperature followed by a return to the original transformation temperature
will then produce appreciable further transformation at this temperature.
As discussed by Hehemann (3), these effects are all consistent with a model
of strain-assisted autocatalytic nucleation similar to that used in martensite formation; treatment above $B_S$ may relax these strains and so remove
the additional nuclei. Stepped transformation experiments in which the
second temperature is lower than the first, lead to a retardation of the
reaction at the second temperature (92), and this effect may perhaps be
compared with thermal stabilisation in martensite formation.

The analogy with martensite is further strengthened by early
experiments which showed that the transformation is accelerated by an
externally applied tensile stress  (93-97). The stress also apparently
acts to stimulate nucleation rather than to increase the growth rate (3).
Strangely, there seems to have been little recent work on this effect,
although it has been much investigated for martensite and might prove very
helpful in differentiating between various proposed mechanisms of bainite
formation.

The overall reaction rates, whilst usefully distinguishing the
macroscopic kinetic behaviour of the various decomposition products of
austenite, do not differentiate between nucleation and growth rates. There
have been several measurements of the growth kinetics of individual bainite
plates using hot stage optical microscopy or thermionic emission
microscopy (1-3, 98-101). These in-situ experiments are facilitated by
the surface relief of the growing plate, but are subject to stereological
errors and surface effects; it is not, however, possible to use the
sequential heat treatment procedures which are customary for pearlitic
growth. All investigators agree that lengthwise (or edgewise) growth of
both upper and lower bainite occurs at a constant (linear) rate, increasing
with increasing temperature; the apparent activation energy is smaller for
upper bainite than it is for lower bainite.

Following Kaufman et al (102), one popular model for edgewise growth
has been to assume that the necessary driving force requires segregation of
carbon, so that the growth rate is limited by the diffusion of carbon away
from the ferrite-austenite interface into the surrounding .austenite. They
used the Zener-Hillert model (103) to calculate the growth rate with the
results shown in Figure 12. For a plain carbon steel, this model gives
growth rates within an order of magnitude of the measured rates, also shown
in the figure, but for alloy steels, the growth is much slower than is
permitted by the carbon diffusion model, although the temperature
dependence is reasonably well predicted. In view of the known deficiencies
of the Zener-Hillert model too much emphasis should perhaps not be laid on
the quantitative disagreement, even though it is very large, but the significant factor is that carbon diffusion alone is unlikely to be able to
account for the difference between the measured growth rates of plain
carbon and alloy steels. Later measurements of lengthening kinetics by
Simonen et al (100) gave comparable results but were compared with the
dendrite growth theory of Trivedi (103, 104) which has different boundary
conditions. The activation energy for edgewise growth of lower bainite for
a wide range of steels is in the range 63-80 kJ mol$^{-1}$; this is consistent

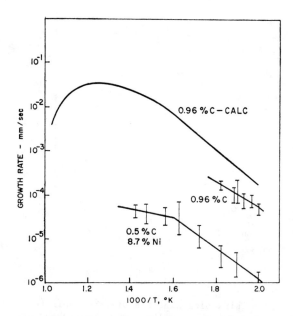

Figure 12 - Comparison of calculated and measured edgewise growth rates of bainite.  After Kaufman, Radcliffe
and Cohen (102).

with a model based on the diffusion of carbon in ferrite, but Figure 12
shows that it could be expected also on the basis of the Zener-Hillert model.
However, this model would imply carbide precipitation in the austenite and
thus cannot easily be reconciled with the morphology of lower bainite.

For upper bainite, growth measurements give activation energies of only
12-35 kJ mol$^{-1}$, and as seen in Figure 12 this is appreciably lower than is
predicted from the Zener-Hillert model.  There is also an appreciable
difference between these activation energies for growth and those measured
from the overall kinetics of the reaction.  Whilst for lower bainite a
value of about 75 kJ mol$^{-1}$ has again been obtained, for upper bainite, the
overall activation energy is about 125 kJ mol$^{-1}$.  This presumably reflects
the strong effect of the nucleation rate on the reaction kinetics.

Very few measurements of thickening rates of bainitic ferrite plates
have been made, but there is some evidence of a discontinuous thickening
rate, as shown by Figure 13, and this has been claimed to give support to
the ledge mechanism (2).  Measurements at relatively high temperatures
have been made for Widmanstätten side-plates (1, 2, 68, 75) which show
discontinuous growth with an average rate below that given by the volume
diffusion solution for a planar boundary.

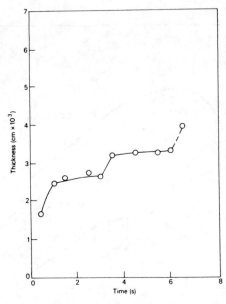

Figure 13 - Thickening kinetics of upper bainite in an
Fe-1%C-6%Mn-1%Mo alloy isothermally transformed at
371°C. After Hehemann, Kinsman and Aaronson (2).

## Possible Mechanisms of Bainite Formation

In discussing mechanisms, we take as our starting point the observation
of Ko and Cottrell (62) of surface relief effects associated with bainite
formation; many of the models suggested from the original identification of
bainite as a separate decomposition product of austenite (105) up to 1950
are listed in their paper. Attention since the work of Ko and Cottrell has
focussed on the fact that the bainitic transformation occurs in a temperature region between that of the diffusional ferrite/pearlite reactions and
the diffusionless martensitic transformation. Furthermore, bainite has the
acicular appearance of martensite, exhibiting similar crystallographic and
surface relief effects, but reaction kinetics, which, apart from the
existence of incomplete transformation between $B_s$ and $B_f$, are those of
nucleation and growth reactions. These facts have led to proposals which
attempt to incorporate features common to both types of transformation.

Bainite is generally considered to be an aggregate of ferrite and
carbide, and Hillert (106) introduced the idea of the cooperative growth of
these two phases from austenite, distinguishing between pearlite which
forms with a high degree of cooperation and bainite where the cooperation is
much less. In a later interpretation of Hillert's ideas Aaronson (75)
has described bainitic growth as being non-cooperative and he supposes that
the two phases grow alternately rather than mutually, although the interface
between one product phase and the parent phase does provide a favourable
site for nucleation of the other phase. Aaronson's description may possibly
be related to the extensive work of Honeycombe and his collaborators (41)
on 'interphase precipitation' of carbides (generally alloy carbides, not
cementite) at austenite-ferrite boundaries in alloy steels. In this
relatively high temperature precipitation process, carbide precipitates are
nucleated on the faces of grain boundary ferrite allotriomorphs which appear

to thicken by a step or ledge mechanism, thereby incorporating the precipitates into the ferrite in bands parallel to the interface.  The possibility then exists that a similar process might account for the banded precipitate morphology of lower bainite although here the bands are at an
angle of 55-65° to the habit plane interface.

Some of the results described above, especially the existence of
several precipitate variants within a single ferrite plate (23), are at
variance with this concept of interface precipitation, and whilst we do not
think there is sufficient evidence to eliminate the possibility at present,
we consider that the balance of the evidence suggests that lower bainite
forms first as supersaturated ferrite, and that the precipitation of
cementite (or $\varepsilon$-carbide) within this ferrite is then a subsequent stage of
the reaction.  This view is strengthened by the appreciable evidence that
ferrite formation is the first stage in upper bainite formation; indeed
in some silicon steels, no other phase is produced.

If we accept that the important stage of the reaction is initially
the formation of ferrite, what then is the distinction between upper and
lower bainite?  There appear to be two possibilities.  The older description, originating from Ko and Cottrell, was that in order to provide
sufficient driving force for the growth process in upper bainite, carbon
must diffuse out of the ferrite plate into the surrounding austenite.  The
growth rate is then limited by the diffusion field, as for example, in the
Zener-Hillert model which has been applied to bainitic growth.  Ferrite in
lower bainite, in contrast, originally forms with a full supersaturation,
which is subsequently relieved by precipitation; the growth rate may be
initially very fast but the plate may cease to grow at an early stage until
the precipitation releases more free energy to drive the process.  A more
recent description supposes that the initial stage in both upper and lower
bainite is the formation of regions of ferrite which are supersaturated with
carbon.  Individual regions are assumed to reach a limiting size very
rapidly, and the apparent finite growth rate of an upper bainite sheaf is
caused by the isothermal nucleation of subsequent laths or sub-units.  (The
evidence for sub-units in lower bainitic ferrite plates is not as clear, so
that the finite growth rate of such a plate may not be describable in this
way).  Once a sub-unit has ceased to grow, the excess carbon diffuses into
the surrounding austenite in the upper bainitic range, and there precipitates as carbide unless, as in silicon steels, the austenite is very stable.
In contrast, the increased driving force and lower mobility of the carbon
leads to in-situ precipitation in the ferrite at temperatures in the lower
bainitic range.

We have yet to mention the shape change.  The implication of the sub-
unit model of Hehemann (3, 22) which we have just described is that
bainite begins, quite simply, as a martensitic transformation, and the
subsequent carbide precipitation might then be regarded as a mode of
tempering, so that we have a possible connection of bainite with auto-
tempered martensite.  The similarity between upper bainite and lath martensite is quite marked and in a recent paper, Thomas and Sarikaya (107) have
reversed this identification and propose that lath martensite should be
regarded as untransformed (incompletely transformed might be better) upper
bainite.  According to the Ko and Cottrell model for upper bainite or the
models for lower bainite which require precipitation during growth, the
transformation is not martensitic in the normal sense, but because of the
low mobility of the iron atoms, the change of crystal structure has to be
accomplished without extensive atomic migration, so that the transformation
possesses  the crystallographic and morphological characteristics of a
'military' transformation.  There are now several pieces of evidence which

indicate that the Hehemann description may be correct, although none of these is conclusive. The evidence is:

1) The $B_S$ temperature is almost always below $T_O$ (or $T_O'$) as is required by the Hehemann description but not by the Ko and Cottrell model.

2) The results of Bhadeshia and his co-workers on the incomplete reaction phenomenon are reasonably consistent with the assumption that transformation ceases when the carbon content in the austenite reaches the $T_O$-composition curve.

3) The known facts about the formation of pro-eutectoid Widmanstätten ferrite with a marked shape change (76) suggests that the Ko and Cottrell type model should be applied to this process. However, the detailed crystallography and shape change measurements as well as the morphology and internal substructure are very different for the two products, and this strongly suggests that the details of the reaction mechanism are different.

For many years a dispute has existed between the proponents of martensite-like descriptions of the formation of bainitic ferrite and those to whom such descriptions are anathema. Following a general theory of morphology developed to explain anisotropic precipitate shapes, Kinsman and Aaronson (51, 68, 75) have advocated strongly the so-called ledge theory of growth of proeutectoid Widmanstätten ferrite and of bainite, which they claim is an alternative to the martensitic type descriptions. They suggest that the plate-shape of ferrite develops because of a substantial barrier to growth at one orientation of the interphase boundary where the austenite and ferrite lattices match sufficiently well to form a partly coherent misfit-dislocation interface. This boundary will be immobile in a direction normal to itself, but it is claimed that it is able to advance in this direction by the formation and migration of ledges across it. These ledges are supposed to have a disordered structure which enables them to migrate at rates governed by long range diffusion. In so doing, they produce 'geometrical surface relief' which, it is implied, is somehow different from the shear of a martensitic transformation.

From these concepts, a continuing argument has developed about whether the growth is martensitic or utilises the ledge mechanism; we believe, however, that the dispute makes no sense. As discussed above, a systematic shape change necessarily implies a lattice correspondence, and all the interfaces surrounding an enclosed particle must then be coherent in the sense that displacement of any element of interface produces the shape change in the swept volume. Continuity of structure requires that the lattice correspondence is preserved across a ledge, so that it cannot be strictly correct to regard its structure as disordered. The lattice correspondence and associated shape change arise because the iron atoms are not able to migrate freely during the transformation; the question of whether the habit plane interface as a whole is glissile or whether the atoms are able more readily to make their small relative displacements at a ledge is a secondary matter. Both atomic height step and non-step mechanisms are believed to operate in different types of martensite, and the same may apply to bainite. It should be noted, however, that the habit plane is probably irrational (i.e. already continuously stepped on an atomic scale), that no microscopical evidence for the postulated steps has ever been found in the case of bainite, and that the nucleation of ledges of many atom layers in height seems most improbable since this requires an element of the supposedly immobile interface effectively to be displaced through these layers.

We conclude that incoherent ledge mechanisms of growth are not probable for bainite and are likely to be important only for transformations where low energy planar interfaces are formed; in any event, they do not affect the conclusion that bainite formation, since it involves a shape change, is closely analogous to martensite. We also think it probable that the lath or plate shapes of the ferritic regions arise, as in martensite, because of the large strain energy of the constrained reaction, rather than from an anisotropy of interface mobility.

The crystallographic theory of martensite has been successfully applied to lower bainite in several investigations (34, 37, 43); in all cases a non-twinning mode of lattice invariant deformation had to be used to account for the experimental results. This implies a difference between the initial formation of bainitic ferrite and of martensite in the same alloy, as was directly verified by Srinivasan and Wayman (37). This difference may arise at the nucleation stage; not only is the structure different (cubic ferrite in bainite, tetragonal in martensite) but there is also the possibility (46) that the bainite nucleus has the equilibrium carbon composition, even though it may grow into a region supersaturated with respect to carbon. This is indeed one of several possible reasons for the relatively slow nucleation of sub-units which is required by the Hehemann description of bainite formation.

Watson and McDougall (76) have also successfully applied the crystallographic theory of martensite to explain their results on the formation of Widmanstätten ferrite in an iron-0.45% carbon alloy. The shear component of the shape deformation accompanying this transformation is rather large, and Bhadeshia (46, 108) has suggested that in order that the available driving force can overcome the resulting strain energy, the plates take advantage of a near degeneracy in the crystallography and form, for the most part, as self-accommodating pairs. Experimental evidence in favour of this back-to-back formation has also been obtained by Aaronson et al (109); the slightly different habit planes of the two accommodating plates lead to a thin wedge morphology, with a 'roof-top' surface relief.

Bhadeshia (46) has also attempted a rationalisation of the description of the various transformations in steels which are accompanied by a shape change and so involve a lattice correspondence. Widmanstätten ferrite which is nucleated and grows with the equilibrium carbon concentration begins to form at a temperature ($W_S$) below Ae' where the chemical driving force for the reaction ($\gamma \rightarrow \alpha + \gamma$) exceeds the free energy required for a finite nucleation rate; this condition is believed automatically to include the condition that the stored strain energy shall be supplied, i.e. that $W_S$ is below Ae$_3''$. Bainite is assumed to nucleate also with the equilibrium concentration but to grow as supersaturated ferrite; this means that the nucleation condition is now probably redundant and $B_S$ corresponds to $T_o'$. Finally the $M_S$ temperature requires sufficient cooling below $T_o$ to give a driving force ($\gamma \rightarrow \alpha$) sufficient to enforce athermal nucleation. With reasonable values of these energies, it is possible to show that in low, medium and high alloy steels all three reaction products, only bainite and martensite, and only martensite, respectively, may be formed.

## Conclusions

In this paper, we have attempted to present a broad survey of established and more recent experimental information on the morphology, crystallography and kinetics of the bainite reaction in steels and on possible composition changes in the participating phases. Our tentative conclusions are:

1) The balance of the evidence is that bainite formation should be
   regarded as a separate reaction from those which produce proeutectoid
   ferrite and pearlite.  When attention is paid to substructure and
   composition of the ferrite, the microstructural and kinetic $B_s$
   temperatures are equivalent.  The bainite bay in alloy steels is
   probably due to separate C-curves for the different reactions rather
   than to solute drag on a ferrite-austenite interface.

2) The observed shape change implies a lattice correspondence during
   growth, the necessity for which arises from the low mobility of the
   iron atoms at the transformation temperatures.  This conclusion is
   independent of the argument about whether or not the growth rate is
   controlled by the diffusion of carbon, and of whether a ledge mechanism
   is involved.  The same conclusion applies to those Widmanstätten
   plates of proeutectoid ferrite which show surface relief, but this
   reaction is distinguished from bainite by its different crystallography
   and substructure.

3) Although the Ko and Cottrell model of (carbon) diffusion-controlled
   growth for upper bainite cannot be finally eliminated, it now seems
   more probable that this description applies to proeutectoid ferrite.
   The growth of bainite below $B_s$ is more likely to involve the concept
   of successive nucleation and rapid (martensitic) growth of lath-like
   sub-units, as suggested by Hehemann, followed in upper bainite by
   diffusion of carbon into the surrounding austenite where it ultimately
   precipitates as cementite.  In lower bainite, it is probable that
   cementite, or in some cases $\varepsilon$-carbide, precipitates directly from the
   supersaturated ferrite; interphase precipitation during the growth of
   the ferrite is improbable.

## References

1. H. I. Aaronson, The Mechanism of Phase Transformations in Crystalline
   Solids, Inst. of Metals, London, Mono. No. 33, 1969, pp. 207-281.

2. R. F. Hehemann, K. R. Kinsman and H. I. Aaronson, Met. Trans., 3(1972)
   pp. 1077-1094.

3. R. F. Hehemann, Phase Transformations, ASM, 1970, pp. 397-432.

4. A. R. Troiano, Trans. ASM, 41 (1949) pp. 1093-1112.

5. K. R. Kinsman and H. I. Aaronson, Transformation and Hardenability in
   Steels, Climax Molybdenum Co., Ann Arbor, Michigan, 1967, pp. 39-53.

6. K. J. Irving and F. B. Pickering, J. Iron Steel Inst., 188 (1958)
   pp. 101-112.

7. K. Shimizu and Z. Nishiyama, Mem. Inst. Sci. Ind. Res., Osaka Univ.,
   20 (1963) pp. 43-46.

8. K. J. Irving and F. B. Pickering, The Physical Properties of Martensite
   and Bainite, Iron and Steel Inst., London, Spec. Rep. No. 93, 1965,
   pp. 110-125.

9.   D. N. Shackleton and P. M. Kelly, The Physical Properties of Martensite and Bainite, Iron and Steel Inst., London, Spec. Rep. No. 93, 1965, pp. 126-134.

10.  R. F. Mehl, Hardenbility of Alloy Steels, ASM, 1939, pp. 1-54.

11.  P. Vasudevan, L. W. Graham and H. J. Axon, J. Iron Steel Inst., 190 (1958) pp. 386-391.

12.  J. Barford, J. Iron Steel Inst., 204 (1966) pp. 609-614.

13.  L. J. Habraken and M. Economopoulos, Transformation and Hardenability in Steels, Climax Molybdenum Co., Ann Arbor, Michigan, 1967, pp. 69-107.

14.  M. Hillert, Jernkontorets Ann., 141 (1957) pp. 757-789.

15.  R. W. K. Honeycombe and F. B. Pickering, Met. Trans., 3 (1972) pp. 1099-1112.

16.  J. W. Christian, The Theory of Transformations in Metals and Alloys, 1st ed., pp. 824-831; Pergamon, Oxford, 1965.

17.  G. A. Chadwick, Metallography of Phase Transformations, 1st ed., pp. 251-268; Butterworths, London, 1972.

18.  R. W. K. Honeycombe, Steels: Microstructure and Properties, 1st ed., pp. 106-120; Edward Arnold, London, 1981.

19.  H. I. Aaronson, Decomposition of Austenite by Diffusional Processes, Interscience, New York, 1962, pp. 387-546.

20.  F. B. Pickering, Transformation and Hardenability in Steels, Climax Molybdenum Co., Ann Arbor, Michigan, 1967, pp. 109-129.

21.  K. R. Kinsman and H. I. Aaronson, Met. Trans., 1 (1970) pp. 1485-1488.

22.  J. M. Oblak and R. F. Hehemann, Transformation and Hardenability in Steels, Climax Molybdenum Co., Ann Arbor, Michigan, 1967, pp. 15-30.

23.  H. K. D. H. Bhadeshia and D. V. Edmonds, Met. Trans., 10A (1979) pp. 895-907.

24.  G. V. Smith and R. F. Mehl, Trans. AIME, 150 (1942) pp. 211-226.

25.  Z. Nishiyama, Sci. Rep. Tohoku Univ., 23 (1934) pp. 637-664.

26.  G. Wassermann, Arch. Eisenhüttenw., 6 (1933) p. 347-351.

27.  G. Kurdjumov and G. Sachs, Z. Physik, 64 (1930) pp. 325-343.

28.  B. P. J. Sandvik, Met. Trans., 13A (1982) pp. 777-787.

29.  W. Pitsch, Acta Met., 10 (1962) pp. 897-900.

30.  D. N. Shackleton and P. M. Kelly, Acta Met., 15 (1967) pp. 979-992.

31.  Yu. A. Bagaryatskii, Doklady Akad. Nauk SSSR, 73 (1950) pp. 1161-1164.

32. Y. Ohmori and R. W. K. Honeycombe, Trans. Iron Steel Inst. Japan (Suppl.), 11 (1971) pp. 1160-1164.

33. I. V. Isaichev, Zhur. Tekhn. Fiziki, 17 (1947) pp. 835-838.

34. Y. Ohmori, Trans. Iron Steel Inst. Japan, 11 (1971) pp. 95-101.

35. Private communication, A. T. Davenport, Republic Steel Corporation Research Centre, Cleveland, Ohio, Feb., 1974.

36. S. Hoekstra, Acta Met., 28 (1980) pp. 507-517.

37. G. R. Srinivasan and C. M. Wayman, Acta Met., 16 (1968) pp 609-620, 621-636.

38. H. K. D. H. Bhadeshia, Acta Met., 28 (1980) pp. 1103-1114.

39. D. Kalish and M. Cohen, Mater. Sci. Eng., 6 (1970) pp. 156-166.

40. D-H. Huang and G. Thomas, Met. Trans., 8A (1977) pp. 1661-1674.

41. R. W. K. Honeycombe, Met. Trans., 7A (1976) pp. 915-936.

42. K. H. Jack, J. Iron Steel Inst., 169 (1951) pp. 26-36.

43. J. S. Bowles and N. F. Kennon, J. Austral. Inst. Met., 5 (1960) pp. 106-113.

44. C. Zener, Trans. AIME, 167 (1946) pp. 550-595.

45. J. C. Fisher, Trans. AIME, 185 (1949) pp. 688-690.

46. H. K. D. H. Bhadeshia, Acta Met., 29 (1981) pp. 1117-1130.

47. J. W. Christian, Proc. Int. Conf. on Martensitic Transformations, ICOMAT 1979, Cambridge, MA., pp. 220-234.

48. H. I. Aaronson, H. A. Domian and G. M. Pound, Trans. AIME, 236 (1966) pp. 753-767.

49. G. J. Shiflet, J. R. Bradley and H. I. Aaronson, Met. Trans., 9A (1978) pp. 999-1008.

50. H. K. D. H. Bhadeshia, Metal Sci., 15 (1981) pp. 175-177.

51. K. R. Kinsman and H. I. Aaronson, Discussion of Reference 22, pp. 33-38.

52. W. Steven and A. G. Haynes, J. Iron Steel Inst., 183 (1956) pp. 349-359.

53. H. K. D. H. Bhadeshia and A. R. Waugh, Acta Met., 30 (1982) pp. 775-784.

54. H. K. D. H. Bhadeshia and D. V. Edmonds, Acta Met., 28 (1980) pp. 1265-1273.

55.    H. K. D. H. Bhadeshia and A. R. Waugh, Solid-Solid Phase
       Transformations, H. I. Aaronson et al., ed.; AIME, New York, N. Y.,
       1983, pp. 993-997.

56.    H. K. D. H. Bhadeshia, Solid-Solid Phase Transformations,
       H. I. Aaronson, et al., ed.; AIME, New York, N. Y., 1983, pp. 1041-
       1045.

57.    D. J. Hall, H. K. D. H. Bhadeshia and W. M. Stobbs, Proc. ICOMAT -
       1982, J. de Phys. C-4 (Suppl.) pp. 449-454.

58.    K. R. Kinsman, E. Eichen and H. I. Aaronson, Met. Trans., 2 (1971)
       pp. 346-348.

59.    F. B. Pickering, Discussion of Reference 20, p. 132.

60.    B. A. Bilby and J. W. Christian, The Mechanism of Phase
       Transformations in Metals, Inst. of Metals, London, Mono. No. 18,
       1956, pp. 121-172.

61.    D. Hull, Bull. Inst. Metals, 2 (1954) pp. 134-139.

62.    T. Ko and S. A. Cottrell, J. Iron Steel Inst., 172 (1952) pp.
       307-313.

63.    T. Ko, J. Iron Steel Inst., 175 (1953) pp. 16-18.

64.    H. M. Clark and C. M. Wayman, Phase Transformations, ASM, 1970,
       pp. 59-114.

65.    J. W. Christian, Decomposition of Austenite by Diffusional Processes,
       Interscience, New York, 1962, pp. 371-386.

66.    B. A. Bilby and J. W. Christian, J. Iron Steel Inst., 197 (1961)
       pp. 122-131.

67.    C. Laird and H. I. Aaronson, Acta Met., 15 (1967) pp. 73-103.

68.    H. I. Aaronson, C. Laird and K. R. Kinsman, Phase Transformations,
       ASM, 1970, pp. 313-396.

69.    J. W. Christian, The Physical Properties of Martensite and Bainite,
       Iron and Steel Inst., London, Spec. Rep. No. 93, 1965, pp. 1-19.

70.    J. W. Christian, The Mechanism of Phase Transformations in
       Crystalline Solids, Inst. of Metals, London, Mono. No. 33, 1969,
       pp. 129-142.

71.    J. D. Eshelby, Proc. Roy. Soc., A 241 (1957) pp. 376-396.

72.    J. W. Christian, Proc. 1st. Int. Conf. on New Aspects of Martensitic
       Transformations, Trans. Jap. Inst. Met. (Suppl.) 17 (1976) pp. 21-33.

73.    J. W. Christian, The Theory of Transformations in Metals and Alloys,
       2nd. ed., Pergamon, Oxford, 1975.

74.    M. J. Buerger, Phase Transformations in Solids, R. Smoluchowski et al.,
       ed., (Natnl. Res. Council), Wiley, New York, N. Y., 1951, pp. 183-209.

75.  H. I. Aaronson, Phase Transformations, Institn. Metall., London, 1979, Vol. 1, pp. II 1 - II 18.

76.  J. D. Watson and P. G. McDougall, Acta Met., 21 (1973) pp. 961-973.

77.  J. M. Rigsbee and H. I. Aaronson, Acta Met., 27 (1979) pp. 365-376.

78.  J. M. Rigsbee and H. I. Aaronson, Acta Met., 27 (1979) pp. 351-363.

79.  M. S. Wechsler, D. S. Lieberman and T. A. Read, Trans. AIME, 197 (1953) pp. 1503-1515.

80.  J. S. Bowles and J. K. Mackenzie, Acta Met., 2 (1954) pp. 129-137.

81.  J. K. Mackenzie and J. S. Bowles, Acta Met., 2 (1954) pp. 138-147.

82.  N. F. Kennon and N. A. Kaye, Met. Trans., 13A (1982) pp. 975-978.

83.  S. V. Radcliffe and E. C. Rollason, J. Iron Steel Inst., 191 (1959) pp. 56-65.

84.  R. H. Edwards and N. F. Kennon, J. Austral. Inst. Met., 19 (1974) pp. 45-50.

85.  J. P. Sheehan, C. A. Julien and A. R. Troiano, Trans. ASM, 41 (1949) pp. 1165-1184.

86.  T. Lyman and A. R. Troiano, Trans. ASM, 37 (1946) pp. 402-448.

87.  T. Lyman and A. R. Troiano, Trans. AIME, 162 (1945) pp. 196-222.

88.  A. R. Troiano and J. E. DeMoss, Trans. ASM, 39 (1947) pp. 788-800.

89.  E. Houdremont, W. Koch and J. J. Wiester, Arch. Eisenhuttenw., 18 (1945) pp. 147-154.

90.  W. Jellinghaus, Arch. Eisenhuttenw., 23 (1952) pp. 459-470.

91.  E. P. Klier  and T. Lyman, Trans. AIME, 158 (1944) pp. 394-422.

92.  R. F. Hehemann and A. R. Troiano, Trans. AIME, 200 (1954) pp. 1272-1280.

93.  A. H. Cottrell, J. Iron Steel Inst., 151 (1945) pp. 93-104.

94.  M. D. Jepson and F. C. Thompson, J. Iron Steel Inst., 162 (1949) pp. 49-56.

95.  K. Winterton, J. Iron Steel Inst., 151 (1945) pp. 87-91.

96.  G. J. Guarnieri and J. J. Kanter, Trans. ASM, 40 (1948) pp. 1147-1164.

97.  S. Battacharyya and G. L. Kehl, Trans. ASM., 47 (1955) pp. 351-371.

98.  G. R. Speich and M. Cohen, Trans. AIME, 218 (1960) pp. 1050-1059.

99.   R. H. Goodenow,   R. H. Barkalow and R. F. Hehemann, The Physical
      Properties of Martensite and Bainite, Iron and Steel Inst., London,
      Spec. Rep. No. 93, 1965, pp. 135-141.

100.  E. P. Simonen, H. I. Aaronson and R. Trivedi, Met. Trans., 4 (1973)
      pp. 1239-1245.

101.  R. D. Townsend and J. S. Kirkaldy, Trans. ASM, 61 (1968) pp. 605-619.

102.  L. Kaufman, S. V. Radcliffe and M. Cohen, Decomposition of Austenite
      by Diffusional Processes, Interscience, New York, 1962, pp. 313-352.

103.  J. W. Christian, The Theory of Transformations in Metals and Alloys,
      2nd  ed., pp. 494-500; Pergamon, Oxford, 1975.

104.  R. Trivedi, Met. Trans., 1 (1970) pp. 921-927.

105.  E. S. Davenport and E. C. Bain, Trans. AIME, 90 (1930) pp. 117-154.

106.  M. Hillert, Decomposition of Austenite by Diffusional Processes,
      Interscience, New York, 1962, pp. 197-237.

107.  G. Thomas and M. Sarikaya, Solid-Solid Phase Transformations,
      H. I. Aaronson et al., ed.; AIME, New York, N. Y., 1983, pp.
      999-1003.

108.  H. K. D. H. Bhadeshia, Scripta Met., 14 (1980) pp. 821-824.

109.  K. R. Kinsman, E. Eichen and H. I. Aaronson, Met. Trans.,
      6A (1975). pp. 303-317.

# THERMODYNAMICS OF THE BAINITIC TRANSFORMATION IN Fe-C

T. Y. Hsu (Xu Zuyao) and Mou Yiwen

Department of Materials Science and
Engineering, Shanghai Jiao Tong
University, Shanghai 200030
The People's Republic of China

The total driving force and the driving force for growth (nucleation) associated with probable mechanisms for the bainitic transformation in Fe-C, i.e. $\gamma \rightarrow \alpha + \gamma_1$, $\gamma \rightarrow \alpha + Fe_3C$ and $\gamma \rightarrow \alpha'$ and $\alpha' \rightarrow \alpha + Fe_3C$, are calculated with the KRC and LFG models. Comparison of these results indicates that the formation and growth of bainitic ferrite with the martensitic (shear) mechanism raises difficulties .

## Thermodynamics Approach

### Total Driving Force

The total change in free energy attending the reactions: $\gamma \rightarrow \alpha + \gamma_1$, $\gamma \rightarrow \alpha + Fe_3C$, and $\gamma \rightarrow \alpha'$ with the same composition and $\alpha' \rightarrow \alpha_B + Fe_3C$ ( $\alpha_B$ denotes the composition of the bainitic ferrite) are calculated through the KRC (1) and LFG (2,3) models whose activity expressions were found in reference (4). The expressions of activity of iron in ferrite, $a_{Fe}$, however, are derived by differentiating activity of carbon with respect to carbon concentration and applying the Gibbs-Duhem equation. We have the following results:

$$\ln a^{\alpha}_{Fe} = \frac{3}{Z_{\alpha}-3} \ln \frac{3-Z_{\alpha}X_{\alpha}}{3(1-X_{\alpha})} \tag{1}$$

for the KRC model, and

$$\ln a^{\alpha}_{Fe} = 9\ln \frac{3(1-X_{\alpha})}{3-4X_{\alpha}} + 12\ln \frac{3(1-2J_{\alpha})+(8J_{\alpha}-3)X_{\alpha}-\delta_{\alpha}}{2J_{\alpha}(4X_{\alpha}-3)} \tag{2}$$

for the LFG model, where

$$Z_{\alpha} = 12-8e^{-w_{\alpha}/RT} \tag{1a}$$

$$J_{\alpha} = 1-e^{-w_{\alpha}/RT} \qquad \delta_{\alpha}=[9-6(2J_{\alpha}+3)X_{\alpha}+(9+16J_{\alpha})X^2_{\alpha}]^{1/2} \tag{2a,b}$$

In both KRC and LFG models we have taken the interaction energy between carbon atoms in austenite, $w_{\gamma}= 8054$ Jmol$^{-1}$, the partial molar enthalpy and non-configurational entropy of carbon in austenite, $\Delta\bar{H}_{\gamma}=38573$ Jmol$^{-1}$, $\Delta\bar{S}^{xs}_{\gamma}=$ 13.48 JK$^{-1}$mol$^{-1}$ from Shiflet et al. (4), and equivalents in ferrite, $w_{\alpha}=$ 48570 Jmol$^{-1}$ from Bhadeshia (5), $\Delta\bar{H}_{\alpha}= 112206$ Jmol$^{-1}$ and $\Delta\bar{S}^{xs}_{\alpha} = 51.46$ JK$^{-1}$mol$^{-1}$ from Lobo et al. (6). The change in free energy attending $\gamma \rightarrow \alpha$ of the pure iron, $\Delta G^{\gamma\rightarrow\alpha}_{Fe}$ values were taken from Kaufman et al., Mogutnov and Orr and Chipman.

The free energy change accompanying the reaction $\gamma \rightarrow \alpha + \gamma_1$ is calculated through eqs. (19) and (20) of reference (4). The free energy change of $\gamma \rightarrow \alpha + Fe_3C$ may be written as (1):

$$\Delta G^{\gamma\rightarrow\alpha+Fe_3C} = (1-X_{\gamma})G^{\alpha}_{Fe}+X_{\gamma}G^{G}_{C}+X_{\gamma}\Delta G^{Fe_3C}-G^{\gamma} \tag{3}$$

where $G^{\alpha}_{Fe}$, $G^{G}_{C}$ and $G^{\gamma}$ are the free energies of pure $\alpha$ iron, graphite and austenite, the formation free energy change of cementite, $\Delta G^{Fe_3C}$ can be obtained from table 5 in reference (1). Applying the KRC and LFG models to eq. (3) obtains:

$$\Delta G^{\gamma\rightarrow\alpha+Fe_3C}=(1-X_{\gamma})\Delta G^{\gamma\rightarrow\alpha}_{Fe}+X_{\gamma}(\Delta G^{Fe_3C}-\Delta\bar{H}_{\gamma}+\Delta\bar{S}^{xs}_{\gamma}T)-\frac{RT}{Z_{\gamma}-1}[(1-Z_{\gamma}X_{\gamma})\ln(1-Z_{\gamma}X_{\gamma})$$

$$-(1-X_{\gamma})\ln(1-X_{\gamma})+X_{\gamma}(Z_{\gamma}-1)\ln X_{\gamma}] \tag{4}$$

and

$$\Delta G^{\gamma\rightarrow\alpha+Fe_3C}=(1-X_{\gamma})\Delta G^{\gamma\rightarrow\alpha}_{Fe}+X_{\gamma}(\Delta G^{Fe_3C}-6w_{\gamma}-\Delta\bar{H}_{\gamma}+\Delta\bar{S}^{xs}_{\gamma}T)-5RT[(1-X_{\gamma})\ln(1-X_{\gamma})$$

$$-(1-2X_{\gamma})\ln(1-2X_{\gamma})-X_{\gamma}\ln X_{\gamma}]-6RT[X_{\gamma}\ln\frac{\delta_{\gamma}-1+3X}{\delta_{\gamma}+1-3X}+(1-X_{\gamma})\ln\frac{1-2J_{\gamma}+(4J_{\gamma}-1)X_{\gamma}-\delta_{\gamma}}{2J_{\gamma}(2X_{\gamma}-1)}] \tag{5}$$

where $\delta_{\gamma}=[1-2(1+2J_{\gamma})X_{\gamma}+(1+8J_{\gamma})X^2_{\gamma}]^{1/2}$ $\qquad J_{\gamma} = 1-e^{-w_{\gamma}/RT}$ (5a,b)
The free energy change of $\gamma \rightarrow \alpha'$ ( the same composition ) would be expressed as

$$\Delta G^{\gamma\rightarrow\alpha'} = (1-X_{\gamma})\Delta G^{\gamma\rightarrow\alpha}_{Fe}+RT[X_{\gamma}\ln\frac{a^{\alpha}_{C}}{a^{\gamma}_{C}}+(1-X_{\gamma})\ln\frac{a^{\alpha}_{Fe}}{a^{\gamma}_{Fe}}] \tag{6}$$

Substituting the KRC and LFG models into eq. (6) obtains:

$$\Delta G^{\gamma\rightarrow\alpha'} = \frac{RT}{(Z_{\alpha}-3)(Z_{\gamma}-1)}[(Z_{\gamma}-1)(3-Z_{\alpha}X_{\gamma})\ln(3-Z_{\alpha}X_{\gamma})-(Z_{\alpha}-3)(1-Z_{\gamma}X_{\gamma})\ln(1-Z_{\gamma}X_{\gamma})$$

$$+(Z_{\alpha}-3Z_{\gamma})(1-X_{\gamma})\ln(1-X_{\gamma})-3(Z_{\gamma}-1)(1-X_{\gamma})\ln 3]+(1-X_{\gamma})\Delta G^{\gamma\rightarrow\alpha}_{Fe}+[\Delta\bar{H}_{\alpha}-\Delta\bar{H}_{\gamma}-(\Delta\bar{S}^{xs}_{\alpha}-\Delta\bar{S}^{xs}_{\gamma})T]X_{\gamma}$$

and

$$\tag{7}$$

$$\Delta G^{\gamma\to\alpha'} = RT\Big\{ 2X_\gamma\ln X_\gamma + 4(1-X_\gamma)\ln(1-X_\gamma) + 5(1-2X_\gamma)\ln(1-2X_\gamma) - 3(3-4X_\gamma)\ln(3-4X_\gamma)$$

$$9(1-X_\gamma)\ln 3 + 4X_\gamma\ln\frac{\delta_\alpha-3+5X_\gamma}{\delta_\alpha+3-5X_\gamma} - 6X_\gamma\ln\frac{\delta_\gamma-1+3X_\gamma}{\delta_\gamma+1-3X_\gamma} + 12(1-X_\gamma)\ln\frac{3(1-2J_\alpha)+(8J_\alpha-3)X_\gamma-\delta_\alpha}{2J_\alpha(4X_\gamma-3)}$$

$$-6(1-X_\gamma)\ln\frac{1-2J_\gamma+(4J_\gamma-1)X_\gamma-\delta_\gamma}{2J_\gamma(2X_\gamma-1)}\Big\} + (1-X_\gamma)\Delta G_{Fe}^{\gamma\to\alpha} + X_\gamma[\Delta\bar{H}_\alpha - \Delta\bar{H}_\gamma - (\Delta\bar{S}_\alpha^{xs} - \Delta\bar{S}_\gamma^{xs})T + 4w_\alpha - 6w_\gamma] \quad (8)$$

The free energy change of $\alpha' \to \alpha'' + Fe_3C$ ($\alpha''$ being with lower supersaturation than $\alpha'$) would be written as

$$\Delta G^{\alpha'\to\alpha''+Fe_3C} = RT\Big[ X_{\alpha'}\ln\frac{a_C^{\alpha''}}{a_C^{\alpha'}} + (1-X_{\alpha'})\ln\frac{a_{Fe}^{\alpha''}}{a_{Fe}^{\alpha'}} \Big] \quad (9)$$

Substituting the KRC and LFG models into eq. (9) obtains:

$$\Delta G^{\alpha'\to\alpha''+Fe_3C} = RT\Big\{ X_{\alpha'}\ln\frac{(3-Z_\alpha X_{\alpha'})X_{\alpha''}}{(3-Z_\alpha X_{\alpha''})X_{\alpha'}} + \frac{3(1-X_{\alpha'})}{(Z_\alpha-3)}\ln\frac{(3-Z_\alpha X_{\alpha''})(1-X_{\alpha'})}{(3-Z_\alpha X_{\alpha'})(1-X_{\alpha''})} \Big\} \quad (10)$$

and

$$\Delta G^{\alpha'\to\alpha''+Fe_3C} = RT\Big\{ X_{\alpha'}\Big[ 3\ln\frac{(3-4X_{\alpha''})X_{\alpha'}}{(3-4X_{\alpha'})X_{\alpha''}} + 4\ln\frac{(\delta_{\alpha''}-3+5X_{\alpha''})(\delta_{\alpha'}+3-5X_{\alpha'})}{(\delta_{\alpha'}-3+5X_{\alpha'})(\delta_{\alpha''}+3-5X_{\alpha''})} + (1-X_{\alpha'})\cdot$$

$$\cdot\Big[ 9\ln\frac{(1-X_{\alpha''})(3-4X_{\alpha'})}{(1-X_{\alpha'})(3-4X_{\alpha''})} + 12\ln\frac{[3(1-2J_\alpha)+(8J_\alpha-3)X_{\alpha''}-\delta_{\alpha''}](4X_{\alpha'}-3)}{[3(1-2J_\alpha)+(8J_\alpha-3)X_{\alpha'}-\delta_{\alpha'}](4X_{\alpha''}-3)} \Big]\Big\} \quad (11)$$

The free energy changes of the reactions: $\gamma \to \alpha + \gamma_1$, $\gamma \to \alpha + Fe_3C$ and $\gamma \to \alpha'$ are shown in Figs. 1 to 6, respectively.

Driving Force for Growth (Nucleation)

The maximum driving force for nucleation (growth) of the proeutectoid ferrite in austenite may be calculated through the following equations (7):

$$\Delta G_{Fe}^{\gamma\to\alpha} + RT\ln\frac{a_{Fe}^\alpha\{X_m\}}{a_{Fe}^\gamma\{X_\gamma\}} - RT\ln\frac{a_C^\alpha\{X_m\}}{a_C^\gamma\{X_\gamma\}} = 0 \quad (12)$$

where $X_m$ represents the composition of the nucleus of ferrite, and

$$\Delta G_{Nm}^{\gamma\to\alpha+\gamma_1} = RT\ln\frac{a_C^\alpha\{X_m\}}{a_C^\gamma\{X_\gamma\}} \quad (13)$$

The expressions of the KRC model would be written as

$$[Z_\alpha\ln(3-Z_\alpha X_m)-3\ln(3(1-X_m))]/(Z_\alpha-3)+\ln(X_\gamma/X_m)+[\ln(1-X_\gamma)-Z_\gamma\ln(1-Z_\gamma X_\gamma)]/$$

$$(Z_\gamma-1) = [\Delta\bar{H}_\alpha - \Delta\bar{H}_\gamma - (\Delta\bar{S}_\alpha^{xs}-\Delta\bar{S}_\gamma^{xs})T - \Delta G_{Fe}^{\gamma\to\alpha}]/RT \quad (14)$$

$$\Delta G_{Nm}^{\gamma\to\alpha+\gamma_1} = RT\ln\frac{X_m(1-Z_\gamma X_\gamma)}{X_\gamma(3-Z_\alpha X_m)} + \Delta\bar{H}_\alpha - \Delta\bar{H}_\gamma - (\Delta\bar{S}_\alpha^{xs}-\Delta\bar{S}_\gamma^{xs})T \quad (15)$$

For the nucleus of ferrite with any given composition in austenite, we have

$$\Delta G_N^{\gamma\to\alpha+\gamma_1} = (1-X_\alpha)\Delta G_{Fe}^{\gamma\to\alpha} + RT\Big\{ X_\alpha\ln\frac{a_C^\alpha\{X_\alpha\}}{a_C^\gamma\{X_\gamma\}} + (1-X_\alpha)\ln\frac{a_{Fe}^\alpha\{X_\alpha\}}{a_{Fe}^\gamma\{X_\gamma\}} \Big\} \quad (16)$$

The driving force for growth (nucleation) of the proeutectoid cementite may be expressed as

$$\Delta G_N^{\gamma\to Fe_3C+\gamma_1} = X_{Fe_3C}G^{Fe_3C} - G^\gamma\big|_{X_\gamma} - (X_{Fe_3C}-X_\gamma)\frac{dG^\gamma}{dX}\Big|_{X_\gamma} \quad (17)$$

We have upon arrangement:

$$\Delta G_N^{\gamma\to Fe_3C+\gamma_1} = 0.25(\Delta G^{Fe_3C} - RT\ln a_C^\gamma) + 0.75(\Delta G_{Fe}^{\gamma\to\alpha} - RT\ln a_{Fe}^\gamma) \quad (18)$$

Substituting the KRC and LFG models into eq. (18) obtains:

$$\Delta G_N^{\gamma\to Fe_3C+\gamma_1} = 0.25(\Delta G^{Fe_3C} - \Delta\bar{H}_\gamma + \Delta\bar{S}_\gamma^{xs}T) + 0.75\Delta G_{Fe}^{\gamma\to\alpha} - \frac{RT}{4(Z_\gamma-1)}[(4-Z_\gamma)\cdot$$

$$\cdot\ln(1-Z_\gamma X_\gamma) + (Z_\gamma-1)\ln X_\gamma - 3\ln(1-X_\gamma)] \quad (19)$$

and

$$\Delta G_N^{\gamma\to Fe_3C+\gamma_1} = 0.25(\Delta G^{Fe_3C} - \Delta\bar{H}_\gamma + \Delta\bar{S}_\gamma^{xs}T - 6w_\gamma) + 0.75\Delta G_{Fe}^{\gamma\to\alpha} - 0.25RT[15\ln(1-X_\gamma)$$

$$-10\ln(1-2X_\gamma)-5\ln X_\gamma+6\ln\frac{\delta_\gamma-1+3X_\gamma}{\delta_\gamma+1-3X_\gamma} + 18\ln\frac{1-2J_\gamma+(4J_\gamma-1)X_\gamma-\delta_\gamma}{2J_\gamma(2X_\gamma-1)} \Big] \quad (20)$$

The calculation values of the driving force for the growth of the proeutectoid ferrite are shown in Fig. 7 and those of the proeutectoid cementite are shown in Figs. 8 and 9. The driving force for the growth (nucleation) of the bainitic ferrite with the same composition is equal to that for the transformation of austenite into ferrite with the same composition (let $X_\alpha = X_\gamma$ in eq. (16)), i.e., as shown in Figs. 5 and 6.

## Discussion and Conclusion

The different results are obtained when the different models and various $\Delta G^{\gamma \to \alpha}$ values are used. The magnitude of the driving force with various reactions is in the following order: $\Delta G^{\gamma \to \alpha + Fe_3C}$, $\Delta G^{\gamma \to \alpha + \gamma_1}$ and $\Delta G^{\gamma \to \alpha'}$ at the temperature range of the bainite formation, as shown in Figs. 10 and 11 in which the results of Kaufman et al. (1) are also shown. With the carbon content of 0.1 - 0.55 wt%, the driving force of $\gamma \to \alpha + \gamma_1$, where $\alpha$ is the carbide-free bainite or the Widmanstätten ferrite, is about 45 J mol$^{-1}$ from the KRC and 60 J mol$^{-1}$ from the LFG model. The values seem too low to drive the shear transformation.

For 0.8 wt% C alloy ($X_c = 0.036$), the starting temperature of the upper bainite being about 823 K, the carbon content of the extrapolated $\alpha/(\alpha+\gamma)$ phase boundary is $X_c = 0.0016$ as calculated. Taking this carbon content as that of the concentration of the bainitic ferrite, we get the change in free energy attending the cementite precipitation is -552 Jmol$^{-1}$ from both KRC and LFG models. The free energy difference between austenite and ferrite of the same composition at 823K is 207 Jmol$^{-1}$ from the KRC and 212 Jmol$^{-1}$ from the LFG model. So we have the change in free energy $\Delta G$ attending the reactions: $\gamma \to \alpha'$ and $\alpha' \to \alpha''(X_c = 0.0016) + Fe_3C$, $\Delta G = 207 - 552 = -345$ J mol$^{-1}$ from the KRC and $\Delta G = 212 - 552 = -340$ Jmol$^{-1}$ from the LFG model. The Zener ordering energy, if calculated through Fisher's method (8), is about $- 26$ Jmol$^{-1}$. However, the driving force for martensitic shear transformation in Fe-C of at least 1170 Jmol$^{-1}$ (280 cal mol$^{-1}$) (7) is required.

Referencing Fig. 7 to 9, we can verify thermodynamically that the leading phase of the decomposition of austenite must be ferrite in hypoeutectoid steel because of the driving force for the nucleation of $\gamma \to \alpha + \gamma_1$ is larger than that of $\gamma \to Fe_3C + \gamma_1$ with $X_c < 0.035$, while in hypereutectoid one, the leading phase is cementite because of $\Delta G_N^{\gamma \to Fe_3C + \gamma_1} > \Delta G_N^{\gamma \to \alpha + \gamma_1}$.

As the driving force for the growth of $\alpha'$, being same as the total driving force of the transformation, is much smaller than that of the proeutectoid ferrite in the temperature range of the bainitic transformation (see Fig. 12), the conclusion may be drawn that the formation and growth of the bainitic ferrite with   martensitic (shear) mechanism takes difficulties.

## References

1.  Larry Kaufman, S. V. Radcliffe, and Morris Cohen, "Thermodynamics of the Bainite Reaction", pp. 313-352 in Decomposition of Austenite by Diffusional Processes, V. F. Zackey and H. I. Aaronson ed.; Interscience, New York, N. Y., 1962.
2.  J. R. Lacher, "The Statistics of the Hydrogen-Palladium System", Proceedings Cambridge Philosophic Society, 33 (1937) pp. 518-523, Cambridge University Press, 1937.

3.  Ralph H. Fowler and E. A. Guggenheim, Statistical Thermodynamics, p. 442; Cambridge University Press, New York, 1939.
4.  G. J. Shiflet, J. R. Bradley, and H. I. Aaronson, "A Re-examination of the Thermodynamics of the Proeutectoid Ferrite Transformation in Fe-C Alloys", Metallurgical Transactions, 9A (7) (1978) pp. 999-1008.
5.  H. K. D. H. Bhadeshia, "Thermodynamics of Steels: Carbon-carbon Interaction Energy", Metal Science, 14 (6) (1980) pp. 230-232.
6.  Joseph A. Lobo and Gordon H. Geiger, "Thermodynamics and Solubility of Carbon in Ferrite and Ferritic Fe-Mo Alloys", Metallurgical Transactions, 7A (9) (1976), pp. 1347-1357.
7.  H. K. D. H. Bhadeshia, "Thermodynamic Analysis of Isothermal Transformation Diagrams", Metal Science, 16 (3) (1982) pp. 159-165.
8.  J. C. Fisher, Transactions AIME, 185 (1949) p. 688.
9.  T. Y. Hsu (Xu Zuyao), "Thermodynamics of Martensitic Transformation in Fe-C", Acta Metallurgica Sinica, 15 (1979) pp. 329-338.

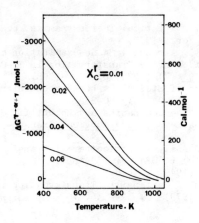

Figure 1 - The driving force of the proeutectoid ferrite, calculated through the KRC with $\Delta G^{\gamma \to \alpha}_{Fe}$ of Kaufman et al.

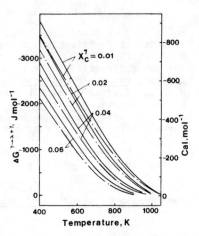

Figure 2 - The driving force of the proeutectoid ferrite, calculated through the LFG with $\Delta G^{\gamma \to \alpha}_{Fe}$ of Mogutnov (———) and Kaufman et al ( — — —).

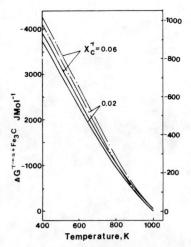

Figure 3 - The driving force of $\gamma \to \alpha + Fe_3C$, calculated through the KRC with $\Delta G^{\gamma \to \alpha}_{Fe}$ of Kaufman et al (———) and Mogutnov (— — —).

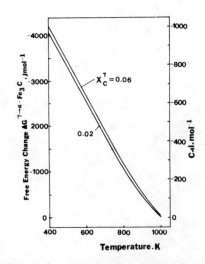

Figure 4 - The driving force of $\gamma \to \alpha + Fe_3C$, calculated through the LFG with $\Delta G^{\gamma \to \alpha}_{Fe}$ of Mogutnov.

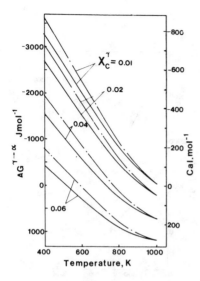

Figure 5 - The change in free energy associated with the transformation of austenite to ferrite with the same composition through the KRC with $\Delta G_{Fe}^{\gamma \to \alpha}$ of Kaufman et al (——) and Mogutnov (– – –).

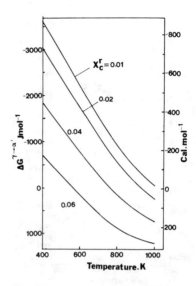

Figure 6 - The change in free energy associated with the transformation of austenite to ferrite with the same composition through the LFG with the $\Delta G_{Fe}^{\gamma \to \alpha}$ of Mogutnov.

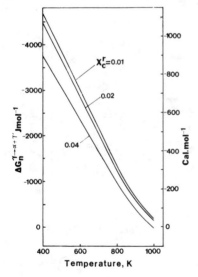

Figure 7 - The driving force for the growth of the proeutectoid ferrite calculated through the KRC with the $\Delta G_{Fe}^{\gamma \to \alpha}$ of Kaufman et al.

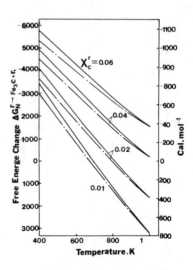

Figure 8 - The driving force for the growth of the proeutectoid cementite, calculated through the KRC with the $\Delta G_{Fe}^{\gamma \to \alpha}$ of Orr et al (——) and Kaufman et al.(– – –).

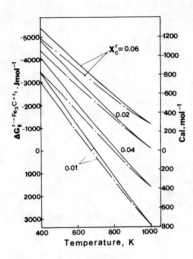

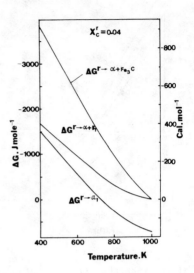

Figure 9 - The driving force for the growth of proeutectoid cementite, calculated through the LFG with the $\Delta G_{Fe}^{\gamma \rightarrow \alpha}$ of Orr et al.(———) and Kaufman et al (— — —).

Figure 10 - Comparison of the driving force associated with the various reactions, calculated through the KRC with the $\Delta G_{Fe}^{\gamma \rightarrow \alpha}$ of Kaufman et al.

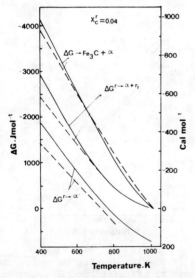

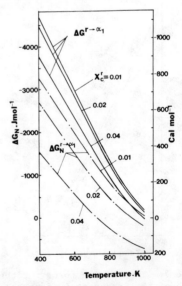

Figure 11 - Comparison of the driving force associated with the various reactions, calculated through the LFG with the $\Delta G_{Fe}^{\gamma \rightarrow \alpha}$ of Mogutnov. The results of Kaufman et al.(1) are also shown as borken lines.

Figure 12 - The driving force for the growth (nucleation) of $\gamma \rightarrow \alpha + \gamma_1$ and $\gamma \rightarrow \alpha'$, calculated through the KRC with the $\Delta G_{Fe}^{\gamma \rightarrow \alpha}$ of Kaufman et al.

# SOLUTE-DRAG, KINETICS AND THE MECHANISM OF THE BAINITE REACTION IN STEELS

H. K. D. H. Bhadeshia

University of Cambridge
Department of Metallurgy and Materials Science
Pembroke Street
Cambridge CB2 3QZ, England.

Various models for the growth of bainite in steels are compared with experimental results and it is suggested that neither the local equilibrium nor the paraequilibrium mechanisms of transformation are appropriate for bainite growth. It is also demonstrated that segregation induced solute drag cannot be claimed to influence the growth of bainite. The growth rate of an individual bainite sub-unit is found to be many orders of magnitude greater than that expected under conditions of carbon diffusion control.

## Introduction

The purpose of this work is to critically examine the ways in which bainite might grow and to compare these growth models with information obtained from high-resolution experiments. Attention is focussed primarily on the kinetic aspects, but the results should be considered in the context of other recent work (see ref.1 for details) on the mechanism of the bainite reaction in steels. Before discussing the results in detail, we note that it is now well established that long range redistribution of substitutional (X) alloying elements does not occur during bainite growth, so that the X/Fe atom ratio is the same in the parent and product lattices; this is expected since the transformation is accompanied by a surface relief effect which has the characteristics of an invariant-plane strain with a significant shear component. On the other hand, short range X atom concentration fluctuations have sometimes been suggested to exist the vicinity of the transformation interface although this would seem inconsistent with the observed surface relief effects. Any short-range X atom concentration fluctuations at the transformation interface may reduce the mobility of the interface by means of various solute-drag effects.

## Theory, Results and Discussion

### Local Equilibrium

In the local equilibrium model (2), both the X atoms and C atoms partition (between the $\alpha$ and $\gamma$ lattices) during transformation, such that each element has the same partial molar free energy in the $\alpha$ and $\gamma$ lattices at the interface; local equilibrium therefore exists at the interface. At low supersaturations, where equilibrium partitioning is a thermodynamic necessity, it is the long range diffusion of X atoms that is meant to control interface motion. The parent and product lattices thus differ substantially in X element concentration. At high supersaturations, bigger departures from equilibrium can be tolerated and the $\gamma \to \alpha$ transformation is supposed to occur in such a way that although local equilibrium is maintained at the interface, only a very narrow X atom spike is allowed to exist in the $\gamma$ immediately adjacent to the interface. The growth of $\alpha$ is then controlled by the diffusion of carbon in the $\gamma$ and the parent and product lattices have a negligible difference in X/Fe atom ratio.

From the point of view of bainite, we are concerned with the high supersaturation case. Hence, in addition to implying a carbon diffusion controlled growth rate, the local equilibrium model also requires the existence of a narrow X atom spike in the $\gamma$ at the interface, such that the concentration of X at the interface amounts to the equilibrium concentration. Fig.1 illustrates an imaging-atom-probe study of the austenite/bainitic-ferrite interface. The bainite was formed by isothermally transforming a Fe-0.43C-2.02Si-3.0Mn (wt.pct.) alloy at $350^{\circ}$C; the details are given in (1). The imaging atom probe has both spatial and compositional resolution on an atomic scale and the micrographs illustrate a uniform distribution of substitutional alloying elements despite the presence of the interface. The required composition spike of the local equilibrium model does not exist.

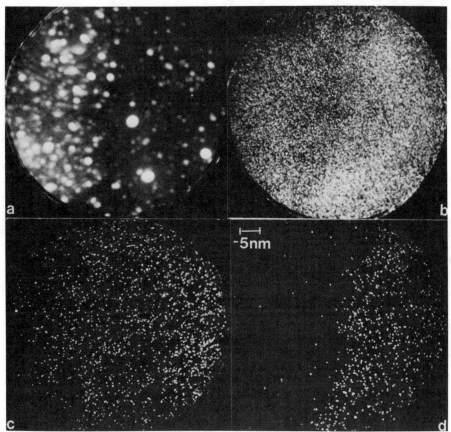

Figure 1: Imaging atom-probe micrographs (1) taken across and austenite-bainitic ferrite interface. (a) Field ion image, the $\alpha/\gamma$ interface being vertical. (b) Corresponding iron map. (c) Corresponding Si map. (d) Corresponding C map.

## Solute Drag

The theory of solute segregation induced drag at transformation interfaces is not well established (3) and the experimental evidence in this area is all the more difficult to interpret. There have been a number of suggestions implying the existence of significant interactions between substitutional alloying elements and austenite-ferrite interfaces, but there is no clear evidence for such effects as far as the bainite reaction is concerned.

All of the recent solute drag models rely entirely on the segregation of solute elements to the austenite-ferrite interface. Fig.1 provides direct evidence (on an atomic scale) that there is no solute segregation at the bainitic-ferrite/austenite interface. Solute drag cannot therefore be claimed to affect the growth of bainite.

Paraequilibrium Transformation

Paraequilibrium transformation (1) refers to the formation of α from alloyed austenite without any redistribution of substitutional alloying elements, even on the finest conceivable scale. Carbon is supposed to partition during transformation, such that its partial molar free energy is equal in both the phases at the interface. The X/Fe atom ratio is therefore constant throughout the transforming sample, even at the interface so that X elements are not in equilibrium at the interface. Ferrite growth occurs at a rate controlled by the diffusion of carbon in the austenite ahead of the interface.

The possibility of bainite growing by a paraequilibrium mechanism was examined with hot-stage experiments in a Photo-Emission electron microscope, so that the individual sub-units of bainite could be directly resolved during transformation. The present work is a preliminary report on these experiments; many more experiments and other details will be published elsewhere. Fig.2 illustrates a sequence of Photo-Emission electron micrographs, taken at 1 second intervals, showing the growth of a bainite sub-unit at 380$^{\circ}$C in the Fe-Mn-Si-C alloy described earlier. The measured growth rate of the arrowed sub-unit is about $7.5 \times 10^{-5} \text{ms}^{-1}$.

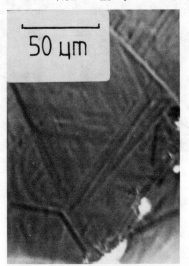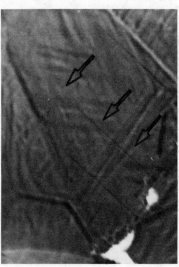

Figure 2: Photoemission electron micrographs showing the growth of an individual bainite sub-unit, as discussed in the text.

These results were compared with the theoretically expected carbon diffusion controlled paraequilibrium growth rate using the theory due to Trivedi (4). This theory has been presented in detail elsewhere (4), but the following crucial points should be noted:
(a) The paraequilibrium phase diagram was calculated as in (5), giving the paraequilibrium carbon concentration in the austenite at 380$^{\circ}$C as 0.1164 mol.frac.
(b) The calculations assume that the plate tip radius adjusts itself to a value consistent with the occurrence of a maximum growth velocity. Hence, the equilibrium concentration of carbon in the austenite at the plate tip is

0.096869 mol. frac. when the $\alpha/\gamma$ interface energy is taken to be $0.2Jm^{-2}$.
(c) Strain energy effects are ignored; these can only reduce the calculated growth velocity.
(d) The calculations are based on a carbon diffusion coefficient evaluated as a weighted average of the composition range in the matrix, as discussed by Trivedi and Pound (4). The relevant composition range which delimits the carbon concentration profile ahead of the growing edge is thus 0.0196-0.096869, giving the effective diffusivity as $0.1205 \times 10^{-15} m^2 s^{-1}$. The diffusivity was expressed as a function of carbon content using the method of Bhadeshia (6).

All the above conditions should give an accurate estimate of the <u>maximum</u> possible growth rate under carbon diffusion control, and this was calculated to be $8.34 \times 10^{-8} ms^{-1}$. This is clearly many orders of magnitude lower than that observed experimentally, despite the fact that the calculations allow the <u>maximum</u> possible growth rate to be evaluated. The growth of bainite clearly occurs at a rate much faster than expected from carbon diffusion control. Recent work (1) has indicated that this is only to be expected since the experimental evidence (1) implies that bainite grows by a shear mechanism with a supersaturation of carbon in the bainitic ferrite.

## Conclusions

It has been demonstrated that bainite does not grow under conditions of local- or para- equilibrium; the growth rate as measured experimentally is many orders of magnitude higher than that expected from C diffusion control. Atom-probe experiments also indicate that the segregation necessary for solute drag does not exist.

## Acknowledgements

The author is grateful to Professor R.W.K. Honeycombe FRS for the provision of laboratory facilities and to Professor W. Form and Dr. D.V. Edmonds for arranging the Photo-Emission experiments. Thanks are also due to Dr. R. Waugh for help with the atom-probe work.

## References

(1) H.K.D.H. Bhadeshia and A.R. Waugh, "Bainite: An Atom-Probe Study of the Incomplete Reaction Phenomenon", Acta Metall., 30 (1982) pp.775-784.
(2) D.E. Coates, "Diffusional Growth Limitations to Hardenability", Metall. Trans., 4 (1973) pp.2313-2325.
(3) H.K.D.H. Bhadeshia, "Considerations of Solute Drag in relation to Transformations in Steels", J. Mat. Sci., 18 (1983) pp.1473-1481.
(4) J.W. Christian, "Theory of Tranformations in Metals and Alloys", Part 1, 2nd Edition, p.472, Pergamon Press, Oxford, 1975.
(5) H.K.D.H. Bhadeshia and D.V. Edmonds, "The Mechanism of Bainite Formation in Steels", Acta Metall., 28 (1980) pp.1265-1273.
(6) H.K.D.H. Bhadeshia, "Diffusion of Carbon in Austenite", Metal Science, 15 (1981) pp.477-478.

# FORMATION OF FERRITE FROM CONTROL-ROLLED AUSTENITE

G. R. Speich
Department of Metallurgical and Materials Engineering
Illinois Institute of Technology
Chicago, Illinois  60616

L. J. Cuddy
United States Steel Corporation
Technical Center
Monroeville, PA  15146

C. R. Gordon and A. J. DeArdo
Department of Metallurgical and Materials Engineering
University of Pittsburgh
Pittsburgh, PA  15261

Ultrafine-grained plate steels with higher strength and lower ductile-to-brittle transition temperatures than conventional plate steels are produced by careful control of the time-temperature-deformation sequence during hot rolling. The mechanism of achieving ultrafine ferrite grain sizes by such controlled-rolling processes have been intensively studied over the last 25 years. It has been generally recognized that lower finishing temperatures gave the finest ferrite grain sizes and that the presence of microalloying additions (V, Ti, Cb) resulted in further grain size refinement. More recently, it has been shown that the low temperature rolling resulted in unrecrystallized austenite, and that the deformation caused the unrecrystallized austenite to become deformed into flattened elongated grains with a very high grain-boundary surface area/unit volume. Transformation of these flattened austenite grains resulted in a very fine ferrite grain size, because of the high nucleation rate at the austenite grain boundaries. The role of microalloying additions was to delay the recrystallization of the austenite.

Many variations on this simple controlled-rolling practice have been suggested including: use of low reheating temperatures; preliminary rolling, cooling below $A_1$, and then reheating into the austenite range followed by finish rolling; forming fine-grained ferrite from recrystallized austenite; and use of accelerated cooling following controlled rolling.

In the present paper, we review the basic aspects of the formation of ferrite from controlled-rolled austenite including a discussion of the effect of deformation and cooling rate on (1) the nucleation and growth of ferrite, (2) overall reaction kinetics, (3) the structural state of the austenite, (4) the austenite-to-ferrite grain size relationships, and (5) application of these concepts to various controlled-rolling practices.

## Introduction

Ultrafine-grained ferrite plate steels with higher strength and lower ductile-to-brittle transition temperatures than conventional plate steels have been produced for the last 25 years by careful control of the time-temperature-deformation sequence during hot rolling. Early research revealed that there was a direct relationship between the rolling sequence and the final properties; i.e., lower finishing temperature resulted in higher yield strengths and lower ductile-to-brittle transition temperatures, Figure 1 (1). Both of these property improvements were found to be related to the finer ferrite grain size exhibited by the steels. In addition, it was found that when this low temperature deformation was applied to steels which contained small additions of microalloying elements such as Nb, V, or Ti, further improvements in grain refinement were realized.

Over the last 10 years, much research has been conducted to explain how the low-temperature rolling caused the observed grain refinement and why the grain refinement was magnified by the presence of microalloying additions. Much of this information has been summarized in recent conference proceedings and review articles (2-5). It was shown that the low-temperature deformation resulted in unrecrystallized austenite in which the grain shape was altered from equiaxed to flattened, elongated grains. Kozasu and co-workers (6) demonstrated that the increased surface area/unit volume of such grains gave a higher ferrite nucleation rate than equiaxed recrystallized austenite. The role of microalloying elements was to delay the recrystallization of the austenite so that such flattened, deformed austenite grains could be formed and would remain until the steel cooled to the transformation temperature, Figure 2 (7).

This research led to the development of sophisticated controlled-rolling practices in which a series of high-temperature roughing passes are used to refine the austenite by sequential recrystallization, followed by a delay to allow the plate to cool to lower temperature, and then a series of low-temperature finishing operations are used to flatten the austenite grains, Figure 3. Extremely fine-grained ferrite is produced from such flattened austenite grains, Figure 4 (8).

Many variations on this simple controlled-rolling practice have been suggested including: use of low reheating temperatures; preliminary rolling, cooling below $A_1$, and reheating into the austenite range followed by finish rolling; forming fine-grained ferrite from recrystallized austenite; and use of accelerated cooling following controlled rolling (9-13).

Because of the intense interest in this field, it seemed desirable to review recent research on the transformation of ferrite from control-rolled austenite. In the first part of this paper, we briefly review the basic elements of nucleation and growth theory and classify the high energy sites in both recrystallized and deformed austenite. In the second part, we discuss the major processing parameters that affect the structural state of the austenite. In the third part, we consider the relationships between the structural state of the austenite, the cooling rate, and the final ferrite grain size. In the fourth part, we discuss some applications of these principles to commercial controlled-rolling practices.

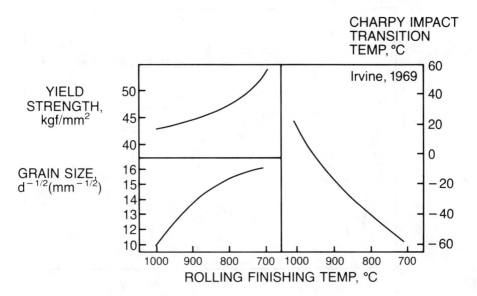

Figure 1 - Mechanical properties of ultrafine-grained plate steels (1).

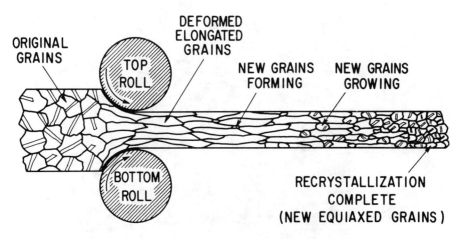

Figure 2 - Formation of flattened, deformed austenite grains and their
subsequent recrystallization during controlled rolling (7).

TEMPERATURE

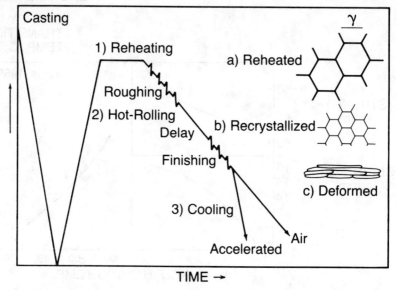

Figure 3 - Controlled-rolling process.

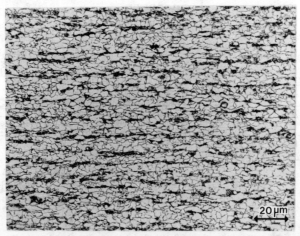

Figure 4 - Ultrafine-grained ferrite produced from flattened,
unrecrystallized austenite (8).

## Nucleation of Ferrite

### Nucleation Kinetics

The overall nucleation rate of a new phase in the solid state is given by the general nucleation equation

$$I = Z\beta N exp\left(-\frac{\tau}{t}\right) exp\left(-\frac{\Delta G^*}{kT}\right) \tag{1}$$

where I is the number of nuclei formed per unit time per unit volume, Z is the Zeldovich factor, $\beta$ is a frequency factor related to the volume diffusivity, N is the number of active sites per unit volume, $\tau$ is the incubation time, t is time, $\Delta G^*$ is the critical free energy (or activation energy) for nucleation, k is Boltzmann's constant, and T is the absolute temperature (14). In general, the nucleation rate can be separated into a transient and a steady-state period. In the transient period, the nucleation rate increases continuously once the incubation time is exceeded. The nucleation rate is nearly constant in the steady-state period.

In general, two distinct types of nucleation are possible—homogeneous and heterogeneous. In homogeneous nucleation, the nucleus forms within a grain and is independent of boundaries, dislocations, etc. In heterogeneous nucleation, the nucleus forms at a high energy site such as a grain boundary where the available boundary free energy can be used to lower the critical activation energy for nucleation. Although the number of sites for homogeneous nucleation are much greater, heterogeneous nucleation dominates most phase transformations because of the much lower activation energy for nucleation (note I depends on $\Delta G^*$ in an exponential manner and on N in a linear manner, Equation 1).

In most steel transformations, values of Z, $\beta$, and $\tau$ are not easy to calculate from first principles. As a result, changes in these parameters because of controlled rolling are not generally considered. However, changes in N and $\Delta G^*$ caused by changes in processing parameters are considered to be significant. For the special case of heterogeneous nucleation of ferrite at austenite grain boundaries, we can simplify Equation 1 to

$$I_B = K_B N_B exp\left(-\frac{\Delta G_B^*}{kT}\right) \tag{2}$$

where $I_B$, $N_B$, $\Delta G_B^*$, and $K_B = Z\beta exp(-\tau/t)$ refer to boundary nucleation. Similar equations can be derived for heterogeneous nucleation at grain corners, grain edges, twin interfaces, deformation bands, inclusions, etc. In general, more than one type of heterogeneous site may be activated and then

$$I_T = \sum_i K_i N_i exp\left(-\frac{\Delta G_i^*}{kT}\right) \tag{3}$$

where $I_T$ is the total nucleation rate from all heterogeneous sites of type i.

Heterogeneous Nucleation Sites

In recrystallized austenite, heterogeneous nucleation of ferrite can occur at grain boundaries, edges, or corners, at twin interfaces, or at inclusions, Figure 5 (14). The effectiveness of each site is dependent on both the value of $N_i$ and $\Delta G_i^*$ for each site.

Cahn (15) has shown that the value of N for grain boundaries ($N_B$), grain edges ($N_E$), and grain corners ($N_C$) are given approximately by

$$N_B = N_A \, \delta \left( \frac{1}{d} \right) \tag{4}$$

$$N_E = N_A \, \delta^2 \left( \frac{1}{d^2} \right) \tag{5}$$

$$N_C = N_A \, \delta^3 \left( \frac{1}{d^3} \right) \tag{6}$$

where $N_A$ is the number of atomic sites per unit volume, $\delta$ is the grain boundary thickness, and d is the austenite grain diameter.

Because of the ease of measuring the surface area of boundaries per unit volume, $S_V$, by quantitative metallographic techniques, it is common to replace $1/d$ in Equations 4-6 by a term involving $S_V$ because the austenite grain boundary area is related to the austenite grain diameter by the expression

$$\frac{1}{d} = C \, S_V \tag{7}$$

where C is a constant which is dependent on the shape of the individual grains (C = 1/3 for cubes or spheres, C = 1/2 for normal grain shapes) (16).

The values of N for annealing twins or inclusions can also be derived from their size, shape, and number of density. In the case of annealing twins, it is believed that only the incoherent, high energy ends of the twins would be an effective nucleation site because the broad faces of the twins are coherent, low energy interfaces, Figure 5D. In the case of inclusions, both the effect of the additional interfacial free energy of the metal/inclusion interface as well as the effect of increasing the chemical free energy change by locally depleting the surrounding austenite of alloy content, e.g. Mn for MnS, has to be considered.

The value of $\Delta G_i^*$ is sensitive to the models chosen for each site. In the simplest case for nucleation at a boundary, a double hemispherical cap is assumed, strain energy is ignored, and the $\alpha/\gamma$ interface is assumed to be isotropic and incoherent, Figure 6. Russell (14), following the work of Cahn (15), has shown that the critical free energy for nucleation of such a nucleus is given by

$$\Delta G_B^* = \frac{32\pi \left( \sigma^{\alpha\gamma} \right)^3}{3 \left( \Delta G_V^{\gamma \to \alpha} \right)^2} \, S(\theta) \tag{8}$$

where $\sigma^{\alpha\gamma}$ is the interfacial energy of the ferrite ($\alpha$)/austenite ($\gamma$) boundary, $\Delta G_V^{\gamma \to \alpha}$ is the free energy change per unit volume for the austenite to ferrite transformation, and $S(\theta)$ is a geometric factor that is

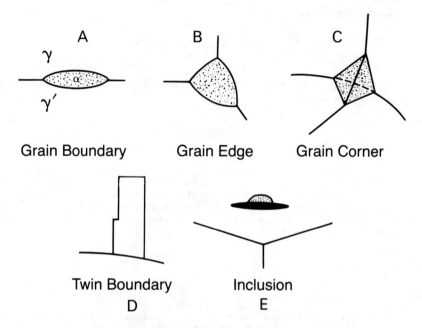

Figure 5 - Heterogeneous nucleation sites in recrystallized austenite:
(A) grain boundary, (B) grain edge, (C) grain corner,
(D) annealing twin boundary, and (E) inclusion (14).

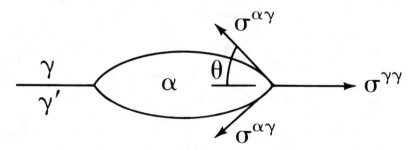

Figure 6 - Double hemispherical cap model for nucleation of grain-
boundary allotriomorph (14).

dependent on the dihedral angle $\theta$ of the cap. For the geometry shown in Figure 6, $S(\theta)$ is given by

$$S(\theta) = \frac{2-3\cos\theta + \cos^3\theta}{4} \qquad (9)$$

The dihedral angle $\theta$ is defined by the condition that the surface tensions be balanced at the $\gamma/\gamma/\alpha$ junction

$$\sigma^{\gamma\gamma} = 2\sigma^{\alpha\gamma}\cos\theta \qquad (10)$$

For nucleation at grain edges, or corners, Figure 5B and 5C, equations can be derived in a similar manner for $\Delta G_E^*$ or $\Delta G_C^*$, but the geometric term $S(\theta)$ is much more complicated. Cahn (15) has conveniently plotted the values of $\Delta G_E^*$ and $\Delta G_C^*$ in the form of a ratio $\Delta G_E^*/\Delta G_H^*$ and $\Delta G_C^*/\Delta G_H^*$ where $\Delta G_H^*$ is the critical free energy for homogeneous nucleation given by Russell (14) as

$$\Delta G_H^* = \frac{16\pi \left(\sigma^{\alpha\gamma}\right)^3}{3 \left(\Delta G_V^{\gamma\to\alpha}\right)^2} \qquad (11)$$

The results for various values of $\cos\theta$ indicate that in general $\Delta G_B^* > \Delta G_E^* > \Delta G_C^*$, Figure 7. This result is a necessary consequence of the geometry because more boundary free energy/unit volume can be utilized at a corner site than at an edge or a boundary site.

Because $\Delta G_B^*$ is extremely sensitive to the value of $\sigma^{\alpha\gamma}$, other possible nucleation models involving low energy, semicoherent $\alpha/\gamma$ interfaces have been considered, especially by Aaronson and co-workers, Figure 8 (17-19). In the simplest case of a semi-coherent allotriomorph, Figure 8A, Russell (14) shows that the critical activation free energy is lowered with respect to the incoherent case (equation 8) to a value of

$$\Delta G_{B(coh)}^* = \frac{16\pi \left(\sigma_{coh}^{\alpha\gamma}\right)^3}{3 \left(\Delta G_V^{\gamma\to\alpha} + \varepsilon\right)^2} S(\theta) \qquad (12)$$

where $\sigma_{coh}^{\alpha\gamma}$ is the coherent $\alpha/\gamma$ interfacial free energy and $\varepsilon$ is the strain energy required to maintain coherency. In general, completely coherent nuclei are not formed; dislocations at the interface make the interface semicoherent and lower the value of $\varepsilon$ which results in a lower $\Delta G_B^*$ (coh). (Note $\varepsilon$ is positive and $\Delta G^{\gamma\to\alpha}$ is negative.) Coherency or semicoherency at the $\alpha/\gamma$ interface requires that the ferrite adopt the special Kurdjumov-Sachs orientation relationship with respect to austenite. As subsequently discussed, the resulting semicoherent interface may have low mobility because it migrates by ledge formation whereas the incoherent interface migrates by normal diffusional processes.

The most recent work by Lange and Aaronson (18) suggests that experimentally observed nucleation rates are close to those calculated from Equation 1 if a semicoherent pillbox nucleation model, Figure 8C, is assumed. In this model, the broad top and bottom faces are low energy

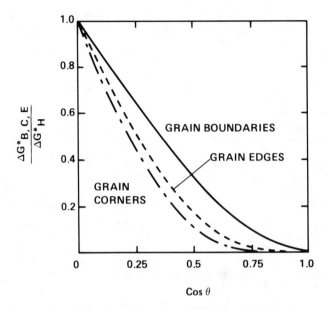

Figure 7 - Activation energy for nucleation at grain boundaries, faces, and edges (15).

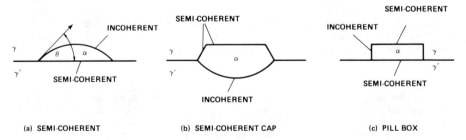

(a) SEMI-COHERENT

(b) SEMI-COHERENT CAP

(c) PILL BOX

Figure 8 - Models for semicoherent nucleation of ferrite at austenite grain boundaries (14).

semicoherent interfaces, and the thin circular edge is a high energy incoherent interface.

The value of $\Delta G_B^*$ can also be lowered by increasing the value of $\Delta G_V^{\gamma \to \alpha}$. The value of $\Delta G_V^{\gamma \to \alpha}$ is given approximately by the expression

$$\Delta G_V^{\gamma \to \alpha} = \frac{RT}{V_m} \ln\left(\frac{1 - X_O}{1 - X_\gamma^{\gamma \alpha}}\right) \tag{13}$$

where $X_O$ is the mol fraction of carbon in the steel, $X_\gamma^{\gamma \alpha}$ is the mol fraction of carbon in the austenite at the ferrite/austenite phase boundary, and $V_m$ is the molar volume (20). Because the value of $X_\gamma^{\gamma \alpha}$ increases with decreasing temperature, Figure 9, higher nucleation rates (and finer ferrite grain sizes) are achieved by increasing the cooling rate or adding alloying elements to force the transformation of austenite to occur at lower temperatures.

Of course, as Cahn (15) and Christian (21) have both shown, the relative nucleation rates do not necessarily increase in the same order as the values of $\Delta G_i^*$ decrease because the density of sites $N_i$ also decreases as the mode of nucleation changes from boundary to edge or corner. From Equations 3 and 4 to 6, it is easy to show that

$$\ln\left(\frac{I_E}{I_B}\right) = kT\ln\left(\frac{\delta}{d}\right) - \left(\Delta G_E^* - \Delta G_B^*\right) \tag{14}$$

with a similar equation for $\ln I_C/I_E$. From Equation 14, $I_E > I_B$ if $kT\ln \delta/d > \Delta G_E^* - \Delta G_B^*$. Similarly, $I_C > I_E$ if $kT\ln \delta/d > \Delta G_C^* - \Delta G_E^*$. With a value of $\sigma^{\alpha\gamma} = \sigma^{\alpha\alpha} = 500$ ergs/cm$^2$, a value of $\Delta G_V^{\gamma \to \alpha} = -15.5$ J/cm$^3$ (Equation 13 for a 1.4Mn; 0.11C steel transformed at 750°C) (20), we calculate $\Delta G_H^* = 8.6 \times 10^{-15}$ J from Equation 11. From Figure 5 we obtain $\Delta G_C^* = 8.6 \times 10^{-16}$ J, $\Delta G_E^* = 1.3 \times 10^{-15}$ J, and $\Delta G_B^* = 2.4 \times 10^{-15}$ J. For $\delta = 5 \times 10^{-8}$ cm and $d = 50 \times 10^{-4}$ cm, $kT\ln \delta/d = -1.6 \times 10^{-19}$ J. Therefore $I_C > I_E > I_B$ and grain-corner nucleation will occur first, followed by edge, and then boundary nucleation.

Of course, these arguments are valid for incoherent nucleation. It is highly unlikely that semicoherent nucleation of ferrite can occur at corners or edges as it does at boundaries, because the need to satisfy orientation relationships between three or four grains rather than with two. Also, unless sufficient growth can occur from the corner or edges to satisfy the required equilibrium volume fraction of ferrite upon cooling, then nucleation of ferrite at boundaries will be dominant. This could be the case for coarser austenite grain sizes where the distance between corner and edge sites is very large, as shown in Figure 10.

## Effect of Deformation

Upon deformation below the recrystallization temperature, (but above the transformation) a number of changes in the structural state of the austenite occur that can increase the rate of nucleation of the ferrite. First, increases in $S_V$ occur because deformation changes the austenite grain shape, creates deformation bands, changes the low-energy broad interfaces at annealing twins into high-energy interfaces, and changes the

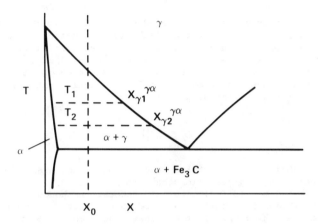

Figure 9 - Schematic Fe-C phase diagram.

Figure 10 - Nucleation of ferrite at undeformed austenite
grain boundaries (75X) (24).

inclusion shape. Second, incipient recrystallization at deformed
austenite grain boundaries may cause bulges to develop which greatly
increase the nucleation potency of the grain boundaries. Third, the
increased $\Delta G_V^{\gamma \to \alpha}$ caused by the stored energy in the austenite phase may
lead to a higher nucleation rate (Equation 8). Finally, strain-induced
precipitation of microalloy carbides, usually on subboundary networks in
the austenite phase, may drain microalloying elements from the austenite
phase and thus promote nucleation of ferrite by increasing the value of
$\Delta G_V^{\gamma \to \alpha}$.

The most frequently considered effect of deformation on nucleation is
the increased $S_V$ caused by the change in the austenite grain shape. For
hot rolling of sheet and plate, which may be regarded as plane strain
deformation, the austenite grain shape is changed as shown schematically
in Figure 11. For the simple example of an array of cubical grains, the
value of $S_V$ is dependent on the rolling reduction ratio R (original
thickness/final thickness) according to

$$S_V = \frac{1}{d} \left( 1 + R + \frac{1}{R} \right) \tag{15}$$

Equation 15 indicates that the value of $S_V$ is not greatly affected until
substantial rolling reductions have been achieved, Figure 12. More
sophisticated models (22) take into account the noncubical shape and the
anisotropy caused by rolling and use quantitative metallographic relationships to determine $S_V$. In the planar-linear model, $S_V$ is given by

$$S_V = 0.429 \; N_{L_{||}} + 2.571 \; N_{L_{\perp}} - N_{L_{|}} \tag{16}$$

where $N_L$ is the number of intercepts per unit length on the planes and in
the directions shown in Figure 13.

The plastic incompatibility of adjacent austenite grains, may require
the formation of deformation bands (regions of intense local shear) in
addition to general slip. Such deformation bands, Figure 14 region A
(24), provide additional nucleation sites for ferrite and increase the
effective value of $S_V$. Ouchi et al. (23) have shown that the density of
deformation bands, $S_{V(DB)}$, is linearly related to the percent reduction
(% Red) and that the minimum rolling reduction to form deformation bands
is about 30 percent. This leads to an expression for $S_{V(DB)}$ of the form .

$$S_{V(DB)} = 0.63 \; (\% \; Red - 30) \; mm^{-1} \tag{17}$$

The overall $S_V$, $S_{V(TOTAL)}$, is obtained by summing $S_V$ for grain boundaries,
$S_{V(GB)}$, and $S_{V(DB)}$, i.e.,

$$S_{V(TOTAL)} = S_{V(GB)} + S_{V(DB)} \tag{18}$$

as shown in Figure 15 (23,24). In many cases, deformation bands in as-
rolled and quenched austenite are difficult to identify. Partial isothermal transformation of the austenite to ferrite can then be used to
decorate the deformation bands and make them easily identified, Figure 14,

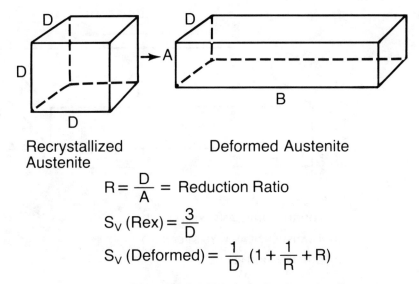

Recrystallized
Austenite                     Deformed Austenite

$$R = \frac{D}{A} = \text{Reduction Ratio}$$

$$S_V(\text{Rex}) = \frac{3}{D}$$

$$S_V(\text{Deformed}) = \frac{1}{D}\left(1 + \frac{1}{R} + R\right)$$

Figure 11 - Plane-strain deformation of cube-shaped austenite grain.

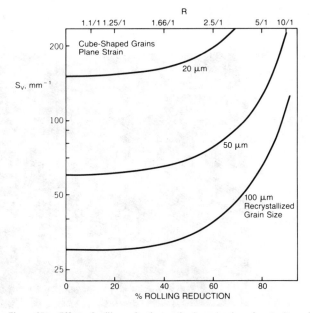

Figure 12 - Effect of rolling reduction on $S_V$ for cube-shaped austenite grains.

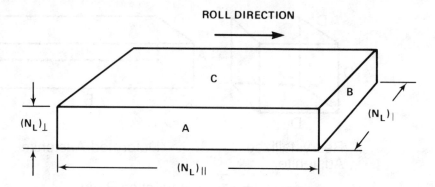

ROLL DIRECTION

A - LONGITUDINAL ANALYSIS PLANE

B - TRANSVERSE ANALYSIS PLANE

C - ORIENTATION PLANE

Figure 13  -  Planar-linear model for $S_V$ (22).

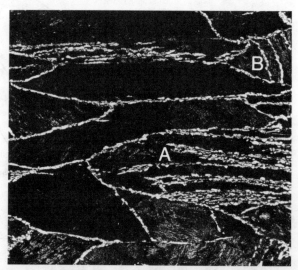

Figure 14  -  Nucleation of ferrite at deformed austenite grain boundaries,
deformation bands A, and annealing twins B (24).

region A (24). The frequency of occurrence of the deformation bands per grain is extremely irregular and related to both the austenite grain size and the orientation difference between adjacent grains (25).

In addition to the additional surface area for nucleation created by changing the shape of the austenite grains and by forming deformation bands, some new surface area for nucleation may be generated by changing the coherent low-energy interface of an annealing twin into a high-energy interface. Examination of region B in Figure 14 (and other areas) indicates that deformation and rotation of the annealing twin has changed the low energy coherent interface into a high-energy incoherent interface, and nucleation of ferrite has occurred there. There is increasing evidence that many so-called "deformation bands" are in fact deformed annealing twins (24,26).

Sandberg and Roberts (20), Roberts et al. (27), Ouchi et al. (23), and Amin and Pickering (28), have indicated that the increased nucleation of ferrite in deformed austenite is more than a simple effect of the change in $S_V$, because when the final ferrite grain size (assumed to be proportional to I and thus to $S_V$) is plotted versus $S_V$ for both recrystallized and deformed austenite, finer ferrite grain sizes are achieved with deformed austenite, Figure 16. Roberts et al. (27) suggests that this is caused by an increased nucleation potency of the deformed austenite grain boundary because of the bulging caused by incipient recrystallization, Figure 17 (27). The bulging leads to a lower activation energy for nucleation than a straight boundary because of geometric effects similar to those observed at grain edges or corners where more grain-boundary energy is available per unit volume of nucleus. Amin and Pickering (28) adopt a different viewpoint and argue that the increased nucleation potency of deformed austenite grain boundaries arises because subboundries have a higher density in the vicinity of grain boundaries than in the grain interior. These subboundaries, possibly aided by precipitation of microalloying carbides, Figure 18 (29) provide convenient sites for nucleation of ferrite.

A more subtle effect of deformation on nucleation of ferrite has been proposed by Ouchi et al. (23). Ouchi et al. found that the $Ar_3$ temperatures of Si-Mn steels was only slightly affected by rolling reduction at various temperatures. In contrast, the $Ar_3$ temperatures of Nb steels increased sharply with increased rolling reduction, especially at lower temperatures, Figure 19. After correction for different $S_V$ values, Ouchi et al. concluded that even at constant $S_V$, the Nb steels with a deformed austenite grain structure had a consistently higher $Ar_3$ temperature than those with an equiaxed grain structure. Ouchi et al. explained these results by arguing that Nb in solution lowers the $Ar_3$ temperature because of the lowering of the value of $\sigma_{\gamma\gamma}$ by segregation of Nb to the $\gamma/\gamma$ grain boundary. The lower value of $\sigma_{\gamma\gamma}$ results in a lower ferrite nucleation rate (Equations 8 to 10). Since deformation results in a lowering of the Nb content in solution by strain-induced precipitation of NbCN, deformation would result in a higher $Ar_3$ temperature because Nb would then no longer be available to segregate to the $\gamma/\gamma$ grain boundary.

Of course, since deformation of the austenite creates a high dislocation density, the stored energy of the austenite phase is increased, and this increased stored energy results in a higher value of $\Delta G_V^{\gamma\to\alpha}$. From Equation 8 this leads to a lower value of $\Delta G_B^*$. However, both Roberts et al. (27) and Ouchi et al. (23) have argued that this is a small effect. Roberts et al. estimate that the additional stored energy from the dislocations, $\Delta G_d$, is given by

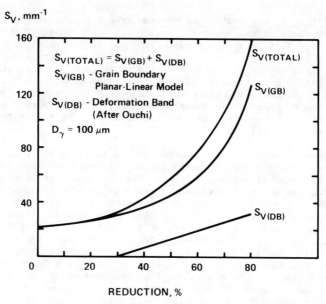

Figure 15 - Increase in $S_V$ because of deformation bands (23,24).

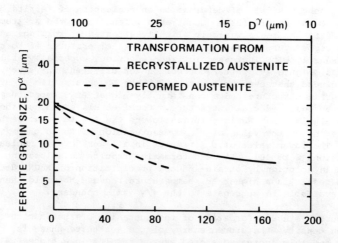

AUSTENITE GRAIN BOUNDARY AREA/UNIT VOLUME $S_V$ [mm$^{-1}$]

Figure 16 - Effect of $S_V$ for recrystallized and deformed austenite
on ferrite grain size (6,27).

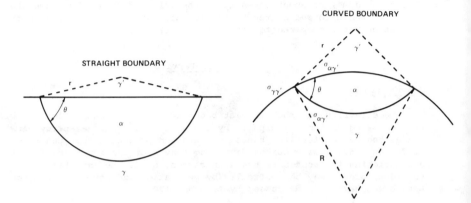

Figure 17 - Nucleation of ferrite at bulges in deformed austenite boundaries (27).

Figure 18 - Precipitation on microalloy carbides at subboundaries in deformed austenite (29).

$$\Delta G_d = - \left( \mu \, \frac{b}{2}^2 \right) \rho_o \tag{19}$$

where $\mu$ is the shear modulus, b is the Burgers vector, and $\rho_o$ is the dislocation density. With $\mu = 8.06 \times 10^4$ MPa, b = 1/2 <111> = 2.47 $\times 10^{-10}$ m, the value of $\Delta G_d$ for a dislocation density $\rho_o = 10^{15}/m^2$ is equal to 2.48 $\times 10^6$ J/m$^3$. The corresponding chemical driving force $\Delta G_V^{\gamma \to \alpha}$ for a 0.11%C, 1.45%Mn steel at 750°C according to Equation 13 is 15.5 $\times 10^6$ J/m$^3$, or about seven times the value of $\Delta G_d$. Thus, the stored energy of the deformed austenite should only have a small effect on the nucleation rate, compared to other effects (23,27).

## Growth of Ferrite

### Growth Models

Once ferrite has nucleated at boundary sites, growth of ferrite will proceed being driven principally by the difference in free energy between the austenite and ferrite phases. Growth can proceed either under diffusion control with carbon diffusing away from the advancing ferrite and enriching the austenite phase in carbon, or if the interface is semi-coherent, growth may be controlled by generation and movement of ledges. Both cases have been discussed by Aaronson (30).

In the case of a grain-boundary allotriomorph, the kinetics can be separated into thickening and lengthening kinetics, Figure 20. Since the interface of the grain-boundary allotriomorph is considered to be incoherent, straightforward diffusion models have been applied to both cases. For thickening, a simple one-dimensional diffusion model predicts a parabolic thickening rate (31,32)

$$S = \alpha \, t^{1/2} \tag{20}$$

where S is the half-thickness, $\alpha$ is a parabolic rate constant, and t is time. The value of $\alpha$ is given by the expression

$$\left( \frac{x_\gamma^{\gamma\alpha} - x_\gamma}{x_\gamma^{\gamma\alpha} - x_\alpha^{\alpha\gamma}} \right) \left( \frac{D_c^\gamma}{\pi} \right)^{1/2} = \frac{\alpha}{2} \exp\left( \frac{\alpha^2}{4D_c^\gamma} \right) \mathrm{erfc}\left( \frac{\alpha}{2\sqrt{D_c^\gamma}} \right) \tag{21}$$

where $x_\gamma^{\gamma\alpha}$ is the mol fraction of carbon in the austenite at the $\gamma/\alpha + \gamma$ phase boundary, $X_\gamma$ is the mol fraction of carbon in the parent austenite, $x_\alpha^{\alpha\gamma}$ is the mole fraction of carbon in the ferrite at the $\alpha/\alpha + \gamma$ phase boundary, and $D_c^\gamma$ is the diffusion coefficient of carbon in the austenite. Since $D_c^\gamma$ varies with carbon content, the Wagner approximation is normally applied. In general, the calculated values of S agree closely with the observed growth kinetics.

Lengthening kinetics have also been considered to be diffusion controlled. Hillert (33) shows that the rate of lengthening, $G_\ell$, is given by:

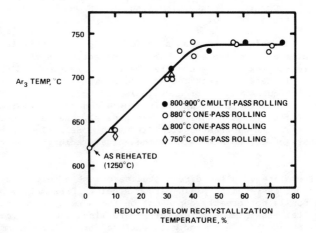

Figure 19 - Increase in $A_{r_3}$ temperature as a result of controlled rolling (23).

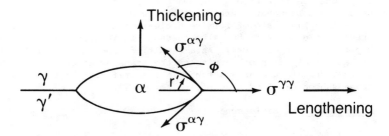

Figure 20 - Thickening and lengthening of grain-boundary allotriomorph (30).

$$G_\ell = \frac{D_c^\gamma \left(x_\gamma^{\gamma\alpha} - x_\gamma\right)}{4r^\prime \left(x_\gamma - x_\alpha^{\alpha\gamma}\right) \sin \phi} \tag{22}$$

where $\phi$ is defined in Figure 20, and $r^\prime$ is the radius of curvature of the allotriomorph in the area immediately adjacent to the junction with the boundary. Observed lengthening kinetics in general agree with Equation 22 within a factor of two.

In the case of Widmanstätten ferrite plates where the broad faces of the plates are considered to be semicoherent, then the thickening kinetics are more complex and are dictated by the nucleation and growth of ledges. In general, the kinetics are slower than for diffusion-controlled growth but quantitative prediction is difficult because neither the nucleation rate of the ledges or their growth can be mathematically treated in terms of measurable quantities. In general, the incoherent tip of the plate can still be treated with diffusion models similar to Equation 22 but where $r^\prime$ is replaced with r the actual radius of the tip of the plate (30).

There are very few direct studies of the effect of deformation on the growth process even in plain carbon steels. Presumably, because diffusion of carbon is not greatly dependent on the defect structure of the austenite, growth models for ferrite forming from deformed austenite in plain carbon steels are not much different from those from undeformed austenite. One recent study of the growth of pearlite in deformed austenite supports this conclusion because austenite deformation had no effect on the diffusion-controlled growth rate (34).

However, in microalloyed steels, the effect of deformation on growth of ferrite can be quite important because of the strain-induced precipitation of microalloy carbides (nitrides) in the austenite phase, and because of the possible effect of deformation on the interface precipitation of microalloy carbides during growth of ferrite (28,35). Some of these interactions could be complex because solute is drained from the austenite when precipitation of microalloying carbides occurs. Crooks et al. (36) show that after nucleation of the ferrite at the deformed austenite grain boundaries in V steels, certain of the ferrite grains grow preferentially out of the fine-grained assembly at the boundary, leading to a selective coarsening of ferrite grains as growth proceeds, Figure 21.

Amin and Pickering (28) have observed structures similar to those shown in Figure 21 in Nb steels, but interpret the results differently. They conclude that strain-induced precipitation of microalloy carbides occurs on subboundary networks that are densest near the boundary. This leads to a high rate of ferrite nucleation near the boundary, as previously discussed, but also leads to inhibition of growth by precipitate pinning effects. As a result, further growth of ferrite requires repeated nucleation of ferrite as the transformation front advances. Coarser ferrite grains are found in the grain interior because of a less dense subboundary and precipitate network.

## Impingement and Coalescence

Because of the high nucleation rate of ferrite at deformed austenite grain boundaries, impingement of ferrite occurs early in the growth process. Once this has occurred, coalescence of ferrite grains will occur to minimize the ferrite grain boundary area, Figure 22. As discussed

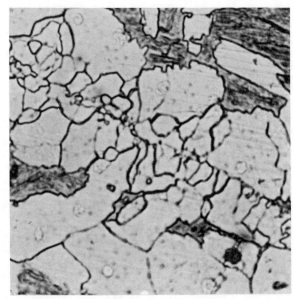

Figure 21 - Selective growth (coarsening) of ferrite grains during isothermal
transformation of a 0.1V steel compressed 30 percent and
subsequently held 1000S at 725°C (36).

## GROWTH SEQUENCE

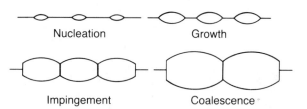

Figure 22 - Nucleation, growth, impingement, and coalescence
steps for formation of ferrite.

earlier, strain-induced precipitation of microalloy carbides may have an important effect on the coalescence process because of their pinning action on grain boundaries.  Such pinning action is dependent on the ferrite grain size, $d_\alpha$, and the volume fraction, $f_v$, and size, a, of the pinning precipitate according to the equation

$$d_\alpha = k_\alpha \frac{a}{f_v} \tag{23}$$

where $k_\alpha$ is a constant (63).  The pinning action of microalloy carbides in the ferrite phase during coalescence is similar to the pinning action of microalloy carbides in the austenite during intial growth of ferrite, which is subsequently discussed.  However, the pinning action is probably more important during coalescence because the driving force is much smaller than during transformation of austenite to ferrite.

Direct studies of the coalescence process of the ferrite are rather scarce.  Webster and Woodhead (38) have studied the coalescence of ferrite grains after transformation is completed in undeformed 0.16C steels with and without Nb additions.  Coalescence of ferrite grains in the steel without Nb occurred more rapidly than in the steels containing Nb, Figure 23.  Coarsening of the ferrite grains also proceeded faster in the direction of the grain boundary than into the grain interior in the plain carbon steel, with slab shaped ferrite grains being produced.  The ferrite grains remained more equiaxed during coalescence in the Nb steel.

## Overall Reaction Kinetics

Because of the rather complex nucleation sequence, followed by growth and coalescence of the ferrite, Figure 22, combined with precipitation of microalloy carbides, no quantitative attempts have been made to relate the volume fraction of ferrite with the relevant nucleation and growth rates of the individual ferrite units in control-rolled microalloyed steels, even when simple isothermal tranformation is considered.  The more complex problem of transformation upon continuous cooling in these steels is even further from quantitative treatment.

For the case of undeformed Fe-C alloys, where ferrite forms as grain-boundary allotriomorphs with parabolic growth rates, Obara et al. (39) have shown that the volume fraction of ferrite for different isothermal reaction times can be satisfactorily calculated from known nucleation and growth parameters using the following relationships.

$$f_v(t) = 1 - \exp\left[-b^{-1/2} f(a)\right]$$

$$a = \left(9\, I_B\, \alpha^2\right)^{1/2} t$$

$$b = \frac{\left(9\, I_B\, t\right)}{4\, S_v^2} \tag{24}$$

$$f(a) = a \int_0^1 \left[1 - \exp\left\{\left(\frac{-\pi a^2}{2}\right)(1 - 2\, m^2 + m^4)\right\}\right] dm$$

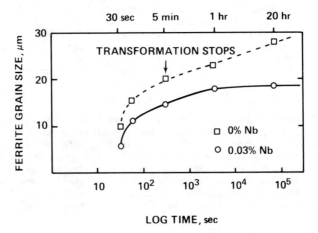

Figure 23 - Coalescence of ferrite grains after transformation in steels with and without Nb additions (38).

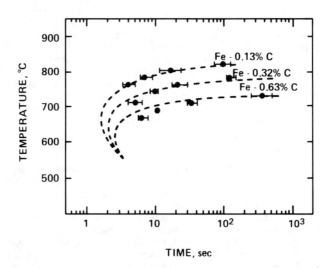

Figure 24 - Comparison of calculated and observed transformation times for isothermal transformation of austenite to ferrite (39).

where $f_v(t)$ is the volume fraction of ferrite at time t, $I_B$ is the steady-state nucleation rate of ferrite at austenite grain boundaries, $\alpha$ is the parabolic rate constant, and $m = Y/\alpha\, t^{1/2}$ where Y is the perpendicular distance between the grain-boundary plane and the plane of polish. This analysis gives reasonable agreement with experimentally determined curves, Figure 24.

Because $\alpha$ is not affected by the deformation of the austenite (the rate of carbon diffusion is not affected by boundaries or dislocations), the main effect of deformation is to increase the nucleation rate by its effect on $S_v$ and by increasing the potency of the deformed boundary through bulging or subboundary formation, as previously discussed. Following nucleation, impingement, coalescence, and inward growth of the ferrite occurs, Figure 22. Typical effects of deformation on the overall reaction kinetics are shown in Figure 25 with the $f_v(t)$ curve being shifted to shorter times (37).

Of course, in practice ferrite forms from deformed austenite during continuous cooling rather than under isothermal conditions. As discussed earlier, the $Ar_3$ temperature is increased by deformation of the austenite at temperatures below the recrystallization temperature of the austenite. Also, it appears that the transformation range is extended over a wider temperature range in deformed austenite compared to re-crystallized (or normalized) austenite, Figure 26 (20). The ferrite nose is also shifted to shorter times by over an order of magnitude, Figure 27 (40).

## Structural State of the Austenite

As previously shown, the kinetics of ferrite formation are sensitive to the structural state of the austenite. In general, depending on the processing route, Figure 3, the austenite may exist in three states (1) reheated austenite, (2) recrystallized austenite, and (3) deformed austenite. Reheated austenite is simply austenite heated to the solution temperature after casting, in preparation for the hot-rolling operation. Recrystallized austenite is formed after hot rolling at high temperatures, an operation sometimes called roughing. The recrystallization process may occur during the hot-rolling operation in the roll gap, in which case it is called dynamic recrystallization; or the recrystallization may occur in the delay period between deformation passes, in which case it is called static recrystallization. Deformed austenite, or unrecrystallized aus-tenite, is formed at low rolling temperatures below the recrystallization temperature of the austenite. The structural state of the austenite for each case is discussed in the following sections.

### Reheated Austenite

The as-reheated austenite ($\gamma$) consists of annealed, equiaxed grains bordered by high-angle grain boundaries and with some grains containing annealing twins, Figure 10. One of the most important microstructural feature of the reheated $\gamma$ is its grain size. The variation of the reheated grain size with temperature has been studied in great detail. These studies have shown that the grain size can vary either continuously or discontinuously with temperature, depending upon the type, size, and volume fraction of the microalloy carbide (nitride). Typical behavior is shown in Figure 28. The continuous variation of the austenite grain size with temperature, illustrated by the plain-carbon steel (silicon-killed) is indicative of "normal" grain growth, i.e., that which occurs in

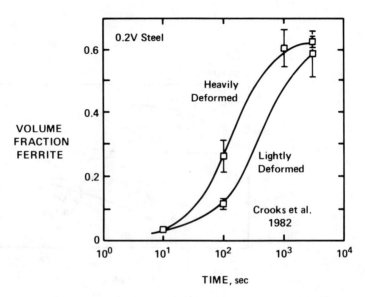

Figure 25 - Effect of deformation on isothermal transformation kinetics (37).

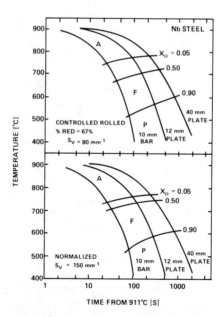

Figure 26 - Continuous-cooling transformation kinetics of control-
rolled and normalized steels (20).

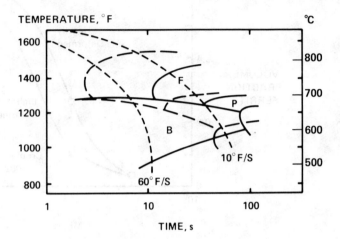

Figure 27 - Effect of deformation on continuous-cooling transformation
behaviour (dashed CCT curves with deformation, solid
curves undeformed) (40).

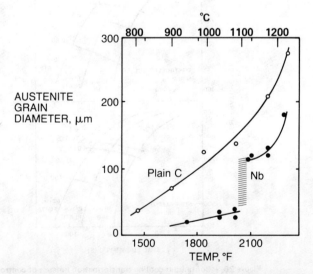

Figure 28 - Austenite grain-coarsening behaviour of C-Mn and Nb steel (41).

the absence of pinning effects. This type of grain coarsening is described by the equation

$$d_\gamma^2 - d_{\gamma o}^2 = k_o \, t \, \exp\left(-\frac{Q}{RT}\right) \qquad (25)$$

where $d_\gamma$ and $d_{\gamma o}$ are the average grain sizes of the austenite at annealing times $t$ and $t = o$, respectively; $k_o$ is a constant; $Q$ is the activation energy for grain growth with a value near 18 kcal/mol°K and R and T have their normal meanings (42).

The discontinuous grain coarsening curves shown by the Nb steel, Figure 28, indicate that "abnormal" grain growth or secondary recrystallization has occurred. Abnormal grain growth can occur only when the first-formed austenite grains are kept to a small size by the retarding force presented by certain pinning precipitates. These pinning precipitates are present in Al-killed steels (AlN) or in steels containing microalloy precipitates (NbCN, TiCN, VCN). In general, for pinning to be effective, the precipitate size must be very small, a~ 100Å, but this is dependent upon the austenite grain size and volume fraction $f_v$ of the precipitate according to an equation analagous to Equation 23

$$d_\gamma = k_\gamma \frac{a}{f_v} \qquad (26)$$

where $d_\gamma$ is the austenite grain size, $k_\gamma$ is a constant, a is the precipitate particle size, and $f_v$ is the volume fraction of precipitate. These and other details of various grain coarsening models have recently been reviewed by Cuddy and Raley (43).

The overall discontinuous grain-coarsening behavior can therefore be divided into three regimes. Below the grain coarsening temperature ($T_{GC}$), the austenite microstructure consists of fine austenite grains pinned by a fine dispersion of precipitate. In the temperature region at and immediately above $T_{GC}$, the austenite microstructure consists of a bimodal size distribution of grains and large coarsened precipitate particles. The austenite grain structure found at temperatures somewhat above $T_{GC}$ is a uniform coarse grained structure with few precipitates, and with the grain size increasing continuously as the temperature is increased.

The value of $T_{GC}$ is dependent roughly on the temperature of dissolution of the pinning precipitate.* Generally, the temperature of dissolution can be calculated in the form of a solubility product, e.g. for NbC (44)

$$\log [\%Nb] [\%C]^{0.87} = -\frac{7530}{T} + 3.11 \qquad (27)$$

where T is the absolute temperature in °K, and the compositions are in wt. %. For a given carbon content, as the Nb content is increased the

---

* Unpinning may occur below the dissolution temperature if Ostwald ripening of the precipitate is sufficient to increase markedly the size of the pinning precipitate (43).

dissolution temperature and $T_{GC}$ increase, Figure 29 (43). Dissolution temperatures are also dependent on the relative solubility of the various precipitates, e.g., $T_{GC}$ for Ti steels is higher than for V steels, Figure 30 (45).

Several studies of the formation of austenite have shown that nucleation occurs on structural features present in the starting material, e.g., carbides in grain boundaries, or at pearlite colonies (46). Thus, the as-reheated austenite structure will also reflect initially the scale of the prior structure, and hence will depend on prior processing.

Because of the time dependency of the precipitate dissolution and grain growth processes, the austenite grain size may also be dependent on process variables such as heating rate.

## Basic Aspects of Hot Deformation of Austenite

When austenite is hot rolled, the behavior of the austenite can be studied by considering its hot flow curve Figure 31 in tension or torsion (47,48). The particular shape of the hot-flow curve exhibited by a given material under a given set of deformation conditions (temperature T and strain rate $\dot{\varepsilon}$) is governed by the relative magnitudes of the hardening and softening processes, and how the relative magnitude of these effects shifts with strain. For single-phase austenite, the hardening process is simply work hardening, and the possible softening processes are dynamic recovery and dynamic recrystallization. When the deformation conditions result in a small amount of dynamic softening, i.e., when T is low and $\dot{\varepsilon}$ is high, then the hardening processes dominate, and the net flow curve has the shape of the top curve, Figure 31. At higher T and lower $\dot{\varepsilon}$, dynamic recovery is possible during deformation, and this leads to a lower flow curve, i.e., the middle flow curve of Figure 31. When the deformation conditions result in a large amount of dynamic softening, i.e., when T is very high and $\dot{\varepsilon}$ is very low, dynamic recrystallization can occur which leads to a substantial lowering of the flow curve, i.e., the lower curve of Figure 31.

The structural state of the austenite at the end of each strain cycle is shown schematically in Figure 32. The onset of each softening process is generally indicated by a critical strain—$\varepsilon_c^r$ for dynamic recovery and $\varepsilon_c^R$ for dynamic recrystallization, Figure 31. The strain at which the flow stress reaches a peak is designated by $\varepsilon_p$. The change in the shape of the flow curve can be most easily represented by the temperature compensated strain rate as proposed by Zener and Hollomon. This parameter is expressed as

$$ Z = \dot{\varepsilon} \, \exp\left(\frac{Q}{RT}\right), \tag{28} $$

where Q is the apparent activation energy for the combined hardening/ softening process. Dynamic recrystallization occurs when Z is very low, and pure work hardening occurs when Z is very high, as indicated in Figure 32.

The structural state of the austenite at the end of a hot deformation process can be understood, therefore, by comparing the applied strain to the various critical strains which exist at the value of Z appropriate to the deformation. Thus, it is important to understand the manner in which the various critical strains vary with both Z and chemical composition.

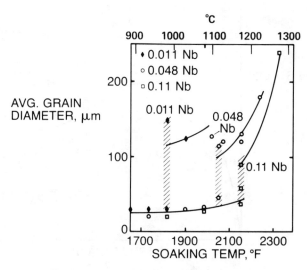

Figure 29 - Austenite grain-growth characteristics in steels containing various Nb additions (43).

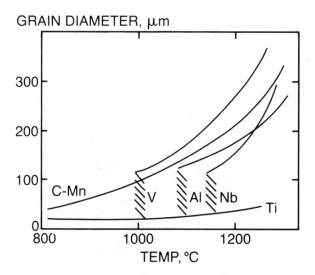

Figure 30 - Austenite grain-growth characteristics in steels containing various microalloying additions (45).

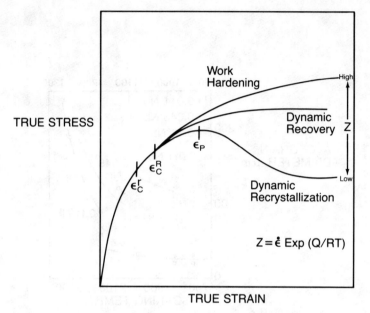

Figure 31 - Hot-flow curve for austenite.

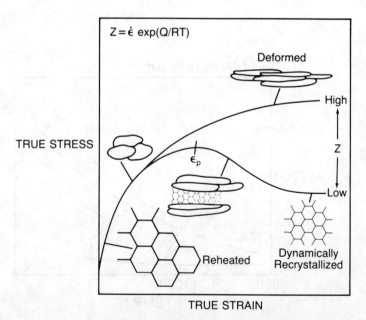

Figure 32 - Structural state of austenite at various stages of deformation.

The influence of Z on $\epsilon_c^r$ and $\epsilon_c^R$ has been the subject of several investigations (47,48) and some typical results are shown in Figure 33 (49). However, these critical strains are also sensitive to chemical composition, especially additions of microalloying elements. In particular, $\epsilon_c^R$ can be greatly increased by the presence of microalloying additions when the deformation occurs at low Z and high supersaturation.

## Recrystallized Austenite

If recrystallization occurs during deformation, it is referred to as dynamic recrystallization. If, after deformation, the steel is held at a sufficiently high temperature for a long enough time, static recrystallization will occur. The time for dynamic or static recrystallization to occur is sensitive to the concentration of microalloying elements, both because of the normal effect of solute elements in delaying recrystallization and because of the pinning action of microalloy precipitates on migrating grain boundaries. Typical results are shown in Figure 34 for the effects of temperature and steel composition on the time for dynamic recrystallization (48).

Some authors have distinguished between the structure of dynamically and statically recrystallized austenite, with the dynamically recrystallized austenite having a higher density of dislocations and annealing twins than statically recrystallized austenite. However, because only static recrystallization occurs for most commercial controlled-rolling processes, we shall not attempt to distinguish between static and dynamic recrystallization.

## Deformed (Unrecrystallized) Austenite

When the austenite is deformed below a temperature where the first evidence of unrecrystallized austenite is observed, this temperature is referred to as the recrystallization stop temperature, $T_{RX}$. This temperature is sensitive to the presence of microalloying elements as shown in Figure 35 (50).

At still lower temperatures, no recrystallization of the austenite occurs, and only pancaking (flattening) of the austenite grains occurs in a manner previously described. This temperature is called the pancaking temperature, $T_P$. Interruption of the rolling sequence between $T_{RX}$ and $T_P$ will result in a mixed austenite grain structure comprised of recrystallized and unrecrystallized austenite.

Pancaked austenite may have several types of internal defect structures, including subboundaries, deformation bands, and deformed annealing twins, all of which may play a role in nucleating ferrite. Also, the increased $S_V$ accompanying pancaking has an important effect on nucleating ferrite. Bulges caused by incipient recrystallization may play a prominent role along with the subboundaries and strain-induced precipitation of microalloy carbides in increasing the potency of nucleation at the grain boundaries. Many of these features were discussed earlier.

## Austenite/Ferrite Grain Size Relationships

As can be seen from the preceding discussion, prediction of the ferrite grain size from the austenite grain size and shape is a difficult task. Not only nucleation but also impingement and coalescence of the ferrite units must be considered, Figure 22. In addition, coalescence may

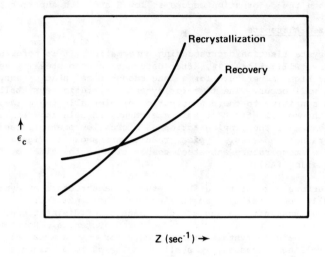

**Figure 33 - Effect of Z on critical strains for recovery and recrystallization (49).**

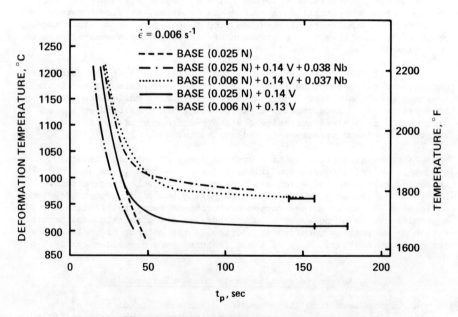

**Figure 34 - Effect of microalloying additions and deformation temperature on time for dynamic recrystallization at fixed strain rate (48).**

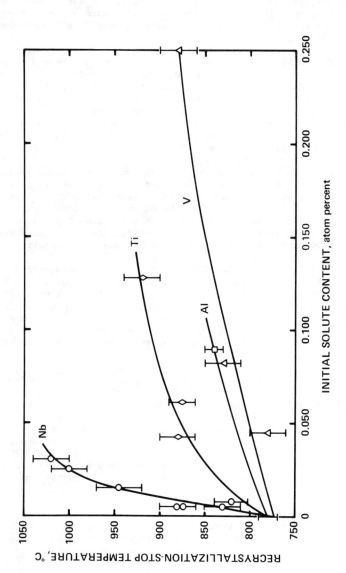

Figure 35 - Effect of microalloying additions on recrystallization-stop temperature (50).

be strongly influenced by strain-induced precipitation of microalloy carbides.  In spite of these difficulties, it is qualitatively correct that
if the nucleation rate of the ferrite is increased, the resulting ferrite
grain size will be reduced.  Therefore, most commercial controlled-rolling
processes for refining the ferrite grain size have concentrated on increasing the nucleation rate of ferrite by increasing the value of $S_V$
either by refining the recrystallized austenite grain size or by flattening unrecrystallized austenite grains to provide a larger number of heterogeneous nucleation sites.

The various changes in the austenite grain size and morphology as the
finishing temperature is lowered through the recrystallization and
austenite grain flattening (no recrystallization) regions are shown in
Figure 36, (51).  The resultant ferrite grain size is gradually refined,
concomittant with the changes in the austenite grain size and shape.  The
finest ferrite grain sizes are achieved when the austenite grains are
flattened.  Bimodal ferrite grain-size distributions result when the austenite is partially recrystallized.

In recrystallized or reheated austenite, the major variables
affecting the ferrite nucleation rate are the austenite grain size and the
temperature at which the austenite phase transforms to ferrite.  The
transformation temperature is lowered when the cooling rate from the hot-
rolling temperature is increased.  This is primarily a function of plate
or sheet thickness for simple air cooling.  In some cases, accelerated
cooling can be employed to further lower the transformation temperature.

In deformed austenite, grain shape must be considered because the
value of $S_V$ depends on the shape of the elongated grain.  Also, the
contribution of the deformation bands and deformed annealing twin to $S_V$
must be considered, Figure 14.  To compare the effect of refining the
recrystallized austenite grain size with flattening of the austenite
grains, comparisons are generally made at constant $S_V$, where the $S_V$ for
unrecrystallized austenite includes the contribution of deformation bands
and deformed annealing twins.

In general, the ferrite grains are smaller than the austenite grains
because of the large number of ferrite grains nucleated in each austenite
grain, Figure 14.  However, the ratio of $d_\gamma/d_\alpha$ never approaches the value
expected from the large number of ferrite nuclei formed per unit volume
because of impingement and coalescence of ferrite grains during air
cooling.  For instance, air cooling of a 12-mm-thickness steel plate
occurs at a rate of about 1°C/s.  To cool the plate from 770°C (beginning
of transformation) to 600°C (lowest temperature for grain coalescence)
requires 170 sec.  During this time period, substantial coalescence of the
initial ferrite nuclei can occur.

A lower limit to the value of $d_\gamma/d_\alpha$ can be obtained from a simple
model that assumes nucleation of ferrite at each austenite boundary and
growth inward.  If growth from each nucleus occurs into only one grain
from each boundary (because the nuclei are semicoherent with the adjoining
grain) then a value of $d_\gamma/d_\alpha \cong 2$ is expected.  If growth at each nucleus
occurs in both directions from the grain boundary into both grains
(because the nuclei are incoherent) then a value of $d_\gamma/d_\alpha \cong 1$ is expected.

Values of $d_\gamma/d_\alpha$ observed for recrystallized austenite generally lie in
the range of 1 to 4 with the value increasing with increasing austenite
grain size and with increasing cooling rate.  Some typical results for
microalloyed steels are shown in Figures 37 and 38.

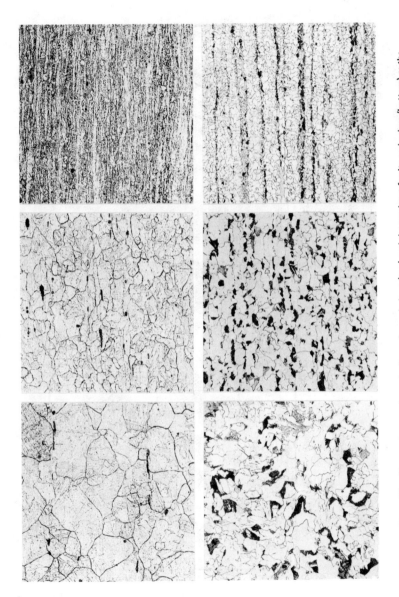

Figure 36 - Changes in austenite grain size and morphology (top) and the resultant ferrite grain sizes (bottom) as the finishing temperature is lowered through the recrystallization and austenite grain flattening (no recrystallization) regions (51).

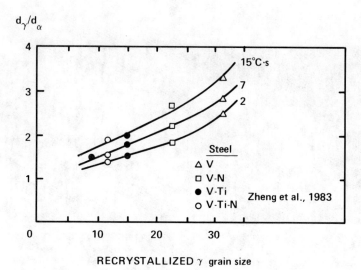

Figure 37 - Values of $d_\gamma/d_\alpha$ for steels with various austenite grain sizes at a cooling of 15°C/s (52).

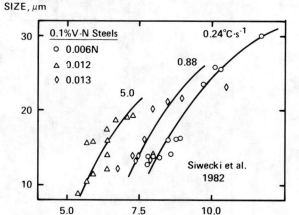

Figure 38 - Relationship between austenite and ferrite grain sizes in microalloyed steels for various cooling rates (53).

In the case of flattened, unrecrystallized austenite grains, the relationship between the ferrite and austenite grain sizes are similar to that for recrystallized austenite grains if the grain height rather than the grain diameter is used, Figure 39 (41). The austenite grain height is, of course, related to the percent rolling reduction because the un-recrystallized austenite grains undergo a simple plane-strain deformation, Figure 11. The ferrite grain diameter is approximately one-half the aus-tenite grain height ($d_\gamma/d_\alpha = 2$) although the ferrite grain diameter becomes nearly equal to the austenite grain height, as the austenite grain height decreases at higher rolling reductions ($d_\gamma/d_\alpha = 1$), Figure 40 (41).

Comparison of ferrite grain sizes from both recrystallized and deformed austenite were discussed earlier. Data from Kosazu, Figure 16, indicated that ferrite grain sizes were finer for deformed austenite than for recrystallized austenite when compared at the same $S_V$. The rate of change of ferrite grain size with $S_V$ is large at small $S_V$ and decreases with increasing $S_V$. More recent work indicates that for higher $S_V$ values there is little difference between recrystallized and pancaked austenite, Figure 41 (51,53). Also, for recrystallized austenite, there appears to be an upper limit of about 200 $mm^{-1}$ for $S_V$, which corresponds to an aus-tenite grain size of 10 $\mu$m, whereas much higher $S_V$ values of 600 $mm^{-1}$ can be obtained by extensive deformation of fine, unrecrystallized austenite, Figure 41. As a result, a lower limit of ferrite grain size of 4 $\mu$m can be obtained from deformed austenite, whereas a lower limit of ferrite grain size of 8 $\mu$m is obtained for recrystallized austenite (51).

Roberts (54) has suggested an empirical relationship between $d_\gamma$ and $d_\alpha$ for recrystallized austenite which embodies both the effect of $d_\gamma$ and cooling rate on the value of $d_\alpha$, i.e.,

$$d_\alpha = 3.75 + 0.18d_\gamma + 1.4 \left(\frac{dT}{dt}\right)^{-1/2} , \ \mu m \qquad (29)$$

where dT/dt is the average cooling rate between 750 and 550°C in °C/s. Very good agreement of the experimental data with Equation 29 for Ti-V-N steels was obtained when ferrite was formed from recrystallized aus-tenite. For flattened, unrecrystallized austenite grains formed during hot rolling the situation is more complex, and Equation 29 is not then applicable.

## Applications

### Normal Controlled Rolling

In normal controlled rolling, the actual deformation cycle that the austenite undergoes before transformation to ferrite can be separated into three steps, Figure 3. In the roughing passes at high temperature, the austenite grain size is continually refined by sequential static re-crystallization of the austenite. This is followed by a time delay to allow the plate to cool to lower temperatures. Finally, the austenite undergoes a series of low temperature rolling passes in which the austen-ite grains are flattened with no recrystallization occurring.

The controlled rolling cycle through the recrystallization stage has been modeled by Sellars (55) and by Roberts et al., (54). In their computer models, static recrystallization is assumed, and the percent recrystallization and grain growth of the material are characterized by

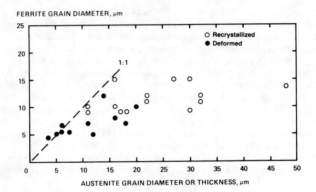

Figure 39 - Comparison of ferrite grain sizes produced from flattened and recrystallized austenite grains (41).

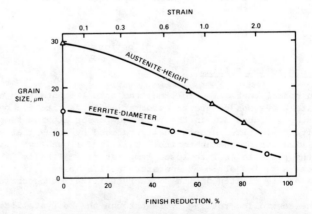

Figure 40 - Effect of finishing reduction below the austenite recrystallization-stop temperature on austenite grain height and resulting ferrite grain size (41).

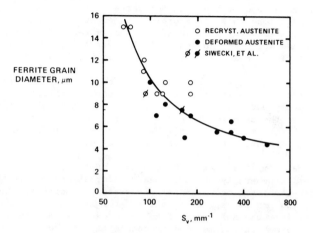

Figure 41 - Ferrite grain sizes produced from recrystallized and unrecrystallized austenite at various $S_V$ values (51).

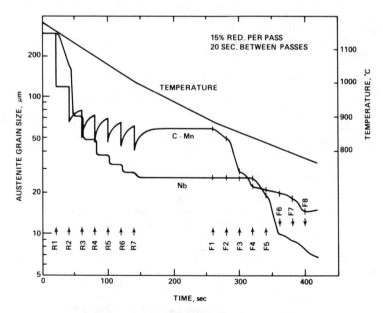

Figure 42 - Calculated response of austenite grain size to controlled-rolling cycle (55).

suitable empirical equations which take into account plastic strain and temperature. A typical calculated controlled-rolling cycle is shown in Figure 42. The calculated cycle gives the changes in the austenite grain size for deformation passes of 15 percent and for a delay time of 20s between passes. The grain growth characteristics of the austenites between passes are seen to be markedly different for the C-Mn steel than for the Nb steel because of the pinning actions of Nb in solution NbCN precipitates. This results in a much finer austenite grain size for the Nb steel at the end of the roughing passes. After a time delay, the finishing passes lead to further grain refinement with grain flattening and only partial recrystallization of the austenite. In the temperature-strain regime where the austenite is recrystallized, the final ferrite grain size can be calculated from the relationship given in Equation 29.

For cases where very low finishing temperatures are used to produce only flattened austenite grains that are unrecrystallized, the modeling is more difficult because of the more complicated mechanism of nucleation of ferrite from the deformed austenite. More modeling work is especially needed in this temperature range because commercial finishing practices to produce the finest ferrite grain sizes are performed here.

Another complication in the modeling of the recrystallization process is that it is generally based on heavy deformations for each pass (15%) that can be achieved only in small laboratory specimens. In commercial plate-rolling practices, much smaller deformations per pass (5 to 10% especially in roughing) are used to avoid overloading the mill. In this smaller deformation regime, the austenite may undergo recovery rather than recrystallization and strain-induced discontinuous grain growth may occur. The resulting coarse austenite grains may persist through the rolling cycle and result in coarse ferrite grains or even bainite upon transformation. Alternately deformations may be such that only partial recrystallization occurs; this also leads to mixed austenite grain sizes which result in mixed ferrite grain sizes. Such steels may have impaired mechanical properties. It is clear that careful control of pass temperature and reduction, and of interpass times are required to assure complete recrystallization of the austenite (56).

Many thousands of tons of plate have been produced by controlled-rolled processes, in particular for use in Arctic grade line-pipe applications where high strength (60 to 70 ksi) and low ductile-to-brittle transition temperatures (< -50°F) are required. However, a severe draw-back to such operations is the decrease in productivity caused by the long delay time needed to allow the plate to cool to lower temperatures, and the high mill loads resulting from the low-temperature rolling.

## Recrystallization Controlled Rolling

To circumvent the problems of decreased productivity and high mill loads characteristic of controlled rolling, it has recently been proposed that ultrafine ferrite grained steels can be produced from recrystallized austenite if the austenite grain size is made small enough (52,54,57). This requires addition of Ti to steels containing sufficient N to form a fine dispersion of TiN which has a high dissolution temperature, and results in a high temperature for austenite grain coarsening, $T_{GC}$. By restricting the reheating temperature to a temperature below $T_{GC}$, but above the temperature for austenite recrystallization, $T_{RX}$, fine-grained equiaxed austenite can be produced. The rolling process may be followed by accelerated cooling to further refine the ferrite grain size. The process is shown schematically in Figure 43.

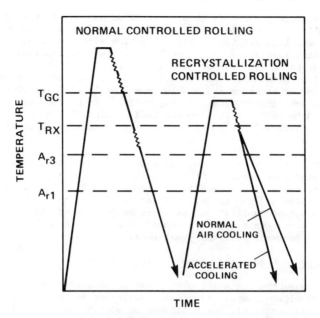

Figure 43 - Comparison of normal controlled rolling and recrystallization
controlled rolling (52,54).

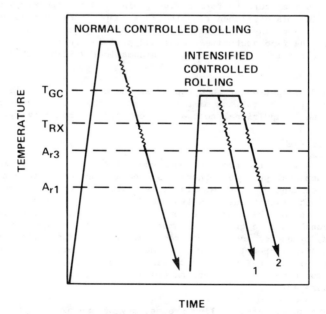

Figure 44 - Intensified controlled rolling (58).

## Intensified Controlled Rolling

To avoid problems with discontinuous austenite grain growth during controlled-rolling operations and to produce the maximum grain refinement, it is best to start the rolling operation with the finest possible austenite grain size. As a result in latest developments, the reheating temperature is kept as low as possible (1150 to 950°C), and heavier controlled deformation is carried out below $T_{RX}$ (see 1 in Figure 44). Additions of titanium and niobium are beneficial in raising the austenite grain-coarsening temperature (Figure 30). Such processes have been referred to as intensified controlled rolling (58) or special controlled rolling (59), and are compared with normal controlled rolling in Figure 44.

If higher strengths and lower ductile-to-brittle transition temperatures are desired, the controlled rolling may be extended to lower temperatures in the intercritical range (see 2 in Figure 44). The ferrite formed below $Ar_3$ now is deformed and develops a texture, resulting in the development of "splits" on the fracture surface of Charpy impact specimens, although this does not appear to markedly deteriorate the overall fracture properties of the steel (60).

## Reheating into Austenite Range During Rolling

Because the austenite grain size is sensitive to the fineness of the starting microstructure (austenite nucleates at carbide/ferrite junctions upon reheating) (46), one modification of the controlled-rolling process involves cooling below $Ar_1$ and reheating into the austenite range after initial rolling. In the initial rolling the austenite grain size is refined by repeated recrystallization. This is followed by an operation in which the plate is cooled by water sprays and air cooling to just below the $Ar_1$ temperature to form a fine grained ferrite/pearlite microstructure. The plate is then reheated to a temperature just above $A_{c_3}$ to form a very fine grained austenite. Finish rolling is then performed to flatten the fine austenite grains and produce a ultrafine-grained ferrite upon air cooling. The process is schematically illustrated in Figure 45 (61).

A disadvantage of the process is the need for a special loop in the conveyor for the rolling mill and an additional reheating furnace resulting in additional cost and a decrease in productivity.

## Accelerated Cooling

Because of the need for improved weldability in many plate steels for constructional purposes, there has been an increasing tendency for the carbon content of plate steels to be lowered to obtain a lower carbon equivalent which results in less tendency for weld-cracking problems. To compensate for the loss in strength caused by the lower carbon content, additional ferrite grain refinement is achieved in these lower carbon steels, by use of accelerated cooling. Typical facilities are shown in Figure 46. Either lamellar flow cooling for intermediate cooling rates or jet spray cooling for faster cooling rates can be used. Typical cooling rates for both types of cooling are shown as a function of plate thickness in Figure 47.

The accelerated cooling process may be employed to produce ultrafine-grained plate steels by combining it with the recrystallization controlled-rolling process, as previously discussed. But even in the case

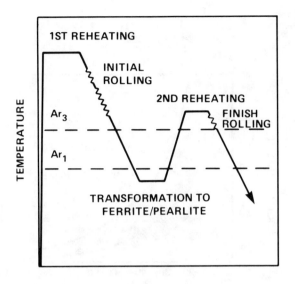

Figure 45 - Reheating into austenite range after initial rolling (SHT process) (61).

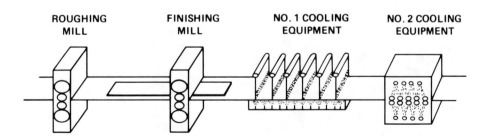

Figure 46 - Accelerated cooling process (62).

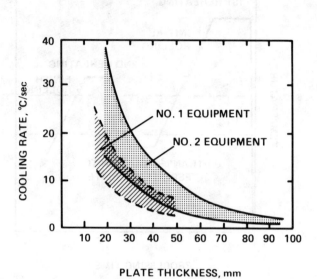

Figure 47 - Cooling rates achieved with accelerated cooling process for different plate thicknesses (62).

of less severe controlled-rolling practices, as used for mild ship plate steels, a considerable refinement in ferrite grain size can be achieved by applying accelerated cooling at the end of the hot-rolling operation.

In higher hardenability steels, such as Mn-Mo-B steels, such accelerated cooling can result in the formation of lower temperature acicular ferrite or fine bainitic transformation products, Figure 27. These acicular transformation products have considerably higher strengths than polygonal ferrite, and thus strengths near 600 MPa (80 ksi) can be achieved even at low carbon contents. The ductile-to-brittle transition temperature of most of these steels do not appear to be equivalent to polygonal ferrite steels (58).

## References

1.  K. J. Irvine, "A Comparison of the Bainite Transformation with Other Strengthening Mechanisms in High-Strength Structural Steel," pp. 55-65 in Steel Strengthening Mechanisms, Climax Molybdenum Co., 1969.

2.  Microalloying 75, Union Carbon and Carbide Corp, New York, NY, 1976.

3.  The Hot Deformation of Austenite, Edited by J. Ballance, Met. Soc. AIME, Warrendale, PA, 1977.

4.  Thermomechanical Processing of Microalloyed Austenite, Edited by A. J. DeArdo, G. A. Ratz, and P. J. Wray, Met. Soc. AIME, Warrendale, PA, 1982.

5.  T. Tanaka, "Controlled Rolling of Steel Plate and Strip," International Metals Reviews, Vol 26, 1981, pp. 185-212.

6.  I. Kozasu, C. Ouchi, T. Sampei, and T. Okita, "Hot-Rolling as a High Temperature Thermomechanical Process," pp. 120-133 in Microalloying 75, Union Carbon and Carbide Corp., 1976.

7.  R. A. Grange, "Microstructural Alterations in Iron and Steel During Hot Working," pp. 299-317 in Fundamentals of Deformation Processing, Proc. of the 9th Sagamore Army Materials Research Conf., Syracuse Univ. Press., Edited by W. A. Backofen et al., 1962.

8.  L. J. Cuddy, "Production of Ultrafine Grained Steels by Controlled Reheating, Rolling and Cooling," in preparation.

9.  E. R. Morgan, T. E. Dancey, and M. Korchynsky, "Improved Steels Through Hot-Strip Mill Controlled Cooling," Journal Metals, Vol. 17, 1965, pp. 829-831.

10. K. Tsukada, K. Matsumoto, K. Hirabe, and K. Takeshige, "Application of On-Line Accelerated Cooling," pp. 347-371 in Mech. Working and Steel Processing Conference, XIX, Iron and Steel Soc. AIME, Warrendale, PA, 1982.

11. C. Ouchi, T. Okita, and S. Yamamoto, "Effects of Interrupted Accelerated Cooling after Controlled Rolling on the Mechanical Properties of Low Alloy Steels," Trans. Iron and Steel Inst. Japan, Vol. 22, 1982, pp. 608-616.

12.  K. Tsukada, T. Ohkita, C. Ouchi, T. Nagamine, K. Hirabe, and K. Yako, "Application of On-Line Accelerated Cooling (OLAC) to Steel Plates—Development of OLAC, Part I," Nippon Kokan Tech. Report, Overseas No. 35, 1982, pp. 24-34.

13.  K. Tsukada, V. Yamazaki, S. Kosuge, T. Tokunaga, K. Hirabe, and K. Arikata, "Development of YS 36 kgf/mm$^2$ Steel with Low Carbon Equivalent Using On-Line Accelerated Cooling (OLAC)—Development of OLAC, Part 2," Nippon Kokan Tech. Report, Overseas No. 35, 1982, pp. 35-46.

14.  K. C. Russell, "Nucleation in Solids," pp. 219-264 in Phase Transformations, American Society for Metals, Metals Park, OH, 1970.

15.  J. W. Cahn, "The Kinetics of Grain Boundary Nucleated Reactions," Acta Metall., Vol. 4, 1956, pp. 449-459.

16.  E. E. Underwood, Quantitative Stereology, Addison-Wesley, New York, NY, 1970.

17.  H. I. Aaronson, "The Proeutectoid Ferrite and Proeutectoid Cementite Reactions," pp. 387-546 in Decomposition of Austenite by Diffusional Processes, Interscience Publishers, New York, NY, 1962.

18.  W. F. Lange, III, and H. I. Aaronson, Met. Trans. A., in press.

19.  H. I. Aaronson and K. C. Russell, "Nucleation—Mostly Homogeneous and in Solids," pp. 371-399 in Solid → Solid Phase Transformations, Met. Soc. AIME, Warrendale, PA, 1982.

20.  A. Sandberg and W. Roberts, "The Influence of Thermomechanical Treatment on the Continuous Cooling Transformation of Austenite in Microalloyed Steels," pp. 405-430 in Thermomechanical Processing of Microalloyed Austenite, Met. Soc. AIME, Warrendale, PA, 1982.

21.  J. W. Christian, The Theory of Transformations in Metals and Alloys, Part I, Pergamon Press, Oxford, 1975.

22.  E. E. Underwood, "Surface Area and Length in Volume," pp. 77-127 in Quantitative Microscopy, Edited by R. T. Dehoff and F. N. Rhines, McGraw Hill, 1968.

23.  C. Ouchi, T. Sampei, and I. Kozasu, "The Effect of Hot-Rolling Condition and Chemical Composition on the Onset Temperature of γ → α Transformation after Hot Rolling," Trans. Iron and Steel Inst. Japan, Vol. 22, 1982, pp. 214-222.

24.  C. R. Gordon, unpublished research, University of Pittsburgh, Pittsburgh, PA, 1983.

25.  T. Tanaka, T. Enami, M. Kimura, Y. Saito, and T. Hatomura, "Formation Mechanism of Mixed Austenite Grain Structure Accompanying Controlled-Rolling of Niobium-Bearing Steel," pp. 195-215 in Thermomechanical Processing of Microalloyed Austenite, Met. Soc. AIME, 1982.

26.  H. Inagaki, "Role of Annealing Twins for Grain Refinement in Controlled Rolling of Low Carbon Microalloyed Steel," Trans. Iron and Steel Inst. Japan, Vol. 23, 1983, pp. 1059-1067.

27.  W. Roberts, H. Lidefelt, and A. Sandberg, "Mechanism of Enhanced Ferrite Nucleation from Deformed Austenite in Microalloyed Steels," pp. 38-42 in Hot-Working and Forming Processes, Metals Society, London, 1980.

28.  R. K. Amin and F. B. Pickering, "Ferrite Formation from Thermomechanically Processed Austenite," pp. 377-403 in Thermomechanical Processing of Microalloyed Austenite, Met. Soc. AIME, 1982.

29.  L. J. Cuddy, unpublished research, United States Steel Research Laboratory, Monroeville, PA, 1983.

30.  H. I. Aaronson, C. Laird, and K. R. Kinsman, "Mechanism of Diffusional Growth of Precipitate Crystals," pp. 313-390 in Phase Transformations, ASM, Metals Park, OH, 1968.

31.  K. R. Kinsman and H. I. Aaronson, "Influence of Molybdenum and Manganese on the Kinetics of the Proeutectoid Ferrite Reaction," pp. 39-53 in Transformation and Hardenability in Steels, Climax Molybdenum Corporation, Ann Arbor, MI, 1967.

32.  C. A. Dubé, PhD Thesis, Carnegie Institute of Technology, 1948 (cited in Reference 30).

33.  M. Hillert, "Role of Interfacial Energy During Solid-State Phase Transformation," Jernkontorets Annaler, Vol. 141, 1957, pp. 757-789.

34.  M. Umemoto, H. Ohtsuka, and I. Tamura, "Transformation to Pearlite from Work Hardened Austenite," Trans. Iron and Steel Inst. Japan, Vol. 23, 1983, pp. 775-784.

35.  R. W. K. Honeycombe, "Transformation from Austenite in Alloy Steels," Met. Trans. A., Vol. 7A, 1976, pp. 915-936.

36.  W. S. Owen and M. J. Crooks, private communication, MIT and Bethlehem Steel Research Laboratory, 1983.

37.  M. J. Crooks, A. J. Garratt-Reed, J. B. Vander Sande, and W. S. Owen, "The Isothermal Austenite-Ferrite Transformation in Some Deformed Vanadium Steels," Met. Trans. A., Vol. 13A, 1982, pp. 1347-1350.

38.  D. Webster and J. H. Woodhead, "Effect of 0.03%Nb on the Ferrite Grain Size of Mild Steel," JISI, Vol. 202, 1964, pp. 987-994.

39.  T. Obara, W. F. Lange, III, H. I. Aaronson, and B. E. Dom, "Prediction of TTT-Curves for Initiation of the Proeutectoid Ferrite Reaction in Fe-C Alloys," pp. 1105-1109 in Solid → Solid Phase Transformations, Met. Soc. AIME, Warrendale, PA, 1982.

40.  R. Kaspar, A. Streisselberger, and O. Pawelski, "Thermomechanical Treatment of Ti- and Nb-Mo-Microalloyed Steels," pp. 555-574 in Thermomechanical Processing of Microalloyed Austenite, TMS-AIME, Warrendale, PA 1982.

41.  L. J. Cuddy, "Microstructures Developed During Thermomechanical Treatment of HSLA Steels," Met. Trans. A., Vol. 12A, 1981, pp. 1313-1320.

42. O. O. Miller, "Influence of Austenitizing Time and Temperature on Austenite Grain Size of Steel," Trans. ASM, Vol. 43, 1951, pp. 260-289.

43. L. J. Cuddy and J. C. Raley, "Austenite Grain Coarsening in Microalloyed Steels," Met. Trans. A., Vol. 14A, 1983, pp. 1989-1995.

44. H. Nordberg and B. Aronsson, "Solubility of Niobium Carbide in Austenite," JISI, Vol. 206, 1968, p. 1263.

45. L. J. Cuddy, "Hot Rolling - A Thermomechanical Treatment," in the Encyclopedia of Materials Science and Engineering, Pergamon, Oxford, to appear.

46. G. R. Speich and A. Szirmae, "Formation of Austenite from Ferrite and Ferrite-Carbide Aggregates," Trans. Met. Soc. AIME, Vol. 245, 1969, pp. 1063-1074.

47. R. A. Petkovic, M. J. Luton, and J. J. Jonas, "Flow Curves and Softening Kinetics in High Strength Low Alloy Steels," pp. 69-112 in Hot Deformation of Austenite, Met. Soc. AIME, Warrendale, PA, 1977.

48. I. Weiss, G. L. Fitzsimons, K. Mielitginen-Titto, and A. J. DeArdo, "The Influence of Niobium, Vanadium, and Nitrogen on the Response of Austenite to Reheating and Hot Deformation in Microalloyed Steels," Thermomechanical Processing of Microalloyed Austenite, Met. Soc. AIME, Warrendale, PA, 1982.

49. H. J. McQueen and J. J. Jonas, "Recovery and Recrystallization During High-Temperature Deformation," pp. 393-493 in Plastic Deformation of Metals, Academic Press, N.Y., 1975.

50. L. J. Cuddy, "The Effect of Microalloy Concentration on the Recrystallization of Austenite During Hot Deformation," ibid, pp. 129-140.

51. L. J. Cuddy, "Grain Refinement of Nb Steels by Control of Recrystallization During Hot Rolling," Met. Trans. A., Vol. 15A, 1984, pp. 87-98.

52. V. Zheng, G. Fitzsimons, and R. J. DeArdo, "Achieving Austenite Grain Refinement Through Recrystallization Controlled Rolling and Controlled Cooling in V-Ti-N Microalloyed Steels," Int. Conf. on Technology and Applications of High Strength Low Alloy Steels, ASM, Philadelphia, PA, 1983.

53. T. Siwecki, A. Sandberg, W. Roberts, and R. Lagneborg, pp. 163-194 in "The Influence of Processing Route and Nitrogen Content on Microstructure Development and Precipitation Hardening in Vanadium Microalloyed Steels," Thermomechanical Processing of Microalloyed Austenite, Met. Soc. AIME, Warrendale, PA, 1982.

54. W. Roberts, A. Sandberg, T. Siwecki, and T. Werletors, "Prediction of the Microstructure Development During Recrystallization Hot-Rolling at Ti-V Steels," Int. Conf. on Technology and Applications of High Strength Low-Alloy Steels, ASM, Philadelphia, PA, 1983.

55.  C.M. Sellars, "The Physical Metallurgy of Hot Working," pp. 3-15 in Hot Working and Forming Processes, Metals Society, London, 1980.

56.  T. Tanaka, T. Enami, M. Kimura, Y. Saito, and T. Hatomura, "Formation Mechanism of Mixed Austenite Grain Structure Accompanying Controlled Rolling of Niobium-Bearing Steel," pp. 195-215 in Thermomechanical Processing of Microalloyed Austenite, Met. Soc. AIME, Warrendale, PA, 1983.

57.  H. Sekine, T. Maruyama, H. Kagryama, and Y. Kawashima, "Grain Refinement Through Hot Rolling and Cooling after Rolling," pp. 141-161 in ibid.

58.  I. Kozasu, "Recent Development of Microalloyed Steel Plates," Int. Conf. on Technology and Applications of High Strength Low Alloy Steels, ASM, Philadelphia, PA, 1983.

59.  The Second United States Steel-Sumitomo Metal Joint Technical Meeting, May 10-11, 1982, U. S. Steel Research Laboratory, Monroeville, PA.

60.  G. R. Speich and D. S. Dabkowski, "Effect of Deformation in the Austenite and Austenite-Ferrite Regions on the Strength and Fracture Behaviour of C, C-Mn, C-Mn-Cb, and C-Mn-Mo-Cb Steels, pp. 557-598 in Hot Deformation of Austenite, Met. Soc. AIME, Warrendale, PA, 1977.

61.  T. Tanaka, N. Nozaki, K. Bessyo, M. Fukuda, and T. Hashimoto, "SHT Plates for Low Temperature Service," The Sumitomo Search No. 19, May 1978.

62.  H. Ohtani, T. Hashimoto, T. Sawamora, K. Bessyo, and T. Kyoguku, "Development of Low $P_cM$ High Grade Line Pipe for Arctic Service and Sour Environment," Proceedings of the International Conference on Technology and Applications of HSLA Steels, ASM, Metals Park, Ohio, to appear.

63.  C. Zener and C. S. Smith, "Grains, Phases, and Interfaces:  An Interpretation of Microstructure," Trans. AIME, Vol. 175, 1948, pp. 47-48.

# EFFECT OF DEFORMATION ON THE $\gamma$- to -$\alpha$ TRANSFORMATION

## IN A PLAIN C AND Mo MODIFIED STEEL.

E. Essadiqi, M.G. Akben and J.J. Jonas

Dept. of Metallurgical Engineering
McGill Unversity
3450 Unversity Street
Montreal, Que., Canada
H3A 2A7

## Introduction

It has been reported by numerous authors (1-9) that the deformation of austenite below the $Ae_3$ temperature accelerates the austenite-to-ferrite transformation. The present work concerns the effect of deformation on the isothermal decomposition of austenite in two steels: (i) a 0.06% C plain carbon steel, and (ii) a 0.30% Mo steel. Dynamic TTT (DTTT) curves were determined for these materials in the temperature range 790 to 840°C.

## Experimental Method and Materials

The chemical compositions of the two steels studied are listed below*. Compression specimens 11.43 mm in height and 7.62 mm in diameter were machined out of the as-received plates, with the longitudinal axis aligned along the rolling direction. Specimens were first normalized at 1000°C to eliminate the rolling texture. They were subsequently annealed at 1060 (Mo steel) and 1030°C (plain C steel) and quenched in order to produce an austenite grain size of about 110 $\mu$m. Each sample was then reheated to 930°C for 5 min and cooled to the test temperature at 1°C/s.

Isothermal compression tests were carried out on an Instron machine which was modified for constant true strain rate testing (10). A prestrain of $\epsilon = 0.10$ was applied at a strain rate of $7.4 \times 10^{-2} s^{-1}$, after which the strain rate was reduced to $7.4 \times 10^{-4} s^{-1}$, without unloading. At the end of each test, the specimen was water quenched. All the tested samples were sectioned parallel to the deformation axis and polished for optical examination. After etching with nital, the volume fraction of ferrite was determined using standard procedures of point counting (11).

## Results and Discussion

The appearance of the flow curves produced by the strain rate change technique can be seen in Fig. 1. After the strain rate decrease, the flow stress continues to rise, due to work hardening, but then drops as a result of the phase

---

* (i) 0.06% C -1.43% Mn -0.24% Si - 0.025% Al:
  (ii) 0.05% C -1.34% Mn -0.20% Si - 0.065% Al - 0.29% Mo.

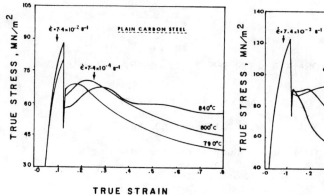

TRUE STRAIN

Fig. 1a. Flow curves produced at 840, 800 and 790°C when the plain carbon steel was tested using the strain rate change technique. The oscillations at 840°C are due to the occurrence of dynamic rerystallization in the austenite.

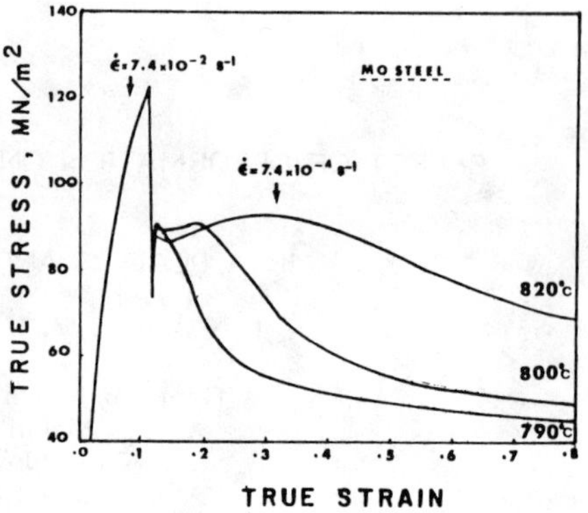

Fig. 1b. Flow curves produced at 820, 800 and 790°C when the molybdenum steel was tested using the strain rate change technique.

transformation. The time to the peak of the flow curve increases with the temperature of the test, indicating that the transformation is initiated more and more slowly as the $Ae_3$ temperature is approached. Before the peak, the highest flow stress is associated with the lowest temperature, in the conventional way. By contrast, at large strains, the lowest flow stress is observed at the lowest temperature. This is because the amount of flow softening after the peak increases as the test temperatue is decreased for the following reasons: (i) the flow stress, at a given temperature, of the $\alpha$ phase is lower than that of the $\gamma$ phase; (ii) the volume fraction of $\alpha$ that is stable increases progressively as the test temperature is reduced. Because of the lower flow stress of the $\alpha$ phase, the rate of flow softening can be related to the rate of the $\gamma$-to-$\alpha$ transformation. The manner in which this is done will be described in a later publication (12).

The dynamic TTT curves presented in Fig. 2 (plain carbon steel) and Fig. 3 (Mo steel) were determined from the metallographic results. The static TTT curves established by standard dilatometric techniques (13) are also reproduced on these diagrams for comparison purposes. The two figures show clearly the extent to which deformation accelerates the decomposition of the austenite. Because of the acceleration, the $PF_s$ temperature associated with a fixed time is appreciably higher for the dynamic TTT than for the conventional TTT curves. The difference between these $PF_s$ temperatures decreases as the time required to initiate the transformation (i.e. the temperature) is increased. For example, for the two steels, the temperatures associated with a given time and the start of transformation are raised about 60, 30 and 10°C respectively at 50,100 and 1000s.

From the structures shown in Fig. 4, it can be seen that the polygonal ferrite is first nucleated at the austenite grain boundaries and that, once transformation has begun, further nucleation takes place largely at the austenite/ferrite interface. Some of the ferrite grains were additionally nucleated within deformation bands, presumably at sub-boundary sites (2,7,14). Although the ferrite grains formed were relatively fine (2 to 16 $\mu$m), they were far from uniform in size, reflecting the influence of the different classes of nucleation sites, their varying nucleus densities, and the different times available for their growth.

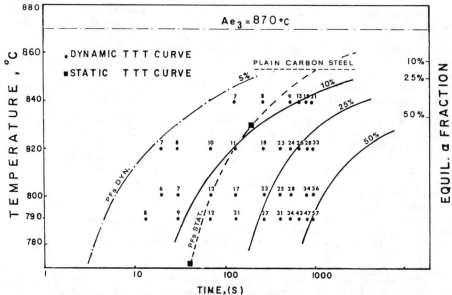

Fig. 2.   Dynamic TTT curve for the plain carbon steel derived from the present metallographic results.   The static TTT curve (broken line) obtained from dilatometer measurements (13) is shown for comparison purposes.

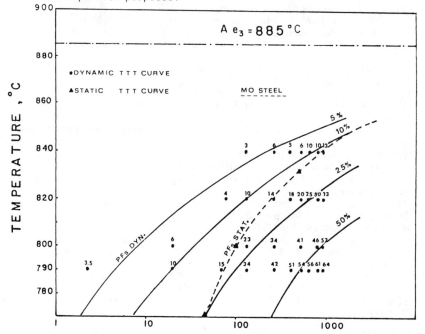

Fig. 3.   Dynamic TTT curve for the molybdenum steel derived from the present results.   The static TTT curve (broken line) obtained from dilatometer measurements (13) is also shown.

The more rapid nucleation and growth in <u>deformed</u> austenite can be attributed to a number of causes. First of all, deformation increases the specific grain boundary area per unit volume, a purely geometric effect. It also changes the <u>structure</u> of the boundaries, and in this way increases the probability of nucleation per unit of grain boundary area (14). Finally, it may also decrease the activation energy that must be provided to form a nucleus (1,3,8). Once nucleation has occurred, the rate of growth is accelerated because the presence of dislocations enables pipe diffusion to play a role in accelerating the rate of solute partitioning.

It is of interest that, at a given temperature, the rates of transformation are approximately equal in the two steels (see Figs. 2 and 3). This is despite the somewhat higher $Ae_3$ temperature of the Mo steel (885°C) compared to that of the plain C steel (870°C), a difference that provides a slightly higher driving force for the transformation in the former material. The higher driving force, however, appears to be offset by a more sluggish intrinsic growth rate in the Mo-modified alloy. The latter can in turn be attributed to the effect of Mo addition in reducing the C activity as well as the C diffusion rate (15). Furthermore, Mo addition also appears to increase the interfacial free energy required to form a ferrite nucleus, as reported by Aaronson and Domain (16). This effect can contribute a further component to the retardation of the transformation.

## Acknowledgements
The authors are indebted to Drs. L.E. Collins and G.E. Ruddle for their assistance, and to the Department of Energy, Mines and Resources, Ottawa for financial support of this work.

## References
1.  W. Roberts, H. Lidefelt and A. Sandberg: "Hot Working and Forming Process", Metals Society, London, 1979, p. 38.
2.  D.J.Walker and R.W. K. Honeycombe: Metal Sci. 1978, Vol. 10, p. 445.
3.  Y. Desalos, A. Lebon, R. Lambry: "Les Traitements thermomecaniques: Aspects theoriques et applications" 24e Colloque de Metallurgie I.N.S.T.N., Saclay (1981).
4.  M.J. Crooks, A.J. Garratt-Reed, J.B. Vaner Sande, W.S. Owen: Metall. Trans. A, 1982, Vol. 13, p. 1347.
5.  J.J. Jonas, R.A. Do Nascimento, I. Weiss and A.B. Rothwell: "Fundamentals of Dual Phase Steels", Ed. R. Kot and, B.L. Bramfitt, TMS-AIME, Warrrendale, Pa., 1981, p. 95.
6.  B. Bacroix, M.G. Akben and J.J. Jonas: "Thermomechanical Processing of Microalloyed Austenite", Eds. A.J. DeArdo and P.J. Wray, AIME, Warrendale, Pa., 1981, p. 293.
7.  R.K. Amin, F.B. Pickering: ibid: p. 455.
8.  A.J. Sandberg, W. Roberts: ibid, p. 405.
9.  R. Priestner: ibid, p. 455.
10. M.J. Luton, J.P. Immarigeon and J.J. Jonas: J. Physics E: Sci. Instr., Vol. 7, 1974, p. 862.
11. ANSI/ASTM Handbook E562-76
12. E. Essadiqi, M.G. Akben and J.J. Jonas: to be published.
13. L.E. Collins and J.R. Barry, CANMET, E.M.R., Ottawa, Private communication.
14. I. Kozasu, C. Ouchi, T. Sampei and T. Okita "Microalloying 75" Ed. M. Korchynsky, Union Carbide Corp., New York, N.Y., 1977, p. 120.
15. T. Wada, H. Wada, J.F. Elliot and J. Chipman: Metall. Trans. A, Vol. 3, 1972, p. 2865.
16. H.I. Aaronsson and H.A. Domain: Trans. AIME, Vol. 236, 1966, p. 781.

Fig. 4    Examples of the ferrite structures produced at 790°C by dynamic transformation.

(a)  Mo steel; $\epsilon$ = 0.15; t= 70s.

(b)  Mo steel; $\epsilon$ = 0.80; t= 953 s.

(c)  Plain C steel; $\epsilon$ = 0.80; t = 953s.

# CCT CHARACTERISTICS OF Nb, V MICROALLOY STEELS

L.E. Collins, J.R. Barry and J.D. Boyd

Physical Metallurgy Research Laboratories
CANMET, Energy, Mines and Resources Canada
Ottawa, Canada  K1A 0G1

Transformation products in steel are kinetically determined and are hence very sensitive to the time-temperature conditions imposed by the processing schedule. In most commercial processes, the austenite ($\gamma$)-ferrite ($\alpha$) transformation occurs over a range of temperatures as the steel cools, and it is thus important to understand the continuous-cooling-transformation (CCT) behaviour of controlled-rolled microalloyed steels. Of particular importance is the effect of austenite deformation on the transformation kinetics [1]. It is known that the $\gamma$-$\alpha$ transformation of deformed, unrecrystallized austenite is accelerated, compared with recrystallized austenite. Although ferrite growth rates may also be enhanced, this effect is attributed primarily to the enhanced ferrite nucleation rate resulting from: (i) increased austenite grain boundary area [2], (ii) increased nucleation site density at grain boundaries [3], and (iii) intragranular nucleation on deformation bands [4]. In addition, deformation of austenite influences the precipitation of Nb, Ti and V carbonitrides, to varying degrees, which alters the solute concentration and hence the transformation kinetics.

Clearly, the CCT behaviour of microalloyed steels is determined by a complex interaction between the deformation, recovery and precipitation in the austenite. In the present study, the effects of deformation temperature, total deformation and cooling rate on the CCT characteristics have been investigated for Nb and Nb + V steels. The compositions (wt %) of the steels studied are:

| Steel | C | Mn | Nb | V | S | P | Si | Al |
|-------|------|------|-------|------|-------|--------|------|------|
| Nb | 0.11 | 1.46 | 0.055 | -- | 0.002 | <0.002 | 0.28 | 0.02 |
| Nb + V | 0.095 | 1.62 | 0.055 | 0.11 | 0.005 | <0.002 | 0.30 | 0.05 |

The transformation studies were carried out using a quench-deformation dilatometer capable of 50% total deformation and cooling rates greater than
$40°C.s^{-1}$. All samples were heated to 1150°C in 5 min, and held 5 min at this
temperature prior to subsequent cooling and deformation. This treatment did
not dissolve all the Nb(C,N) and resulted in a mean austenite grain size of
43 μm. Samples were deformed in compression at a strain rate of approximately
$1 s^{-1}$. Unless otherwise specified, samples were deformed once at a selected
temperature and strain, cooled at a fixed rate, and the start and finish of
the γ-α transformation were determined by dilatometry. Dilatometer results
were reproducible to within ±5°C. The final microstructure was characterized
by optical metallography. For both steels, the austenite remained unrecrystallized prior to transformation after deformation of up to 50% at temperatures less than 900°C.

The effect of deformation temperature was determined by deforming specimens of the Nb steel 50% at temperatures between 775°C and 875°C. The results
are given in Fig. 1 as plots of the fraction transformed versus temperature.
With increasing deformation temperature, the γ-α transformation occurs at
higher temperatures. Since all deformation temperatures were at or below the
$Ac_3$ temperature ($Ac_3$ = 875°C), this effect is due largely to the increased
time allowed for the incubation of ferrite nuclei in samples cooled from higher temperatures. As well, increased precipitation of Nb(C,N) in the austenite
following deformation at temperatures approaching the range of maximum Nb(C,N)
precipitation [≈900-1000°C (5)], reduces the amount of Nb in solution. Since
Nb in solution is known to strongly suppress the ferrite start temperature
[≈10°C/0.01 wt % Nb at cooling rates 1 to 5°C/s (6,7)], this will further
increase the transformation temperature.

Figure 2 shows the effect of varying the total compressive strain for
specimens deformed at 775°C and cooled such that t(775-500°C) = 275 s. As the
deformation increases from 0-50%, the transformation temperatures increase

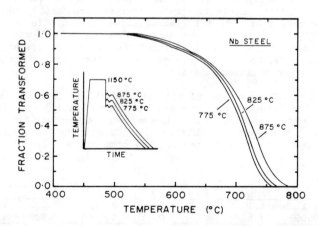

Figure 1 - Effect of deformation temperature on the transformation behaviour
of the Nb steel deformed 50% at the indicated temperatures and cooled such
that $t_{775-500}$ = 275 s.

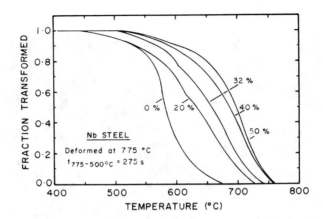

Figure 2 - Effect of total reduction (%) on the transformation behaviour of the Nb steel.

approximately linearly with strain. The final microstructure of an undeformed specimen is predominantly bainite with some grain boundary ferrite (Fig. 3a). With increasing deformation, this structure is gradually replaced by polygonal ferrite and pearlite, until at strains greater than 35% no bainite is evident in the final microstructure (Fig. 3b-3e). The ferrite grain size also becomes finer and more homogeneous with increasing deformation. The observed enhancement of ferrite nucleation with deformation cannot be attributed solely to the increase in grain boundary area of the deformed austenite. At the higher strains, the austenite grain boundaries are increasingly delineated by fine ferrite grains (Fig. 3d, 3e), indicating that deformation introduces highly-effective ferrite nucleation sites at austenite grain boundaries. As deformation is increased, the bainitic structures found in the interior of the prior austenite grains of lightly deformed samples (Fig. 3b) are replaced by a few comparatively coarse ferrite grains. This suggests that, although deformation enhances intragranular ferrite nucleation, the nucleation rate is still much lower than at austenite grain boundaries.

Samples of the Nb steel were cooled at various rates after 50% deformation at 775°C. The ferrite grain size becomes finer and more homogeneous with increasing cooling rate, due to increased nucleation rate and reduced growth rate of ferrite. In addition, since carbon diffusion distances are reduced, a more uniform distribution of smaller pearlite colonies is obtained at the higher cooling rates. The extent to which accelerated cooling can be used to refine the ferrite grain structure in such steels is limited by the onset of bainite formation. After a 50% deformation, a significant fraction of bainite is present in the final microstructure for cooling rates greater than $\approx 3°C.s^{-1}$.

The Nb + V steel exhibited similar transformation behavior to that of the Nb steel (Fig. 4). At t(775-500°C)=275 s the presence of V appears to slightly increase the ferrite start temperature. At slower cooling rates

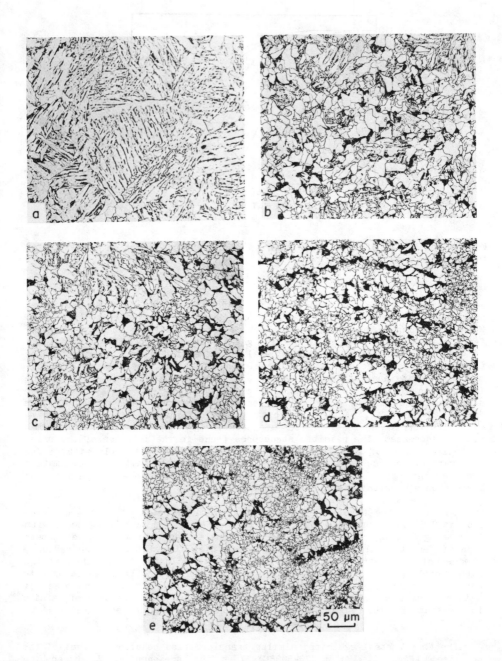

Figure 3 - Cnanges in microstructure with deformation in the Nb steel deformed at 775°C and cooled such that $t_{775-500°C}$ = 275 s. (a) no deformation; (b) 20% reduction; (c) 32% reduction; (d) 40% reduction; (e) 50% reduction.

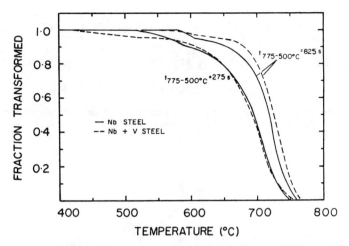

Figure 4 - Comparison of the transformation behaviour of Nb and Nb + V steels. Samples were deformed 25% at 1050°C, held 15 s, quenched to 775°C and deformed to 50% total reduction prior to cooling.

this effect is more pronounced and at t(775-500°C)=825 s the transformation of the Nb + V steel proceeds at temperatures approximately 10°C higher than that of the Nb steel. This result is consistent with the suggestion (8) that V enhances precipitation of Nb(C,N) in the austenite, thereby reducing the Nb solute content and its effect in suppressing ferrite formation. Vanadium also acts to stabilize the high C austenite remaining in the final stages of transformation (X>0.95) resulting in increased M/A content as compared to the Nb steel at some cooling rates.

The response of the Nb + V steel to deformation temperature, total deformation and cooling rate was similar to that of the Nb steel. Regardless of the processing schedule, the hardness of the Nb + V steel was always greater than that of the Nb steel likely due to the precipitation of V(C,N) in the ferrite.

### References

1.  L.E. Collins, M.J. Godden and J.D. Boyd, "Microstructures of Linepipe Steels," Can. Met. Quart., 22 (2) (1983), pp. 169-179.

2.  I. Kozasu, C. Ouchi, T. Sampei and T. Okita, Microalloying '75, p. 120; Union Carbide Corp., New York, NY, 1977.

3.  W. Roberts, H. Lidefelt and A. Sandberg, Hot Working and Forming Processes, p. 38; The Metals Society, London, 1979.

4.  C. Ouchi, T. Sampei, T. Okita and I. Kozasu, The Hot Deformation of Austenite, p. 316; AIME, New York, NY, 1977.

5.  M.G. Akben, I. Weiss and J.J. Jonas, <u>Acta Met.</u>, 29(1), (1981), pp. 111-121.

6.  G.L. Fisher and R.H. Geils, <u>Trans. AIME</u>, 245, (1969), pp. 2405-2412.

7.  W.B. Morrison, <u>J. Iron and Steel Inst</u>., 201, (1963), p. 317.

8.  R.C. Cochrane and W.B. Morrison, Proceedings of Conference on Steels for Linepipe and Pipeline Fittings, London, 1981, to be published by the Metals Society.

# FERROUS PHASE TRANSFORMATIONS – SOME COMMENTS AND DIRECTIONS

Morris Cohen

Department of Materials Science and Engineering
Massachusetts Institute of Technology
Cambridge, MA 02139

This Symposium on "Phase Transformations in Ferrous Alloys" provides an
excellent companion-piece to the Peter G. Winchell Symposium on "The
Tempering of Steel," published in 1983[1]. As a result, many aspects
of ferrous phase transformations have now been brought up-to-date, usefully referenced, and nicely summarized. An extraordinary amount of
information has been collected from the literature, stemming largely
from refined experimentation and critical analysis. "Understanding"
has surely been advanced through the acquisition of more definitive
knowledge, but complexity of findings continues to mount, and general
predictability from the simple to the complex continues to recede. The
situation at hand for physical metallurgy bears some resemblance to that
of high-energy physics in which increasingly powerful (and expensive)
accelerators are accessing finer and finer subnuclear particles deep in
the subquantum recesses of matter. In the latter domain, there is at
least the hope that a fundamental unit of matter will eventually be
revealed which cannot be further subdivided! However, that kind of
expectation is denied to us in the field of physical metallurgy; our
basic building blocks and unit processes for the evolution of microstructures are already extremely complex and, correspondingly, structure/
property relationships are even more so by orders of magnitude. The
present Symposium offers a fresh demonstration of this plight as well as
challenge in physical metallurgy; it is also broadly true in materials
science. We shall have to distinguish increasingly between <u>essence</u> and
<u>detail</u> in this context and to think more carefully about what we mean
by the overworked claim of "understanding."

In dealing specifically with ferrous phase transformations, physical
metallurgists are torn between certain major objectives: (1) we want
to identify and quantify the operative fundamental phenomena (at least
in terms that we find satisfying and useful) leading up to and during a
transformation; (2) we want to apply the resulting microstructures to
account for properties of interest; and (3) we want to establish theoreti-

cal and reliable connections between the above two objectives. This is obviously a tall order. For example, a prevailing paradigm for solid-state phase transformations relies on the concept of nucleation and growth, originally carried over from vapor/liquid and liquid/solid phase changes. For simplicity, one would like to concentrate on the formation of a single unit of a transformation product; but for structure-property relationships, we have to consider the <u>final</u> microstructure, whose history may depend on autocatalysis, changing growth conditions, impingement, competitive reactions, and incompleteness of the transformation. In addition, all these interplays produce various morphological entities, such as the pearlitic lamellae of ferrite and cementite which assemble into colonies, colonies which aggregate into nodules, and nodules which adopt a pattern based on the parent-phase grain boundaries. Lath martensites likewise display a hard-to-predict hierarchy of microstructural elements: sublaths, laths, and packets of laths, all connected back somehow to the parent grain from which they have evolved.

It becomes painfully clear that predictability of final microstructures from the central ideas of nucleation and growth is fraught with enormous difficulties, and may even be unattainable. And then there is the further problem of trying to deduce which aspects of a microstructural hierarchy are important in controlling these properties which are so vital to engineering applications. In some instances, such correlations have been sorted out by systematic experimentation, and quantitative relationships have been established through empiricism or modeling. This process has proved to be very useful for designing experiments and for property improvements, but it cannot be regarded as "predictable from basic principles." In fact, at the present stage, properties of ferrous alloys furnish just as many clues about structure as structure does about properties; it is this reciprocity -- not predictability -- that continues to contribute so much to the success of physical metallurgy and materials science[2].

There are several highlights emerging from this Symposium which are worthy of special attention in future research:

1. As in the aforementioned analogy to elementary particle physics, state-of-the-art high-powered microanalytical instrumentation, including scanning electron transmission microscopy, high-resolution electron microscopy, field-ion and atom-probe microscopy, lattice-fringe imaging, etc., can be expected to disclose important features of the nature of ferrous phase transformations and also to show the way that the resulting microstructures relate to properties. The greatest advances in this connection are likely to come from "surprises" rather than from a filling-in of current knowledge and traditional ideas.

2. It is evident that interfaces hold the key to essential aspects of phase transformations: nucleation, growth, state of coherency, degree of lattice correspondence, mobility, solute drag, defect drag, diffusion control, mechanical behavior, etc. The high-resolution techniques mentioned above are now reaching a capability for shedding new light on these interfacial phenomena. Grain-boundary segregation of solute atoms is now well-documented in embrittlement effects. Subtle segregation of carbon may also be significant in the heterogeneous nucleation of ferrous phase transformations, even the side-by-side nucleation of martensitic and bainitic laths.

3.  One can now visualize a spectrum of phase transformations, starting with plate and lath martensites, extending through the bainites, and even including the Widmanstatten ferrites.  These transformations may be different one from the other in certain respects, but they may also be basically related -- as with individual colors in the optical spectrum.  The connecting link is the transformational shape change, which signals an underlying lattice deformation and a lattice correspondence operating through coherency across the interface.  Depending on temperature and other factors, the shape change may be accommodated elastically, as in shape-memory martensites, or plastically by dislocation multiplication and movements.  These dislocations may be glissile (conservative motions) at low temperatures to involve climb processes (nonconservative motions) at higher temperatures.  The resulting increase in dislocation density, or work hardening of the parent phase, will impose a drag on the moving interface at low temperatures, but is subject to dynamic relaxation at higher temperatures.  Such lattice-invariant deformation and relaxation processes may affect the observed shape change, but not necessarily the basic lattice deformation which accomplishes the intrinsic structural change.  Carbon diffusion can come into play and thereby increase the transformational driving force so that the displacive structural change can ensue at temperatures above where diffusionless martensitic transformations normally occur.  Implicit in this overall viewpoint is a natural coupling between lattice-distortive and diffusion-related phenomena; they are not mutually exclusive and may, indeed, represent a broad class of solid-state phase transformations in which martensitic transformations constitute only a special, diffusionless case.  Moreover, this paradigm, if it can be called such, places full emphasis on the solid-state nature and concomitant stress fields of the transformations, and is not merely a modified version of vapor/liquid and liquid/solid transformations.  Even a simple diffusion-controlled precipitation reaction in the solid state is likely to have significant displacive features, at least in the nucleation and early growth stages.

4.  The many similarities between lath martensite and lower bainite are providing new insights regarding both modes of transformation.  The character of their comparable subunits is specially illuminating.  A reasonable hypothesis is that both transformations start from the same pre-existing nucleation sites in the parent phase, but that time-dependent short-range carbon movement permit the bainitic transformation to nucleate at higher temperatures than is possible for the diffusionless type of transformation.  It also appears that, in both instances, each sublath can propagate rapidly with a shape change and with carbon entrapment until it is decelerated and arrested, perhaps by defect accumulation due to plastic accommodation ahead of the interface.  Autocatalytic nucleation of the next sublath can then take place with local rearrangement of the dislocation arrays (for the case of lath martensite) and of local carbon redistribution (for the case of bainite).  Processes of this sort would be entirely compatible with the spectrum of displacive-type transformations suggested above.

5.  There is another spectrum of ferrous transformations which produce ferrite-carbide aggregates directly.  These comprise lamellar pearlites, fibrous growth of carbides which trail an advancing $\gamma \rightarrow \alpha$ interface, and periodic sheets of carbide precipitates which mark the

advance of a $\gamma/\alpha$ interface during the transformation.  It has been hypothesized that these microstructural differences may be attributed to differences in the nature of the $\gamma/\alpha$ interface, i.e., whether coherent, semicoherent, or incoherent.  There is another instance in which it is desirable not to rely on loose descriptions of interphase interfaces; instead, it is now mandatory to determine the actual interfacial structures which prevail under various transformational conditions and then to establish how such fine-scale features can control the resulting carbide morphology.

6.  In the early 1960's, the Office of Naval Research sponsored several panel studies on "Perspectives in Materials Research," including one on "Phase Transformations in the Solid State."[3]  It would be instructive and useful to undertake a similar study now, particularly from the vantage point of the present Symposium, to discern the kinds of real progress that have been made during the past quarter century, to re-examine terminology and classification schemes, to delineate the crucial research areas for both theoretical and experimental research, and to seek new perspectives for viewing solid-state phase transformations in general as well as ferrous transformations in particular.  The present Symposium has furnished a valuable resource-base for new assessment and, it is hoped, enlightenment.

## References

1.  Peter G. Winchell Symposium on "The Tempering of Steel," Metall. Trans. A 14A (1983) pp. 991-1145.

2.  Morris Cohen, "Unknowables in the Essence of Materials Science and Engineering," Anniversary Volume, Mat. Sci. and Eng. 25 (Sept./Oct. 1976) pp. 3-4.

3.  "Phase Transformations in the Solid State," Part VI of Perspectives in Materials Research, L. Himmel, J. J. Harwood, and W. J. Harris, Jr., Eds., Office of Naval Research, Washington, DC (February 1963) pp. 309-382,

# SUBJECT INDEX

410

# AUTHOR INDEX